This is a volume in the Arno Press collection

HISTORY OF ECOLOGY

See last pages of this volume for a complete list of titles.

PLANT COMPETITION

PLANT COMPETITION

AN ANALYSIS OF COMMUNITY FUNCTIONS

FREDERIC E. CLEMENTS, JOHN E. WEAVER

AND

HERBERT C. HANSON

ARNO PRESS

A New York Times Company

New York / 1977

Editorial Supervision: LUCILLE MAIORCA

Reprint Edition 1977 by Arno Press Inc.

Reprinted from a copy in
The University of Michigan Library

HISTORY OF ECOLOGY
ISBN for complete set: 0-405-10369-7
See last pages of this volume for titles.

Manufactured in the United States of America

Library of Congress Cataloging in Publication Data

Clements, Frederic Edward, 1874-1945.
Plant competition.

(History of ecology)
Reprint of the 1929 ed. published by Carnegie Institution of Washington, Washington, which was issued as its Publication no. 398.
Bibliography: p.
1. Plant competition. 2. Plant communities.
I. Weaver, John Ernest, 1884-1966, joint author.
II. Hanson, Herbert Christian, 1891-1962, joint author.
III. Title. IV. Series. V. Series: Carnegie Institution of Washington. Publication ; no. 398.
QK911.C58 1977 581.5'24 77-74209
ISBN 0-405-10380-8

CARNEGIE INSTITUTION OF WASHINGTON

PUBLICATION NO. 398

1929

PRESS OF
W. F. ROBERTS COMPANY
LANMAN ENGRAVING COMPANY
WASHINGTON, D. C.

PLANT COMPETITION

AN ANALYSIS OF COMMUNITY FUNCTIONS

BY

FREDERIC E. CLEMENTS, JOHN E. WEAVER

AND

HERBERT C. HANSON

PUBLISHED BY CARNEGIE INSTITUTION OF WASHINGTON

SEPTEMBER, 1929

ABSTRACT

An experimental analysis of competition as the paramount function of the community, together with reaction, which is its inevitable concomitant. A number of new methods have been devised and the technique perfected for the comprehensive study of this process under both natural and cultural conditions, involving the utilization of climax and seral communities in nature, field crops, and greenhouse control. Four chapters are devoted to experimental cultures in prairie and woodland, one to field plots of sunflower and wheat, and two to the analysis of factor control and functional adjustment in the greenhouse.

The initial chapter traces the growth of ideas in this field, in especial relation to ecology, forestry and agriculture, while the final one discusses the concepts and principles involved. The latter also indicates the path of future progress in connection with the application of principles and methods to forest, grazing and agronomic research.

Institution Publications by the Authors and Collaborators

242. Clements, F. E. Plant Succession: An Analysis of the Development of Vegetation.

290. Clements, F. E. Plant Indicators: The Relation of Plant Communities to Processes and Practice.

NOTE.—*A combined and condensed edition of "Plant Succession" and "Plant Indicators" entitled PLANT SUCCESSION AND INDICATORS has been issued for Dr. Clements by H. W. Wilson Co., 958 University Avenue, New York City, to whom all inquiries should be made.*

315. Clements, F. E. Aeration and Air-content: The Rôle of Oxygen in Root Activity.

326. Hall, H. M., and F. E. Clements. The Phylogenetic Method in Taxonomy: A Monograph of the North American Species of Artemisia, Chrysothamnus, and Atriplex.

336. Clements, F. E., and F. L. Long. Experimental Pollination. An Outline of the Ecology of Flowers and Insects.

355. Clements, F. E., and J. E. Weaver. Experimental Vegetation: The Relation of Climaxes to Climate.

356. Clements, F. E., and G. W. Goldsmith. The Phytometer Method in Ecology.

398. Clements, F. E., J. E. Weaver, and Herbert C. Hanson. Plant Competition: Analysis of Community Function.

286. Weaver, J. E. The Ecological Relations of Roots.

292. Weaver, J. E. Root Development in the Grassland Formation: A Correlation of the Root Systems of Native Vegetation and Crop Plants.

316. Weaver, J. E., F. C. Jean, and J. W. Crist. Development and Activities of Roots of Crop Plants.

357. Jean, F. C., and J. E. Weaver. Root Behavior and Crop Yield under Irrigation.

PREFACE

The present book is the second of a series dealing with the functions of the plant community as a complex organism. The first was devoted to a consideration of ecesis under climatic and edaphic conditions and prepared the ground for the investigation of competition as the crucial process in the community ("Experimental Vegetation," 1924). Further preparation for this research was afforded by the series of studies on the root systems of plants by Weaver and his colleagues, since these were essential to the understanding of the paramount water relation in grassland ("Ecological Relations of Roots," 1919; "Root Development in the Grassland Formation," 1920; etc.). This series of monographs serves to extend and exemplify the account of community processes and the resulting structures given in "Plant Succession," 1916, and "Plant Indicators," 1920, and in the combined edition, "Plant Succession and Indicators," 1928. The other functions in process of investigation are reaction, coaction, adaptation, and migration and invasion, each of which it is hoped will appear in monograph form.

The experimental organization of the field of competition was begun in 1903 ("Research Methods in Ecology"), but was not resumed on a comprehensive scale until 1924. It has been carried out primarily at Lincoln and Weeping Water, Nebraska, though collateral cultures under garden control have also been maintained at Santa Barbara, California. The major communities concerned have been the true prairie, the subclimax prairie, the subclimax chaparral, and the edge of the oak-hickory association of the deciduous forest climax. While the course and outcome of competition in natural and cultivated communities have furnished the primary objectives, control cultures in the greenhouse at Lincoln have been indispensable in the analysis of the factor complex and the functional response of competitors. It is hoped that the present treatment will constitute an adequate and comprehensive foundation for this important field, but it is fully recognized that the building of the superstructure will require many hands in various climaxes of the globe. In the meanwhile, it is proposed to continue and extend the present series of researches, in accordance with the outline given at the close of chapter nine.

FREDERIC E. CLEMENTS
JOHN E. WEAVER
HERBERT C. HANSON

ALPINE LABORATORY
MANITOU, COLORADO
AUGUST 30, 1928

CONTENTS

ILLUSTRATIONS

Plates

Text-Figures

TEXT-FIGURES—*Continued*

PLANT COMPETITION

AN ANALYSIS OF COMMUNITY FUNCTIONS

BY

FREDERIC E. CLEMENTS, JOHN E. WEAVER

AND

HERBERT C. HANSON

PLANT COMPETITION

1. HISTORY OF THE COMPETITION CONCEPT

Scope of the concept—Like other primary concepts in ecology, the idea of competition has gradually emerged from the general experience of mankind. It must have appeared long before it was recorded, just as the record must have preceded a name for the process itself. As the most striking consequence of the grouping of organisms into communities, it was perhaps first clearly perceived in the case of the forest, but its manifestations were doubtless evident long before in human societies. Certainly the demand for an essential factor in excess of the amount afforded by the environment must have early engaged the attention of thoughtful men, and with most force in human relations. The existence of the same process in the case of plants and animals could hardly have escaped notice, but the formulation of the idea had to await the beginnings of scientific forestry and agriculture.

The appreciation of the importance of competition in community life developed more or less independently in the several fields. For forestry probably the earliest record is that of Petrus de Crescentiis in 1305, while in human society the doctrines of Malthus (1798) were anticipated by Franklin (1751), Hume (1752), and others. The general significance of competition in the plant world was perhaps first clearly perceived by DeCandolle (1820) and Herbert (1837), while its rôle in evolution was first adequately understood by Darwin (1858).

The literature of competition is not extensive, but it is so little known and yet of such importance as to warrant a comprehensive account. An endeavor has been made to render this fairly complete, though the limitations of space and the general nature of many of the studies conspire to make a detailed and exhaustive consideration inadvisable. This treatment falls under five general heads as follows: (1) earlier views; (2) experimental studies; (3) competition and succession; (4) competition in the forest; (5) competition in cultivated fields.

EARLIER VIEWS

During this formative period of about a century, the number of writers was small, but their views were of especial importance in directing the development of the concept in the various fields. These ranged widely, from the rôle and nature of tolerance in the forest to population and the progress of society, the significance of competition in plant communities, the struggle for existence, the response to lime in the soil, and the part played by competition in vegetation, especially in succession.

Tolerance—The effect at least of competition between trees in the forest was understood by Peter de Crescentiis (1305) who directed that cuttings should be made where the trees were too thick, and that where good stems were suppressed by undesirable kinds or by thorns, the latter should be removed (cf. Bühler, 1922:416; Lönnroth, 1925:4). The first clear recognition of its importance in forestry is perhaps to be ascribed to Buffon (1742), who also caught a glimpse at least of its significance for succession (Clements 1916, 1928).

"The best shelter in wet soil is poplar or aspen, and in dry soil, *Rhus,* for the growth of oak. One need not fear that the sumac, aspen or poplar can injure the oak or birch. After the latter have passed the first few years in the shade and shelter of the others, they quickly stretch up and suppress all the surrounding plants" (p. 237-238).

This process received more particular attention at the hands of foresters, such as Duhamel du Monceau (1760, 1764) and Varenne de Fenille (1790), while the first endeavors to distinguish trees as *dominant* and *suppressed* were made in Denmark between 1780 and 1796 (Laschke, 1903:68; Lönnroth, 1925:6). In so far as it has a bearing on competition, the nature and rôle of tolerance are discussed briefly in later sections.

Malthus, 1798—The effect of competition upon mankind, as expressed in terms of population and food-supply, was portrayed so dramatically by Malthus (1798, 1803, 1807) that his name is still the password to many students of population (Thompson, 1915; East, 1926) and is borne by a league devoted to the idea of carrying his doctrines into social effect. For the present purpose, it will suffice to show that Malthus was dealing essentially with aspects of competition in human society and to recall that his views were the inspiration for Darwin's principle of natural selection:

"The cause to which I allude is the constant tendency in all animated life to increase beyond the nourishment prepared for it. It is observed by Dr. Franklin that there is no bound to the prolific nature of plants or animals, but what is made by their crowding and interfering with each other's means of subsistence. Were the face of the earth, he says, vacant of other plants, it might be gradually sowed and overspread with one kind only, as for instance with fennel; and were it empty of other inhabitants, it might in a few ages be replenished from one nation only, as for instance with Englishmen."

"This is incontrovertibly true. Throughout the animal and vegetable kingdoms Nature has scattered the seeds of life abroad with the most profuse and liberal hand; but has been comparatively sparing in the room and the nourishment necessary to rear them. The germs of existence contained in this spot of earth, with ample food and ample room to expand in, would fill millions of worlds in the course of a few thousand years. Necessity, that imperious all-pervading law of Nature, restrains them within the prescribed bounds. The race of plants and the race of animals shrink under this great restrictive law; and man can not by any efforts of reason escape from it."

His basic tenets were expressed as follows:

1. Population is necessarily limited by the means of subsistence.
2. Population invariably increases where the means of subsistence increase, unless prevented by some very powerful and obvious checks.
3. These checks, and the checks which repress the superior power of population and keep its effects on a level with the means of subsistence, are all resolvable into moral restraint, vice, and misery.

Malthus has suffered much, from ardent admirers as well as bitter detractors; his analysis of certain primary factors and processes in human ecology was less brilliant and accurate than assumed by the former; nor did his doctrines and motives deserve the odium cast by the latter.

DeCandolle, 1820—The first definite though necessarily general characterization of plant competition appears to have been given by DeCandolle (1820:26-28). He stated that all the species of a region, all the plants of a given place are in a state of war with respect to each other. The first to establish themselves in any spot tend by the mere fact of occupation to exclude other species from it; the taller suppress the shorter, the more persistent replace those with a short span of life, the more fecund gradually take possession of the space that might be occupied by those of lesser vigor.

To the degree that a habitat is unfavorable to a species it grows less readily; thus, *Carex arenaria* thrives and suppresses all its neighbors in sand, but in compact soil it may be overcome by these very species. In other cases a similar control might be exerted by the temperature, light, water or air. More than this, all the plants of a habitat are engaged in a struggle with each other the success of which varies according to their age. In reclaiming the dunes of the "Landes," *Genista* and *Pinus* are sown broadcast so that the former which grows rapidly will dominate and protect the young pines, even suppressing them when the density is too great. Once it escapes this danger, the pine grows faster, surpasses the broom and finishes by overcoming it in turn.

Even among the species at home in a particular habitat, the most thrifty tend to dominate the area and to exclude the less vigorous. When a particular station favors certain species and hinders others, those that prosper in it end by taking complete possession, and consequently social species are found in special situations. On the contrary, when the conditions favor a large number of species to much the same degree, they compete on more or less equal terms and hence live in a mixed community. It is for this reason that weeds thrive in cultivated fields and that tropical forests present a mixture of many species, while those of temperate regions, less favored by climate, ordinarily exhibit a dominant group.

Darwin, 1859—In connection with the problem of the origin of species, the process of competition has long been popularly known under Darwin's phrase, "the struggle for existence," or Spencer's, "the survival of the fittest." The one emphasizes the dynamic process, the other the outcome of it, as does also Darwin's term, "natural selection," but all of these ideas are included in the word competition. The classic exposition of the rôle of competition in the origin of species was Darwin's, but he himself has pointed out in the sixth edition of his great book that others had anticipated him, in particular Wells (1813) and Matthew (1831). In both cases, the idea of natural selection was briefly developed, as well as incidental to other objectives, and the papers concerned were without influence on the trend of evolutionary thought. Wallace's proposals as to natural selection, read at the same time as Darwin's (1858), suffered also from the eclipse due to the "Origin of Species," as well as from the later books that made the theme of evolution peculiarly Darwin's. Of the galaxy that gathered around the latter, Spencer

(1858, 1864, 1876-1880), Hooker (1859, 1867), and Lyell (1867), in the later editions of his "Principles of Geology," gave effective treatments of the struggle for existence from the Darwinian standpoint.

Darwin's discussion of the struggle for existence is too familiar to warrant an extended summary and hence the following is limited for the most part to those points that bear upon the nature and rôle of competition. He stated that the term, struggle for existence, was used in a large and metaphorical sense to include the dependence of one organism on another, both with respect to the individual and its success in leaving progeny. Carnivorous animals may truly be said to struggle with each other for food in times of dearth, just as a plant may be said to struggle with those of the same and other kinds that already clothe the ground. The mistletoe can not be regarded as a competitor of the tree on which it grows, but several seedling mistletoes growing close together on a branch do compete with each other. Darwin also pointed out that the mistletoe may also compete with other species bearing fleshy fruits in securing the distribution of their seeds by birds, a form of competition that is strikingly exemplified in the carriage of pollen by insects (Clements and Long, 1923).

The doctrine of Malthus was regarded as applying with especial force to plants and animals, so that a struggle for existence follows inevitably from the high rate of increase. This struggle may be that of one individual with others of the same species, with those of distinct species, or with the physical conditions; the latter is obviously not so much competition as a corollary of it.

Since the species of one genus are much alike in habit and constitution, and especially in structure, the struggle will generally be more severe between them than the species of distinct or unrelated genera. It can be dimly seen why the competition should be keenest between allied forms which play much the same rôle in nature, but difficult to say in any particular case why one species emerges victorious. A corollary of the highest importance results from these two facts, to the effect that the structure of every organism is related in the most essential yet often hidden manner to that of all others with which it competes, from which it escapes or on which it preys.

It is interesting to note that Darwin made incidental use of the quadrat and exclosure methods to obtain information as to the results of competition among plants. "If turf which has long been mown, and the case would be the same with turf closely browsed by quadrupeds, be let to grow, the more vigorous plants gradually kill the less vigorous, though fully grown plants; thus out of twenty species growing on a little plot of mown turf (three feet by four) nine species perished, from the other species being allowed to grow up freely." In the case of an extremely barren heath which was enclosed and planted to Scotch fir, the change in the native vegetation was most remarkable, more than is usually seen in passing from one soil to another. Not only were the relative numbers of the heath plants wholly changed, but twelve species of plants (not counting grasses and sedges) flourished in the plantations, which could not be found on the heath.

Nägeli, 1865—The significance of competition in the plant community was broadened by Nägeli (1865:159), who was the first to point out that it furnished the solution of the problems centering about the presence of lime in the soil. He maintained that the chemical composition of the soil as such and by itself alone could not explain the distribution of species, but also stated that it was incorrect to say that it had no part in this. He discussed the classical examples of pairs of calciphile and calciphobe species, such as *Achillea atrata* and *moschata, Rhododendron hirsutum* and *ferrugineum.*

As to the first pair, he adduced the fact that in districts where only one grows, *A. moschata* may also be found in calcareous soil and *A. atrata* in silicious ones, since in each case its competitor is absent. Moreover, in an intermediate soil, such as that derived from calcareous slate, the two may grow side by side, as they are then equal in competitive ability. In the case of *Rhododendron,* the distribution of the two species is in general not determined by the lime-content; throughout the greater part of their area they reveal themselves as indifferent. Where they occur together in considerable number, they become preferent in that they mutually exclude each other. Since they are shrubs, this exclusion is less strict than with herbaceous species, and it is also less decisive where the plants of each are fewer and more scattered.

If the lime-content alone is taken into consideration, it may happen under certain conditions that each species may surpass its competitor on one soil and yield to it on another. Soil preference and indifference are on the whole conditions from which no conclusions can be drawn as to the nature of a species, since they do not determine this. *Achillea atrata* and *A. moschata* in part grow separately and are then indifferent, partly together and are then preferent. If migration had everywhere brought them together, they would have been known only as preferent, while if they had been universally separated, they would have appeared as indifferent. Just as with the chemical conditions, so also is the case with the physical factors of the habitat. The same species may play the rôle of preferent in one, of indifferent in the other. Under the same chemical and physical conditions, it may occur in a wide range of holard or water-content in one and in but a limited one in another.

It is possible under certain conditions for a species to dominate in a soil of a particular holard, but under other to be suppressed or excluded. This is the situation with *Primula officinalis* and *elatior.* When both occur together, they sometimes exclude each other almost completely, the one preferring the drier, the other the moister areas. Each is the more dominant in its own habitat, where it is able to suppress its competitor, but when one alone is present it is much less exacting. *Primula officinalis* is then able to grow in moister, *P. elatior* in drier sites than when both are associated. *Prunella grandiflora* and *vulgaris* behave in similar fashion, while *Rhinanthus alectorolophus* and *minor,* and *Hieracium pilosella* and *hoppeanum* exhibit a comparable response to both nutrient and water content.

Nägeli, 1874—Stimulated by the emphasis placed by Darwin upon the struggle for existence, Nägeli (1874:205) sought to give a mathematical form to the suppression of plants by their competitors. At the outset he stated that competition is not to be regarded as an actual struggle but as a harmless activity, and for the most part as a purely passive relation to the factors of the habitat. In the constant sifting that nature applies to her products, only those capable of existence survive and of these, competition favors those best adapted to the particular habitat. Those less well adapted are thrust aside. At first thought nothing seems more natural than that the stronger of two competitors will completely replace the weaker, and there are indeed many examples of such an outcome. However, related or similar life-forms, between which competition operates most intensely, do not as a rule suppress each other to the degree that one alone persists in the habitat where it is stronger. On the contrary, they tolerate each other in the same situation, and competition operates reciprocally upon the number of individuals. In consequence, suppression is generally partial rather than complete. Nägeli further emphasized the fact that competition takes place only between similar plants under similar conditions; it does not operate between tree and moss, host and parasite.

The major portion of the article was devoted to the mathematical application of probability to the number of individuals of competing species, based upon the primary equation, $z = d \times e$. In this, z represents the number of individuals of a certain species in a definite locality, d the life-period in years, and e the annual increase of new plants. All of the possibilities involved in the reciprocal suppression of two plant forms were passed in review, and the conclusion drawn that their respective numbers were determined by the average life-period and the average annual increase. In the great majority of cases, the coefficients for life-period and increase were positive and of definite value, from which was derived the nearly universal rule that suppression is partial and not total. However, in the case of a very small individual-number, partial suppression may pass into the total condition, owing to the fluctuations brought about by the factors of the habitat. The theoretical probability demands that equal strength (with equal individual-numbers for each form) occur with almost infinite rarity, that dissimilar strength with partial suppression and unequal numbers be the prevailing rule, and inequality with the suppression of one form of fairly rare occurrence.

Macmillan, 1892—The significance of competition between communities as well as within them was clearly recognized by Macmillan (1892:582-611). His field comprised the ecotone between deciduous forest and prairie, which offered especially favorable opportunities for observing the major features of the process. His account gains much in effect without loss of clarity, from the skillful use of metaphor.

"It thus happens that the flora of any region—that is to say the plant society of the region—is in the same condition of mutual interdependence and mutual competition that we discover in human society. Complex interrelations of individual with individual, species with species, formation with formation arise, and the plant population of any area so far from being stable in its composition is in a continual state of battle for soil, light, moisture, heat and useful alliances, both in the physical and biological sense of the word. Thus, in a forest, the pine trees compete with each other for light, each taller one than the rest gaining a distinct advantage; hardwood timber antagonizes the coniferous and along the forest skirmish-line will be found slowly working its way up the streams, gradually isolating the coniferous trees into separate groves, ready at the first sign of misfortune or weakness in the opposing species to seize and occupy its territory. Again forest and prairie—two most notable plant formations of the Minnesota valley, each tenanted by hundreds of species characteristic if not peculiar—carry on a silent warfare with each other and as the chance of battle swings in favor of one, the other is imperceptibly but surely driven back.

"It happens then, to return to the illustration, that we find plants organized much as in human society. The individuals of each species compete with each other for favorable habitats and for the optimum of growth materials and energizing forces. Each species competes with those around it and in this competition the individuals may be said to stand shoulder to shoulder against the common foe, as may be seen in the united efforts of a human tribe or nation against some warring body. And again groups of species, having perhaps a common line of movement or a common need to be supplied, band themselves together and find arrayed against them other united groups of species competing for the same necessity or striving to move in the opposite direction. By the assistance of this fact of organized and stratified competition in the realm of plant society, the dynamic relations of plants to one another are in general to be explained.

"The equatorial region then is a perennial fountain-head from which there is a constant stream of emigration into northern and southern latitudes. With such migration there must, under the stress of natural selection, originate and develop modifications in the rank. What these modifications may be in any particular case depends upon the complicated intermingling of the various conditions of climate, nutrition and competition.

"Before the advancing glacier there must have been, among plants as among animals, a stern race for lower latitudes and more congenial temperatures. In this way periodic returns to the equatorial belt have been characteristic, in a general manner, of plant migration phenomena. Evidently under the competition and struggle of the return, natural selection would operate as before in the development of new characters and the emergence of so-called new species."

Warming, 1895—In the several editions of his pioneer book, Warming considered competition between species and the struggle between communities in general, the treatment being essentially the same in all. After a discussion of the chemical and physical factors in the soil, he stated that plants as a rule are fairly indifferent to the chemical ones so long as they have no competitors, in accordance with the views of Nägeli, Bonnier, and Christ. In the midst of its area a species is often indifferent, but toward the margin competition may restrict it to a particular type.

Commensalism was defined as the relation between species that share with each other the raw materials of soil and air, and in this sense feed at the same table. Commensals were divided into like and unlike, the best examples of the former being communities of a single species. Like commensals make the same demands as to nutrients, water, light, etc., and since these rarely suffice for all, a struggle for existence arises as soon as the space is well occupied. The plants in unfavorable situations and the weaklings are gradually vanquished and then exterminated. In mixed communities the competition is severest between the species that make similar demands, such as the trees of a tropical forest. Success often depends upon the conditions, but as a rule the structural and biological features will determine the outcome. Finally, many communities contain species that differ widely in requirements, and competition then decreases with the disparity in their needs. Thus it is possible for two or more species to be complementary in their occupation and utilization of the same habitat. A few dominant species may hold sway over an area, while others, often present in far greater number, are subordinate or even dependent, the factor of time entering when species develop at different seasons.

Warming further pointed out that a similar battle was taking place between communities, each constantly striving to invade the territory of others. Such competition is necessarily dependent upon that between species. The author suggested that the means by which plants oust each other may arise out of a lack of water or a particular nutrient, of adequate light or heat, or of an excess of some substance, out of mechanical barriers presented by roots and rhizomes, or the depletion of holard and nutrients. The more suitable the climate, the less exacting a species as to soil and the more capable in competition, while the struggle is hardest at the edge of its area. Of great importance also in determining success is the time of arrival, the first-comers seizing the ground and tending to maintain possession against all later invaders.

Clements, 1897, 1904—The rôle of competition in secondary succession was first studied by Clements in such abandoned roadways as those of the old California trail through the mixed prairie of western Nebraska (1897: 968). The successive abandonment of these parallel roads had produced a corresponding series of seral stages marked by narrow but distinct ecotones, in which annuals competed with perennials, the latter with dwarf-shrubs, and these with the climax grasses. The wearing of each new road through the prairie sod gradually destroyed the plants and eliminated their competition, with the result that the bordering *Stipa* became much taller and denser, forming the marginal zones characteristic of all unilateral competition.

In the "Development and Structure of Vegetation" (1904) and "Research Methods in Ecology" (1905), an endeavor was made to analyze the nature and rôle of competition in greater detail and with more accuracy than had been done. This profited from the preliminary results of experimental studies, and the account is here abstracted at some length for the additional reason that both books have long been out of print. As to the precise nature of competition, it is evident that the phrase, "struggle for existence," is true only in the figurative sense. Such a direct relation exists only between parasites, epiphytes and lianes, and the plants that serve to nourish or support them. In the case of plants growing on the same stratum, actual competition between plant and plant does not occur. One individual can affect another only in as much as it changes the physical factors that influence the latter. Competition is a question of the reaction of the plant upon the physical factors that encompass it and of the effect of these modified factors upon the adjacent plants. In the exact sense, two plants do not compete as long as the water-content and nutrients, the heat and light are in excess of the needs of both. The moment, however, that the roots of one enter the area from which the other draws its water supply, or the foliage of one begins to overshade the leaves of the other, the reaction of the former modifies unfavorably the factors controlling the latter, and competition is at once initiated.

The same relation exists throughout the process; the taller, stronger, the more branched or the better rooted plant reacts upon the habitat, and the latter immediately exerts an unfavorable effect upon the weaker, shorter, less branched or more poorly rooted plant. This action of plant upon habitat and of habitat upon plant is cumulative, moreover. An increase in the leaf surface of a plant not merely reduces the amount of light and heat available for those near it or beneath it, but it renders necessary the absorption of more water and nutrient, and still further decreases the amount present. The inevitable result is that the successful individual prospers more and more, and the less successful one loses ground in the same degree. As a consequence, the latter disappears entirely, or is handicapped to such an extent that it fails to produce seeds, or these are reduced in number or vitality.

This exemplifies the simplest type of competition, i.e., that in which the individuals belong to a single species. While these usually show relatively slight differences in height, width, leaf expanse or root surface, still some will have larger, some smaller surfaces for the absorption of water, light and heat. The former will receive the larger share of one or more essential factors, and the consequent reaction will affect the others, expressing itself in reduced stature, branching, inability to flower, etc. This illustrates the primary law of competition, viz., that this is keenest when the individuals are most similar. Such individuals make nearly the same demands on the habitat and adjust themselves less readily to their mutual reactions. The more unlike plants are, the greater the difference in their needs, and hence some adjust themselves to the reactions of others with little or no disadvantage.

In accordance with the above, the competition is closer between species of like form than between those of dissimilar form. This similarity must rest upon vegetation or habitat form, and not merely upon systematic position. Leaf, stem and root characters determine the outcome, and those species most alike in these features will be in close competition, regardless of taxonomic relationship. From this is derived the second principle of competition, viz., the closeness of the competition between plants of different species varies directly with their likeness in vegetation or habitat form. It is obvious that the forms associated in a community are of the first importance in determining the course and result of competition. Identity of vegetation form regularly produces close competition and the consequent numerical reduction or disappearance of one or more species. Dissimilarity tends to eliminate competition and to preserve the advantage of the superior form. Species of trees compete sharply with each other when found together, as do shrubs, herbs, etc. The relation of shrubs to trees or herbs to shrubs is one of subordination rather than of competition.

However, the matter of height and width often enters to such a degree that the tallest herbs compete with bushes and shrubs, and rosettes with grasses. The number and size of leaves are decisive with the same form, except when the leaves are grass-like and erect, or when form and position are very different, as between grasses and forbs.

The position of the competing individuals is of the first importance. The distance between plants affects directly the degree of competition, while their arrangement in groups or singly determines whether the contest shall be between like or unlike forms. The individuals of species with high seed-production and little or no mobility usually occur in dense stands, and the competition is especially severe, for the two reasons of density and similarity. When the seed-production is low or the mobility great, the individuals occur more or less scattered among those of other species, and the degree of competition will depend upon similarity in form. The effect of distance, i.e., of density, upon competition is fundamental; competition increases as the interval diminishes, and the reverse.

The view that competition is purely physical in nature renders untenable a number of conceptions, such as that of vegetation pressure. Masses of vegetation have been assumed to force the weaker species toward the edge, thus initiating an outward or forward pressure. This movement is nothing but simple migration, followed by ecesis, and has no connection with weaker species or the development of a vital pressure. Migration seems to be outward or away from the mass, merely because the ecesis is greater at the edge, where the dissimilarity between plant forms diminishes competition. The actual movement is outward, but it takes place through the normal operation of competition. In this connection, it should also be pointed out that the view that plants require room is inexact, if not erroneous. This is difficult of proof, since it is impossible to distinguish room as such from the direct factors concerned, but the available amounts of these will determine the space occupied by a plant, with little regard to room in the strict sense. Finally, the explanation of competition in terms of physical factors invalidates the view that plants possess spheres of influence, other than the areas within which they exert a demonstrable reaction.

EXPERIMENTAL STUDIES

Experimental studies with competition as the primary objective are still few in number and of recent date. Field experiments to determine the proper rate of planting for various crops, and thinning experiments in the forest have a much longer history, but even of the most recent of these only a very few have taken competition into account definitely. Of the investigations considered in this section, most have dealt directly with competition as a process, either from the standpoint of ecology in general or with special reference to its application to field or forest crops. One or two are not experimental in the strictest sense, but even these have made use of instrumental measurements to secure objective values.

Clements, 1905—The first endeavor to organize a comprehensive experimental attack upon the nature and effects of competition was begun in

1903 and a preliminary account is given in "Research Methods in Ecology" (1905). This study was based upon the competition culture, which was an extension of the quadrat method to include sowing and planting. The complete plan comprised two methods, that of natural and of artificial habitats. The former differed from the usual quadrat only in that attention was focused on competition, especially in the case of the innumerable natural experiments. The method of artificial habitats embraced denuded and transplant quadrats in the field, and sown or planted cultures under control. Most of the experiments were carried out in the greenhouse by means of control cultures.

Competition cultures are naturally more readily controlled in the greenhouse than in the field, though the results are somewhat less readily applicable to cultivated field and native vegetation. Various kinds of cultures were employed, determined by the number, kind and arrangement of species, time of planting, factors concerned, and so forth. *Simple* cultures consisted of a single species, the resulting group being a family and the competition in consequence between like individuals. These permit the simplest and most direct analysis of the process. *Mixed* cultures comprised two or more species planted at the same time; while more complicated, they were of unique value in disclosing the features to which each species owed its effectiveness. *Heterochronous* cultures were those in which one or more species were sown after the first had appeared, and were designed to reveal the working of the process in the case of invasion into an existing community. In *ecad* cultures plants of different response to water and light were grown together in order to evaluate the part played by the adaptation of the plant. *Factor* cultures were employed to afford a closer analysis of reaction, the area being divided into two or more parts which were given different amounts of water or of light. *Biotic* cultures were likewise organized to determine the handicap in competition produced by the parasitism of *Cuscuta*, rusts and other fungi. Finally, *permanent* cultures were established by permitting the plants to ripen and drop their seeds for several generations, just as in nature.

The cultures employed were all a meter square, and were treated essentially like denuded quadrats in the field, with respect to factor readings, charts and photographs. Germination tests were made of a large number of species, and those selected that showed a high percentage. Care was also taken to select species known to be vigorous growers, with the result that nearly all those used were ruderal or subruderal; practically all were forbs, annuals, biennials and perennials being represented. At the time the cultures were started, controls were sown in pots and the seedlings later transplanted singly to large pots, to be grown as closely as possible under the same conditions of water, light and soil as in the competition plots. The cultures and controls were photographed at proper intervals, and the former were charted as quadrats from time to time to show the course of competition.

This investigation was interrupted by removal from Nebraska to Minnesota, and was not resumed until full time could be devoted to research in 1918. In a new series of studies, competition and reaction were investigated in connection with the ecesis of a large number of grasses and forbs in the several prairie associations ("Experimental Vegetation," 1924). With the completion of this series in 1923, a direct attack upon competition was organized from the three-fold approach of native vegetation, field and greenhouse, and this forms the theme of the present volume. The original study had reached the point where it was evident that competition was purely physical in nature, that water and light were the paramount factors in most habitats, and that competition for "room" was to be understood in terms of the amounts available of the direct factors concerned. ("Research Methods," 1905:313; "Plant Physiology and Ecology," 1907:253).

Bates, 1911—The effect of windbreaks on field crops has been studied by Bates in the Middle West, with reference to their value for protection and the damage done by competition. The reaction of the trees upon the physical factors was determined by means of an extensive installation of instruments, and the effects were correlated with the yield of the various crops. The competitive action of windbreaks was found to be localized in a narrow zone adjacent to the trees, where its influence was unfavorable to annual crops. This resulted in the first instance in a serious reduction of the sunlight, the loss varying from 50 to 125 per cent of the light possible on a zone as wide as the height of the trees and being greatest in the case of north-south windbreaks. It was accompanied by a somewhat greater loss of crops in the shadow zone, this increase being due to the inadequacy of light for seed-production, though it may suffice for vegetative growth. The reduction of the water-content may be even more serious, amounting in drouth years to complete failure of the usual crops. This operates over a zone as wide as one to five times the height of the trees, but the effect naturally varies with both species and situation, and it may be reduced by cultivation. There is likewise a temporary reduction of soil fertility, which is a concomitant of the lowered holard.

The protective reaction of the windbreak is felt over a wider zone ranging from 5 to 10 times the height of the trees. All of the effects are due to the influence of the windbreak upon the circulation of the air; they comprise lessened wind movement, reduced evaporation, greater heat during hours of sunshine and less extreme cold at night. The actual value of any of these will depend upon the height and density of the windbreak and upon the direction, velocity and drying power of the prevailing winds.

Montgomery, 1912—The general course and effect of competition has been studied in field plots of wheat, oats, and corn by Montgomery (1912:3). These were repeated twice with two varieties of the first and once with two of the second. The respective distances of planting for wheat were 1.5″, 1″, 0.5″ and 0.25″, giving 112, 168, 336 and 672 seeds per plot of 32″×36″. It was found that the percentage of survivors decreased gradually with the density, the number in the thinnest stand being a third greater. In the case of plants grown from large plump seed in competition with those from small or shrunken ones, the respective decreases were 28% and 38%, by comparison with 35% and 40% when each was planted alone.

In the competition between varieties, Big Frame wheat yielded 15% more than Turkey Red when they were sown singly in 1908, but about 30% less when sown in competition. Three years later, Turkey Red exceeded Big Frame by 18% when grown alone, but was almost twice as productive in competition. This relation was confirmed by a plot of the latter variety grown continuously for 12 years; at the outset this contained a slight admixture of Turkey Red, but the latter gradually became dominant until in 1911 it constituted practically 90% of the stand. In the case of oats, one variety regularly outyielded another when alone, but in competition as regularly fell behind. When the seeds of two varieties were alternated in competition plots, the mixture of the two gave a greater yield than either variety alone.

To throw light upon the possibility of natural selection in corn, a variety was grown in a series of plots planted with 1, 3 and 5 kernels per hill. For six years each lot was planted continuously at the same rate; in 1911, seed from the densest planting yielded 2.5 bushels more than the medium and 7 bushels more than the thin planting.

Kiesselbach, 1917—In following up competition in field crops, Kiesselbach has carried on extensive studies of wheat, oats and corn for more than a decade (1917, 1922, 1926). The most comprehensive of these dealt with the effect of competition between single-row test plots as a source of experimental error in crop-yield tests, the relative yields of two crops planted in blocks with several rows being regarded as the true relative value. In such tests in 1913, Turkey Red wheat was planted at two rates; the thin rate yielded 68% of the thick when grown in single alternate rows, but 90% as much in 5-row blocks. Competition between similar rows of Kherson oats caused the thin rate to yield 20% too low in 1913 and 34% too low in 1914. During 1914, such competition reduced the relative number of stools per plant for the thin rate to about 20% for the oats and 37% for the wheat.

In the case of varieties, single-row competition between Turkey Red and Big Frame wheat caused the latter to yield 10% too high in 1913 and 12% too low in 1914, owing to differences in season. A similar test between Burt and Kherson oats gave 16% and 37% too high a yield for the former for the respective years. When large and small seeds of wheat were planted in competition in the same row, the small seed in consequence of competition gave 15% too little grain and 20% too little straw. In the case of corn, one variety yielded 45% too low when compared in the same hill, and 22% too low in alternate single-row plots.

When surrounded by corn hills with 3 plants each, 2-plant and 3-plant hills produced 10% and 35% more than 1-plant hills in 1914, and 67% and 102% more in 1917. The average grain yield of a 3-plant hill surrounded by similar hills was 466 gm. in 1914. This yield was increased 2.7%, 5.3%, 13.1%, and 43.1% respectively by the presence of (1) one adjacent 2-plant hill (2) one adjacent 1-plant hill (3) one adjacent blank hill and (4) two adjacent blank hills. A repetition of this test three years later gave respectively 2%, 9%, 15%, and 25%. In normal fields, Hogue's Yellow Dent corn yielded 36.6, 44.6 and 40.3 bushels per acre when grown at the rates of 1, 3, and 5 plants per hill, the respective yield per plant being 0.58, 0.23 and 0.13 pounds. At rates of 1, 2, 3, 4, and 5 plants per hill, the yields were 40.7, 49.4, 52.9, 50.7 and 49.3 bushels per acre, or 0.64, 0.39, 0.28, 0.20 and 0.16 pounds per plant.

Further studies of the effect of competition in the improvement of corn varieties disclosed an increase of 0.6 bushels for Hogue's Yellow Dent when the seed had been grown at the rate of 5 plants per hill, while that grown at the rate of one plant per hill yielded 1.8 bushels less per acre than the seed from the customary rate of 3 plants per hill (1922). Significant results were obtained between the density of seeding and the development of spikes in wheat; at the rate of 3, 5 and 8 pecks per acre the number of spikes per plant was 3.5, 2.7 and 1.8, the number of spikelets 14.2, 13.2 and 12.6, and the number of grains per spike 17.1, 15.9 and 14.2, respectively. Other experiments with more or less bearing on competition dealt with the effect of spring pasturing, of good and poor areas in the same field, and of water-content in the development of spikes (1926).

Rayner, 1913, 1921—In several papers on the relation of *Calluna* to the soil, Rayner has sought to determine the nature of the factors that control its distribution and also to throw light on the calcifuge habit. As shown by pot cultures, *Calluna* grows normally in soil from heather areas, but abnormally in that underlaid by chalk. The abnormal behavior consists of reduced and retarded germination, arrest of shoot and root, and small size and reddening of leaves. Intimately connected with these were colonies of bacteria on the roots and a marked diminution of vigor in growth of the mycorhizal fungus. The evidence points to the conclusion that the relation between *Calluna* and the fungus is obligate, successful growth depending upon early infection and the vigor of the fungus. The soil preference of the ling rests upon the maintenance of a balance between the roots and the micro-organisms about them. The bacterial colonies associated with the roots, especially where the fungus is prominent, are to be regarded either as pathogenic or as indicators of soil conditions unfavorable to the fungus (1913:59; cf. 1911:22).

In a later investigation into the causes of the calcifuge habit in *Calluna* (1922:60), Rayner reached the following conclusions. *Calluna vulgaris* will not grow on calcareous soils; in nature, seedlings germinating on such soils would be eliminated immediately by competition. The inimical factor is chemical, and does not depend upon either the physical or biological features of calcareous soils. It does not affect the fungus when growing outside the plant, but probably does in symbiosis by delaying infection and hampering the symbiotic relation. The nature of the unfavorable factor is not known; it may be a toxic substance, the effect of the H-ion concentration, or a difference in the effective concentration of the soil solution. It is consequently uncertain also just how the pathogenic effect upon the plant is exerted, but the unhealthy condition of the root-cells sooner or later leads to parasitism by the endophyte.

Tansley, 1917—In an experimental test of the assumption of Nägeli as to the soil preferences of reciprocal species (p. 7), Tansley has grown *Galium silvestre* and *G. saxatile* in competition on different types of soil. The former grows in nature on limestone soils, the latter on silicious ones; where both soils occur together each species is strictly confined to its own type. *Galium silvestre* was found to germinate on calcareous soil, sandy loam, and acid peat, most freely on the first and least freely on the last. It also established itself on all the soils, but on the acid peat was reduced to a subordinate position. However, some of the plants maintained themselves on peat in competition with *G. saxatile* for at least six years. The latter species likewise germinated on all the soils employed, but to a smaller degree. The rate was lowest in lime soils, where the seedlings became chlorotic and many died. The survivors which became green were vanquished in the competition with *G. silvestre.* Competition appeared to act through the direct suppression of the shoots of one species by those of the other as a result of more vigorous growth on the proper soil. This seems to afford an adequate explanation of the outcome, especially since no evidence of root competition could be found.

It was concluded that both species can establish and maintain themselves in either soil, for several years at least. The calcifuge *Galium saxatile* is heavily handicapped, especially in the seedling stage, as a direct effect of the calcareous soil, and is thus unable to compete successfully with its calcicole congener. On silicious soil the situation is reversed. If the results prove of general application, they would serve to explain the behavior of similar pairs of species, which are preferent where both occur together and indifferent when they grow alone.

Pearsall and Wray, 1927—As a consequence of a study of the calcifuge habit in *Eriophorum angustifolium* (1927:28, 30), Pearsall and Wray accept three of the explanations already offered as furnishing part of the solution of the problem of soil preference. Rayner's conclusion as to the effect of the endophytic fungi of the root is regarded as of the greatest importance in the case of *Calluna* and probably other Ericaceae. The production of chlorosis in many species by highly calcareous media is thought to have the effect of restricting such species to less calcareous soils. Moreover, as Nägeli, Christ and Tansley have shown, competition is an important factor in the calcifuge or calcicole habit, since in its absence a species may grow well on either soil. Their own cultures are considered to afford a parallel case, in as much as *Eriophorum* and *Molinia* both did well in fen and bog waters, although in nature the former is confined to the bog water and the latter to the fen.

It was concluded that moorland plants possess the ability to thrive under a variety of conditions which are separately unfavorable to most normal plants. When these conditions, low calcium, high pH, and peaty toxins are present together, often also with nitrogen and oxygen deficiency and low soil temperatures, the net result may well be to exclude every other type of plant on one or more grounds. It must be obvious from all these considerations that the problem of the calcifuge habit in plants is one of extreme complexity.

Brenchley, 1917-1919—The effect of weeds upon cereal crops has been investigated by Brenchley, with especial reference to direct competition or a toxic exudate from the roots (1917:53). Pot experiments were carried out with *Triticum, Hordeum* and *Fagopyrum* in competition with *Alopecurus agrestis, Brassica alba, Papaver rhoeas,* and *Spergula arvensis,* the three crop plants also being grown alone as controls. The cultures were harvested for the determination of the total dry weight for each set of six pots and the average dry weight of shoots, after which the soil was sifted to remove the rootlets and then returned to the pots for the second crop.

Since these experiments yielded no evidence that the results obtained were due to toxic excretions from the weed roots, nutrient cultures were utilized to further test this possibility. However, from neither series was there either evidence or indication that a direct toxic action was concerned. It was obvious that the competition of plant with plant, irrespective of species, had much to do with the development, and that the time and duration of the competitive check were the chief factors present. If the weed roots actually excreted a toxic substance, the crop should have grown better in the absence than the presence of the weed. On the contrary, the crop in the mixed culture regularly showed better growth, indicating that the vital factor in competition is the mere presence of other plants.

In a later study, Brenchley dealt with the factors in competition and their effects (1919:142). With a limited supply of nutrients, the total growth as measured by dry weight is determined by the amount of available nitrogen. Even with an abundance of nutrients and water, the reduction in light intensity due to overcrowding is a critical factor in competition. In the case of barley, competition for light reduces the number of ears, causes great irregularity in the number of tillers produced, diminishes the amount of dry matter formed, encourages shoot growth at the expense of the roots, and decreases the ability to make use of nutrients.

Aaltonen, 1923—In a comprehensive discussion of the spatial relations of plants, Aaltonen (1923:7) has considered the general question of space in connection with competition and contributed a very useful account of the earlier results and views in the fields of agronomy and forestry. A summary is likewise given of studies of root systems and absorption, as an introduction to his own investigations of roots, and the paper concludes with the description of his studies of competition in maize, both under control and in the field with two different degrees of holard. The criteria taken into account were fresh and dry weight, average and maximum stature, and the osmotic concentration of root cells.

The author is concerned with the problem of space chiefly from the standpoint of the forest, though his citations and experiments deal mostly with field crops. The former were intentionally restricted almost wholly to the literature in German, which necessarily excluded the treatment of competition and root systems from the ecological approach. A brief account of the earlier researches and opinions is made in the later sections on forestry and agronomy, while here are considered two or three of the author's conclusions. He regards light as a subordinate factor in the forest and states that the struggle for room is decided by the underground and not the aerial parts of the trees. Furthermore, it is the quality of the habitat or forest type that determines the result a certain treatment will produce, and not the species of trees with their various light requirements. While the self-pruning of trees has been universally explained on the basis of shading, this appears to stand in closer relation to the quality of the habitat and the absorption of nutrients. In the case of field crops, it is equally important to keep in mind the effect of the underground parts on each other. The most important factor is not the amount of light available, but the soil-mass and the amount of nutrients accessible to each individual.

Clements and Weaver, 1924—In the further organization of the field of experimental vegetation (Clements 1905:306; Clements and Weaver, 1924:4), the method of transplants was applied through a series of four climatic stations from eastern Nebraska to Colorado and of four edaphic ones in the true prairie at Lincoln. This series of investigations was primarily a problem in ecesis, in which competition was the paramount process. The basic methods employed were as follows: (1) sowing seeds; (2) planting seeds or propagules; (3) planting seedlings; (4) transplanting adult plants of various ages; (5) transplanting small communities or portions of communities. The accessory methods involved the use or manipulation of (1) competition; (2) physical factors; (3) protection against animals; (4) instruments; (5) phytometers; (6) seasons and cycles. Various means were utilized for reducing or eliminating competition and modifying the physical factors, but the following proved the simplest and most satisfactory, viz.: (1) sowing on the surface in the midst of natural vegetation; (2) sowing or planting in trenches, by which competition is prevented for a short time; (3) sowing or planting in denuded quadrats, which eliminates competition for a much longer period, but renders the water relations less favorable at the same time that it improves light conditions; (4) transplanting adult plants into living cover or denuded areas, with similar consequences as to competition, water and light; (5) improving the conditions for germination, establishment or survival by watering, shading, thinning, draining, etc. The differences between seasons and between the wet and dry phases of the climatic cycle were definitely taken into account, since they were often greater than the climatic distinctions between the several grassland associations or even between formations.

Yapp, 1925—Yapp has subjected the concepts of association and competition to a stimulating analysis, which though not actually based upon experiment, is in general accord with such results. In discussing the interrelations of plants in vegetation, he considers the reactions of plants upon their habitats, competition, priority, dependence, mutuality, and independence. He emphasizes competition as the dominant function of the community, and treats in turn its incidence, degrees of competition, meaning of the severity of competition, and the exclusion of plants from certain habitats. He defines priority as the relation in which one organism is so situated that it can intercept and retain all it needs of a direct factor before a second can obtain any of it. Priority is regarded as one-sided, competition as mutual though often unequal deprivation. However, he recognizes that the former may be only a special kind of competition (p. 320).

The interrelations from which individuals receive benefit are termed dependence if the latter is one-sided, mutuality if the resulting benefit is mutual. Dependence is defined as the relation in which one form derives some advantage from its association with another, this being one-sided, not reciprocal. It is further distinguished as direct and indirect, the former being typical of so-called obligate, the latter of facultative dependents. However, the experimental transfer of many species from shade to sun and the reverse for more than a decade indicates that perhaps all shade plants are facultative, if the transfer is made with the proper care (Clements, 1927:307). The term mutuality is applied to the cases where mutual benefit arises from the proximity of plants to one another; it is obviously most important in cases of protection against excessive transpiration and evaporation due to insolation and wind (p. 316).

The author's outlook upon the community as a complex organism is also significant, as the following shows:

"The result is that without any central control of a psychological nature, such as is found in a human community, vegetation tends to assume, in accordance with as yet imperfectly known laws, a definite organized structure. * * * The relative independence of the different organs of an 'individual' plant frequently occasions relations between these organs similar to those which obtain between different plants."

Sukatschew, 1928—From a pioneer study of competition between biotypes, Sukatschew has concluded that the forms derived from three original plants of *Taraxacum officinale* differed greatly in competitive ability. It was found that the conditions as to competition were very different in a pure culture of one biotype from those to be found in a mixed one containing several. The biotype that proves most sturdy in the former case may turn out the weaker in the struggle between two or more biotypes, while likewise the form that succeeds best in survival in open cultures may show itself the poorest in dense ones.

In the competition between biotypes from different regions, grown in open stand in pure or mixed culture, those from climates most unlike the local one appeared less sturdy, though this relation was subject to change under greater densities in pure cultures. As measured by survival in dense mixed cultures, two of the foreign biotypes conquered the local, and one other succumbed. When one takes into account both survival and the ability to produce flowers and fruits, it is perceived that the competitive process can greatly alter the relative expression of the different biotypes. The struggle for existence between dense mixed cultures of biotypes is less intense than in pure stands of a single biotype. However, since a variety of competitive conditions may obtain in a restricted area in nature, it is possible for biotypes of very unlike vigor to thrive in close proximity. Finally, it is stated that an intraspecific struggle for existence rarely takes place in nature, since the usual type of competition is between a number of species belonging to widely different genera and families.

COMPETITION AND SUCCESSION

All detailed studies of succession necessarily furnish materials for competition, but a relatively small number have devoted especial attention to this process. The following account is restricted chiefly to these, together with a few others in which the relation of competition to succession appears clearly, even though this was not the main thesis. Tolerance and the "alternation of essences" might equally well be treated here, but it is reserved for the next section.

Clements, 1904, 1910, 1916, 1920, 1928—The rôle of competition as the directive process in succession, together with reaction which is primarily the outcome of it, has been consistently emphasized from the first organization of this great field in the "Development and Structure of Vegetation" (1904:168, 82, 134). It was stated that the closeness of competition between individuals of different species varies directly with their similarity in vegetation or habitat form, and that this principle is of primary importance in the competition that arises between occupants and invaders in the different stages of succession. On the contrary, the species so unlike the occupants that they enter at a clear advantage or disadvantage, establish themselves readily, in one case as a result of the reaction, in the other by taking a subordinate position. This principle lies at the base of the changes in succession that give a peculiar stamp to each stage. A reaction sufficient to bring about the disappearance of one stage can be produced only by the entrance of invaders so different in form as to materially or entirely change the impress of the community. Stabilization results when the entrance of invaders of such form as to exert an efficient reaction is no longer possible. In forests, while vegetation forms can still enter, none of these produce a reaction sufficient to place the trees at a disadvantage, and the climax stage, though it may change in composition, can not be displaced by another naturally, without a change of climate. It was pointed out that the invasion of a community may occur at three different levels, namely, at the level of the dominants, below this level and above it, depending directly upon the height of invaders and occupants. The importance of the invasion level is fundamental, since it determines the outcome of competition and the whole future of the succession. The principles that govern the process of migration, ecesis, competition, reaction and stabilization were formulated and their application pointed out (Cf. also "Research Methods in Ecology," 1905:236, 285; and "Plant Physiology and Ecology," 1907:251, 271).

In the "Life History of Lodgepole Burn Forests" (1910:42), it was concluded that competition in the various stages of the subsere was for water and light, nutrients and soil-air being of little or no effect. Competition was found to occur between the herbaceous cover and the pine seedlings, between seedlings themselves, and between the species and individuals of trees at all stages of their growth. The keenest struggle naturally occurred where the seedlings were closely crowded in a dense ground cover, such as that of *Vaccinium myrtillus*. The average diameter of seedlings in this cover was little more than half and the height about a third of that in bare areas, while the survival of seedlings more than 15 inches was less than a tenth. The major portion of the discussion pertains to tolerance and is touched upon in the next section.

Since competition and reaction are the motive forces in succession, they recur constantly as the central themes of "Plant Succession" (1916), and even the barest abstract of their treatment is out of the question. The following outline will serve to indicate the processes and structures in connection with which the effects of competition are discussed; the numbers refer to the pages in the original edition of "Plant Succession." When there are two sets of numbers, the second designates the pages in "Plant Succession and Indicators" (1928). The headings and pages are as follows: ecesis 68-71; competition 72-74; invasion 75-78; reactions 79-97; stabilization 98-99; life-history stages 100-103; climax and subclimax 105-110; layers and aspects 115-116, 116-117; units of vegetation 124-140, 125-141; quadrat methods 424-437, 187-197.

The major themes of "Plant Indicators" (1920) were likewise succession and climax, considered especially from the standpoint of indicator values. Constantly interwoven with them are discussions of the relation of competition and reaction, especially in connection with the following topics. The second numbers refer in all cases to "Plant Succession and Indicators": associational basis of indicators 47-50, 253-257; successional basis 51-53, 258-260; life-forms 57-70, 263-277; competition forms 71, 278; communities as indicators 72-75, 279-282; light indicators 79-80, 286-287; lime indicators 84, 292; process indicators 91-98, 299-305; grazing indicators 270-324, 347-402; forest indicators 344-362, 418-437.

Woodhead, 1906—In an analysis of the relations between the subdominants of deciduous woodland in England, Woodhead has distinguished two types of societies (1906:333). The controlling species tend to form a biological unit, so that in the case of the meso-pteridetum of *Scilla, Holcus* and *Pteris* a *complementary* society is constituted. In this the underground parts occupy or tend to occupy definite and different layers, i.e., they are edaphically and the aerial parts seasonally, complementary. On the other hand in the xero-pteridetum, the subdominants, *Vaccinium, Calluna, Pteris* and *Deschampsia,* occupy the same layer owing to the soil conditions. After allowance has been made for differences in requirements, these still form a *competitive* society, and sometimes one, sometimes the other species gains the upper hand.

The complementary relations in the first are due to the position of the storage organs and roots on the one hand, and the time of maximum growth and blooming on the other. *Holcus* roots in the loose leaf-mold, the rhizomes of *Pteris* traverse the deeper humus, while the mature bulbs of *Scilla* occur in the firm loam beneath. *Scilla* forms sheets of blue flowers in spring, followed by *Pteris* in summer, while *Holcus* grows rapidly in late winter and early spring. However, the young plants of *Scilla* must meet the competition of its associates as the bulbs are drawn downwards, and wherever *Pteris* increases in density, the tendency is greatly to reduce the flowering activity of *Holcus*, as shown by the early withering of tips, limited period of growth and the eventual restriction of its distribution (cf. p. 326).

Sherff, 1912—In connection with the measurement of evaporation at different levels in a reed-swamp, Sherff has also studied the subterranean organs of the dominants and subdominants (1912:53). He reaches the following conclusions as to the competitive relations between them. Two or more species may live together without competing with each other when (1) their underground parts lie at different depths and their roots are thus produced at different depths; (2) they make unlike demands on the soil, even though the roots lie at the same depth; (3) the aerial shoots have unlike growth-forms; and (4) when these, though similar, grow chiefly at different seasons. In accordance as one or more of these responses control the composition, the community may be called complementary.

The root depth once determined by various factors for the different species, the several root systems then function in a complementary or competitive manner as the case may be. But even when these are complementary, the community may be competitive because of marked competition between the aerial shoots. Moreover, competition between root systems may render competitive a community otherwise complementary. Finally, species that are complementary in one community may be less so in another. Thus, *Polygonum muhlenbergi* and *Sparganium* are complementary in the Scirpo-typhetum, but in the Sio-polygonetum where their rhizomes lie together near or at the surface of the stream-bed, they are "edaphically" competitive, and hence complementary only as to the aerial shoots.

Jefferies, 1915—In connection with a study of *Molinia coerulea*, Jefferies has discussed its competitive relations to *Eriophorum* and likewise to *Calluna vulgaris* (1915:98). *Molinia* owes its success in competition to an unusual array of structural advantages, namely, its compact bunch habit, its adaptations for rapid transpiration and photosynthesis, effective absorption and storage, its evergreen base, and the large amount of woody tissue. However, it appears to be dependent upon good aeration, and hence yields to *Eriophorum* wherever the water of the moorland becomes stagnant and very acid. In competition with *Calluna, Molinia* regularly suppresses the latter; it actively invades groups of bushes, pushing its way up between the stems. The outcome is indicated by the fact that tussocks of the grass cover old stems and roots of *Calluna*, and that the latter frequently occurs in peat underlying wide stretches of the Molinietum. It thus appears that *Molinia* is a common invader of *Calluna* moors and that much of the present grassland of this type was formerly Callunetum.

Jeffreys, 1917—The explanation of the practically complete absence of other plants under *Pteris* has been sought by Jeffreys in the reduction of the light intensity and the effect of the decomposing fronds. However, it was found that the light value was nowhere less than 1/20, while *Vaccinium* and *Deschampsia* were thrifty under *Calluna* in an intensity of 1/100, and *Deschampsia* and *Holcus* under *Ulex* in one of 1/30. However, *Nardus* was not found where the value was less than 1/3, and *Calluna* but once, and the light reaction seemed decisive in the case of both. Under the bracken refuse the light intensity fell as low as 1/40, but even this was inadequate to explain the absence of the most tolerant species. In the hope of discovering the cause for this, fresh fronds were spread over the various species or dug in about their roots and then allowed to decompose.

From the results obtained, it was concluded that the chief reason why no other species are able to grow under a dense stand of *Pteris* is that the refuse of the latter is injurious to them. Whether this injury is due to the dense shade or to the toxic effect of the decomposing fronds depends upon the particular competitor. The following conclusions were reached with respect to these: *Ulex* is destroyed by the shade; *Calluna* is destroyed by the toxic effect, with the influence of shade undetermined; *Nardus* is killed by either factor separately, *Deschampsia* by the toxic substance alone, and *Holcus* is not much affected by either.

Salisbury, 1916, 1924—Salisbury has dealt with the general course of competition in a number of woodland societies in England (1916:14, 83; 1924:1). He emphasizes the fact that species can only grow together without competition where their demands upon the soil are different, their roots situated at different depths, or the leafy shoots produced at different seasons. Thus, *Adoxa, Anemone nemorosa* and *Ficaria verna* root at about the same level and their shoots are likewise competitive, while the root system of *Mercurialis* lies at another level than those of the first and last. *Rubus fruticosus* is an almost exclusive subdominant, even overcoming the persistent *Pteris,* which is at a disadvantage by virtue of the remarkable rapidity with which the former spreads by means of both seeds and offshoots. The *Pteris* society is also poor in species, owing primarily to the heavy shade that it casts.

Competition is regarded as less a matter of the distance from other plants than a question of requirements and capacity for aggression. The plant of low stature is at a disadvantage with tall ones and the one with a light canopy yields to that which casts a dense shade. Competition for light is largely between shoots of the same kind and height, the social habit of woodland species being a means of protection against taller competitors and of aggression against lowly forms. It was suggested that shade species may be better fitted in a positive sense to grow in shaded situations, or that woodland may be a sort of plant slum into which they are driven by the pressure of their more vigorous competitors (1924:4-7).

Farrow, 1917—In his studies of the vegetation of Breckland, Farrow has emphasized the importance of height and rapid growth in connection with success in competition, as well as the influence of the rabbit coaction in determining the outcome (1917:155). The effects of root competition and soil factors, though possibly very great, were regarded as chiefly important because and in so far as they affect the ability of shoots to grow taller and more luxuriant, and thus to overshade their competitors and finally exterminate them. This was illustrated by the suppression of *Festuca ovina* by *Agrostis vulgaris,* owing to the more rapid growth of the latter when both were subjected to an increased water supply.

In analyzing the competition between *Pteris* and *Calluna* or *Carex,* it was observed that the dying and dead plants of the latter usually occurred within the edge of the *Pteris* society. An examination of the respective root systems showed that those of *Pteris* were at an average depth of 25 cm., those of *Calluna* at 5–15 cm. and of *Carex* at 7 cm., thus indicating that they were not directly competitive. However, the light intensity was reduced respectively to 1/60 and 1/96 of its value in the open, and this was in accord with the fact that numerous instances were found in which half of a *Calluna* was killed by the dead fronds, while the uncovered half remained alive. It was also noticed that the bracken fronds even produced this effect before they began to decay, in which cases it is unnecessary to invoke a toxic action. Similar results occurred in the case of *Carex,* the plants on which fronds had fallen becoming etiolated and dying.

By preferring some species to others, rabbits confer an enormous advantage upon the latter, especially *Pteris,* which is only slightly attacked. The existence and dominance of communities of *Carex arenaria* and *Calluna* ultimately depend upon the attack by rabbits injuring the taller species sufficiently to permit the lower ones to suppress them in spite of their potential stature. An intensity of rabbit attack just sufficient to kill the seedling trees allows the relatively dwarf *Calluna* to become dominant, a somewhat greater intensity destroys the *Calluna* and turns the area over to the smaller *Carex,* while still greater attack removes the latter and permits its replacement by the dwarf grass heath. Were it not for the greater damage to the taller species, these would replace the lower and all would finally give way to woodland.

Watt, 1919, 1924, 1925—In his studies of oak and beech woods in Britain, Watt has devoted much attention to the suppression of seedlings and the competition between the trees. Under *Pteris,* seedlings 4 years old were but 6″–9″ tall in comparison with 18″–30″ in the open; these were obviously suppressed by the shade of the fronds, since the decaying fronds remained suspended on the netting and could exert no toxic effect on the seedlings. Seedlings may also be crushed to the ground by the weight of bracken when the latter is bent down by heavy rain or wind (1919).

The presence of ash in exposed situations is dependent upon association with beech. When the two occur in intimate mixture, ash is surpassed at about 35′, so that the transition from scrub to the beech consociation is rapid. With increasing depth of soil, ash competes more vigorously and holds its place until it attains a height of 70′. The accumulation of superficial deposits increases the competitive ability of the oak, but after a certain depth of loam is attained soil texture is more important than depth in favoring the oak. Thus, the retentiveness of the upper layers for water supplies the index to the relative competitive abilities of oak and beech on these soils.

In the absence of animals, the success of sycamore (*Acer*) in competition with ash is determined by its greater tolerance of shade. The relative ability to endure shade is indicated by the fact that *Sambucus nigra* reaches a height of 10′–12′ under ash, of 3′ under mixed ash and sycamore, while it is absent from the pure consocies of the latter. Where beech, ash and sycamore occur together, it is probable that the ash will to a large extent be replaced by sycamore, which will continue as a rival to the beech, largely owing to its tolerance, seed-production and mobility. This association suggests the maple-beech climax of the eastern United States.

Summerhayes and Williams, 1926—The ability of *Epilobium angustifolium* to maintain its position against the competition of *Pteris* has been studied by Summerhayes and Williams (1926:207). The former had covered extensive areas, which were being invaded by the bracken, the ecotone between the two communities being sharply defined. On examining the fern community, many individuals of *Epilobium* were found in it, their great stature making it difficult for *Pteris* to suppress them. Although *Epilobium* starts first, the bracken soon overtakes and shades it until the shoots finally push through the layer of fronds. However, it is probable that reduced nitrification due to the increased acidity under *Pteris* has a greater effect than the shading, the nitrifying process being less active under the bracken than in the *Epilobium* outside.

COMPETITION IN THE FOREST

The direct investigation of competition as the primary process in forest development has been infrequent and incidental. On the other hand, there have been numerous instances of the classification of trees on the basis of the outcome of competition, grounded upon observation for the most part and summed up in the term tolerance. A detailed account of the development of such classifications has been given by Bühler (1922) and by Lönnroth (1925), while Zon and Graves (1911) have contributed a general discussion of tolerance from various angles, and Aaltonen (1923, 1925) has stressed root relations especially.

Light measurements in the forest—The light intensity below the forest canopy is the primary outcome of competition and the resulting dominance of certain individuals or species. It constitutes the most striking reaction of the forest community, though not the only significant one, as has too often been assumed. The measurement of light in plant communities began with the pioneer investigations of Wiesner, and with more or less modification his methods have been in general use. His papers appeared from 1893 to 1907, and their effect is to be found in similar studies by Hesselmann (1904), Wagner (1907), Clements (1910), Pearson (1911), Cieslar (1914), Burns (1916), Bates (1917), and others. Zederbauer (1907, 1908) was the first to question the value of the photographic method, and to attempt to determine the changes in the quality of forest light as a result of absorption by the leaves. Knuchel (1914; cf. Sponsler, 1915) carried out a more extensive study of the transmission and absorption of light by leaves; he concluded that a hardwood canopy modifies the light essentially as do the leaves of the species composing it, while the effect of softwood crowns was negligible (cf. Zon and Graves, 1911:49; Burns, 1916:4; Clements, 1916:92, 1920:79, 349; 1928:92, 286, 423).

Fricke, 1904—The current views as to the nature of tolerance were dramatically challenged by some simple experiments in nature that were devised by Fricke (1904:315). These were all based upon the elimination of root competition on the part of older trees by means of a ditch 10″ deep, which severed the roots of neighboring trees to this depth. The installations were made in a stand of *Pinus silvestris* 70–100 years old, in which there were suppressed groups of young trees 10 years old. The ditches were dug in the spring and the effects were observable the following summer; the terminal shoots grew longer and the needles were twice the length of those of the preceding season. The original ground-cover changed from a meager to a luxuriant stand and in it appeared a number of species not found outside the trenched area, among them seedlings of birch and mountain ash. In another stand of pine on poor soil, natural openings without trees or young growth were enclosed by ditches and then sown with seeds of pine, spruce, beech and oak, without preparation of the ground. The same seeds were also sown in similar areas without ditches. The results were strikingly different, both as to germination and growth, the number and vigor of the seedlings being much greater within the enclosure.

Fricke determined the water-content in the areas with and without the root competition of older trees and found the amount usually 2–3 and occasionally 4–6 times greater in the latter. From this he concluded that deficient holard due to competition was the sole cause of the death of young pines under the parent trees. Zon and Graves (1911:17) point out that this conclusion is not wholly warranted since no measurements were made of light intensity, but readily agree that the importance of water relations in tolerance had been demonstrated (cf. Pearson, 1928, p. 29).

Clements, 1910—In a study of the relation of reproduction to competition in the case of lodgepole pine, it was stated that all the evidence obtained indicated that competition was wholly a question of available water and light (1910:42). In soils actually depleted and those poorly aerated, nutrient or air relations play a part in the process. In natural habitats, however, such as those of the lodgepole forests, where all the nutrient salts taken from the soil are ultimately returned to it and where the loose texture of the gravel permits thorough aeration, these factors are negligible. Competition in all such places is narrowed to the question of the total amount of water and light present, and the amount of each available for any particular individual in consequence of the presence of neighbors.

In many cases of competition there can be no doubt that the amount of both water and light available for a competing group is less than normal development requires. Competition is then for both and in more or less equal degree. In the majority of cases, however, while the total of either factor is inadequate, the deficiency is greater in the case of one of them. The same result is obtained when the size, form, or arrangement of the individuals is such that they react upon one factor more than upon the other.

This is typical of lodgepole communities. The seedlings have little or no effect upon the light available for each until they have attained a height of 1–2 meters, and then only when they are crowded. On the other hand, they compete vigorously for water whenever the roots enter the same area. Owing to the usual density of seeding, such competition is the rule in all recent burns, especially since the water-content is relatively low. While this competition becomes keener as the seedlings develop into adult trees, the inability of lodgepole to endure any material reduction of light intensity seems to give paramount importance to this factor in all stands above 20 years of age.

Burns, 1914, 1927—The most complete and detailed study of tolerance from several points of approach has been made by Burns, much of it under controlled conditions. This has included the development of white pine seedlings in nursery beds (1914), the relation of shade to evaporation and transpiration in nursery beds (1914), discontinuous light in forests (1916), minimum light requirements of forest trees (1923), and leaf efficiency in white pine seedlings (1927). Among his conclusions, the following in particular relate to tolerance. The term tolerance is generally taken to express a light relationship, but it should refer to the total relation of a tree to all the factors of its habitat. The light filtered through the leaves is of little value in photosynthesis; it is the reduced white light that is of primary importance. The times at which light readings are usually taken and the variation in intensity from spot to spot make it evident that much of the data is untrustworthy. Hence, it is concluded that light readings in the forest are of little value, and that the relation of tree growth to light can only be determined by controlled greenhouse and nursery trials.

Tolerance expresses the response of a tree to all the factors of the site and any adequate study of it requires that each factor be given its true value. Light is one of these factors and each species studied has a specific requirement for its minimum light intensity. It seems clear that the development of leaves is fundamentally influenced by the water-content, and the density of crowns should be associated in all probability with this factor rather than with the light intensity. A dense crown does not so much indicate a "tolerant" tree, as it does the amount of water that the latter had for its development.

Pearson, 1923, 1928—The relative effect of the protection and competition arising from the herbaceous cover upon the reproduction of yellow pine (*Pinus ponderosa*) has been discussed by Pearson (1923:39). The handicap due to competition for water and light is offset in some degree by the protection afforded against insolation, wind, and frost in particular. Pine reproduction is strongly handicapped in the densest areas of the bunch-grasses, but where these cover about half the surface, the effects are not critical. Some seedlings are undoubtedly shaded out, but root competition is the vital feature, especially in September, after the rainy season and while the grasses still make heavy demands. The final effect is largely determined by the abundance of the summer rainfall, the mortality being high after dry summers, but rarely severe after a wet season.

Pearson has tested Fricke's results by trenching two plots of yellow pine seedlings generally shaded after 10 a.m.; these were distinctly subnormal in size, color and length of needle. The ditches were dug to 15″ in 1923 and deepened to 18″ in 1924, practically all the tree roots being severed. The soil was replaced to prevent excessive drainage and drying, and the trench was opened in 1924, 1925 and 1927 to cut out new roots. Observations taken in 1924 and 1925 showed no response to the trenching. Each seedling was tagged, and its height and condition recorded during the second season, and measurements were again made in 1928 after height growth had been completed. Although most of the seedlings had died, the survivors were in fairly good condition, but they were growing only half as fast as those of the same age in more open situations.

Hofmann, 1924—In a study of the natural regeneration of Douglas fir (*Pseudotsuga taxifolia*), Hofmann has prepared a list of the common and most important competitors of this species, to the number of forty-four. Practically all of these were trees and shrubs, *Pteris aquilina* and *Epilobium angustifolium* being the significant exceptions. Nearly all these competitors endure more shade than the seedlings of this tree and are able to vanquish them when rapid growth in height does not prevent. A dense young stand of Douglas fir usually succeeds in eliminating all its competitors. The chief advantage of this species over its associates, *Tsuga* and *Thuja,* resides in the ability to establish its seedlings during the first season of growth, owing to its larger seed and greater storage and the consequent ability to develop a much deeper root system early in its life. The seeding age is also an important factor in competition; both the knobcone and lodgepole pine produce seed at 10 years of age or even earlier and hence are able to replace the climax dominants by pure stands, when fires are sufficiently frequent.

Bates and Roeser, 1928—For the purpose of testing the existence of a tolerance of shade and the difference between species in this respect, Bates and Roeser grew seedlings of twelve coniferous species under artificial light for 10 hours a day for a period of 9 months. The seeds of each were sown broadcast over a circular table in such a manner that seedlings would develop at all distances from the central source of light. The total energy below the lamp was about 1/8 that of sunlight, while it fell to 1/160 at the extreme edge of the table. The results showed that there were wide variations in the ability of the different species to utilize weak light of the quality furnished. By far the most efficient was *Sequoia sempervirens* with a minimum of less than three-fourths of 1 per cent, while it increased its original size nearly ten times in light of 10 per cent. *Picea engelmanni* and *Pseudotsuga taxifolia* came next in order, requiring about twice as much light for appreciable growth. The six species of *Pinus,* including such different types as *ponderosa* and *strobus,* were in fairly close agreement in needing four to five times the light intensity for the redwood, while the piñon (*Pinus edulis*) required still higher values. The growth curves for the Colorado and Arizona forms of the yellow pine, which belong to the same climax, were quite similar, but not for the Coast and Colorado Douglas fir, which occur in very different climaxes and climates. A few individuals of several species far excelled the remainder in their ability to grow or survive in weak light, though this was thought to be a matter of greater initial size and vigor rather than of intrinsic differences.

Phillips, 1928—The comprehensive ecological study of the African *Olea laurifolia* by Phillips (1928:169) is valuable in supplying further evidence of the importance of light in forest competition. Cultures of equivalent young seedlings in the same soil were placed under intensities respectively of 1/40, 1/200 and 1/500 of full daylight. The temperature and humidity were practically the same under the three screens employed, and the average water-content was maintained at the same value in each, so that the only effective factor difference was that of light. The mean diameter of the stem increased practically a third from the weak to the medium and from the latter to the stronger light, and the dry weight and ash-content rather more than twice as a rule. The palisade tissue of the leaf was twice as thick in the strong as the weak light, and the mesophyll two and a half times as thick.

The experiment was repeated with the same technique, but the seedlings were placed under dense *Hemitelia capensis* and in the same community but with *Hemitelia* removed. The respective light values were 1/800–1/1000 and 1/80–1/100. The dry weight was again much greater in the stronger illumination, while the ash-content was nearly fifteen times higher.

Toumey, 1928—Tolerance has been defined by Toumey (1928:56) as the capacity of a species or variety for growth and survival under a natural canopy. It is determined by several factors, of which light is often only a subordinate one, and it varies more or less with each of these, especially water-content and temperature. When root competition is eliminated in dense stands of white pine, the bare forest floor develops an abundant ground cover. Trenched quadrats in such areas may exhibit a water-content 2–9 times higher than the checks during critical periods, and they clearly demonstrate that the absence or suppression of a species is due to the reaction of the trees upon the holard and other soil factors, and only in a secondary degree to light. He has also shown that the nearly complete cessation of growth in white pine under a dense canopy of grey birch is not caused primarily by a shortage of light. On the other hand, he regards the poor development of tree seedlings and herbaceous vegetation on the forest floor as an outcome of the weakened and reflected light that reaches them through the crowns, and concludes that we are far from an adequate understanding of the rôle of light in the forest.

He also reports experiments of Grasovsky (1929), in which seedlings of conifers and hardwoods were grown in pots under illuminations of 10,000, 300, 65 and 25 foot-candles. After ten months all the seedlings in the 300 value showed growth and were in good condition; those under 65 foot-candles ranged from good to dead, while those in the minimum intensity were dead, with a single exception. These results indicate that only a narrow margin above the minimum suffices to keep such seedlings growing and that increases in the light value do not produce a proportional effect.

COMPETITION IN CULTIVATED FIELDS

Competition in field and garden crops differs in two important respects from that in natural communities. It is regularly between the individuals of one species or variety, and hence of the type found in simple or pure cultures. In the second instance, the habitat is controlled in as large a degree as possible in favor of the crop. This means not merely maintaining the water-content as nearly as possible at an optimum by means of tillage, irrigating or draining, but also the nutrient-content by rotation or fertilizing, and to some extent the soil-air as well. In spite of cultivation, however, weeds do enter the situation, and in actual practice field crops run the entire gamut from pure or nearly pure stands to those with mixed competition arising from the advent of one or more species of ruderals. Naturally, the complexity of the process may be further increased by the sowing of mixed crops. It is further varied by the method of sowing or planting, the presence or absence of cultivation, the nature of the life-form, whether annual or perennial, herbaceous or woody, etc.

Earlier Studies

Sachs, 1860—The first definite study of competition in crop plants was probably that of Sachs (1860:1; cf. Aaltonen, 1923:18), who sought to determine the effect of soil-mass on yield. He grew buckwheat in flower-pots, planting 6 in one and 12 in the other at equal distances. At the outset both grew with equal rapidity, but a difference soon developed, and at maturity the two cultures were strikingly unlike. The open culture produced large vigorous plants with many branches, large deep-green leaves and many flowers, while the stature was hardly a third as great in the denser stand, the habit pinched and the leaves small and pale. Sachs stated that the better development in the open culture was not a result of stronger illumination, since the larger plants shaded each other more than the smaller ones did. Moreover, he thought the explanation did not lie in the water relations, but rather in the fact that the nutrient-content, while large enough for 6 plants, was inadequate for twelve.

Zoller, 1867—In cultures of dwarf beans, Zoller (1867:193) showed that with respect to yield it was much the same whether 2, 3 or 4 plants were grown in 3.5 l. of soil. In fertile soil the larger number gave the better yield, while the smaller produced a relatively lower one. In ordinary soils larger and smaller numbers were more or less similar in yield, the smaller sometimes giving the better results. In consequence, it was concluded to be profitable to use a larger amount of seed in fertile than in poor fields.

Wollny, 1881—Extensive investigations were carried on by Wollny (1881:25, 217, 493) to determine the relation of density or rate of planting to the amount of seed used and the method of sowing. He likewise found that the yield rose with the soil-mass to a certain limit, progressively at first and then gradually diminishing. With respect to amount of seed and nature of soil, the more fertile the latter, the smaller the amount of seed needed for the maximum yield. With an equal quantity, the greater growth and consequent crowding brought about suppression with its attendant ills. However, the more the crop development is affected by climatic or seasonal extreme, the greater the number of seed to be sown.

As to manner of sowing, the space for each individual should be as equal as possible, and consequently drilling is more advantageous than broadcasting. As to the distance between rows, there is an optimum spacing for each seed quantity in terms of yield, and the latter falls whether the distance is increased or decreased. The same statement holds with reference to yield and spacing in each row. These results likewise indicate that in fruitful soil greater distance between rows and smaller amounts of seed should be employed than in poor.

As to the relatively higher yield with a smaller soil-mass for each individual, Wollny sought the explanation in the better utilization of the soil nutrients and water, since the roots came into more intimate contact with the soil particles. However, in denser stands the reciprocal shading acts disadvantageously, since the roots as well as the leaves develop more poorly in consequence. A very high density also works unfavorably upon the soil temperatures, and especially upon the water-content. The latter effect was confirmed by studies of four crop plants, each grown in three densities. These bore no constant relation among themselves or between the several species, but in spite of this the lowering of the holard was progressive in every case. The greatest reductions were found in the case of lentils and potatoes, a density somewhat more than twice for the first depressing the water-content from 20% to 12% approximately, and one of five times for the second decreasing it from 19.8% to 10.6%.

Hellriegel, 1883—The importance of light in the competition of field crops was so much appreciated by Hellriegel that he regarded them as shade plants to a certain degree. Except for the first stage before the stand is closed, and the last when it becomes more open by the death and drying of leaves, the individuals constantly shade and handicap each other. In consequence, the amount of light falling upon the surface of a field and the best utilization of its energy constitute the first and most important factor to be considered.

Hellriegel also devised a number of pot cultures in order to determine the significance of soil-mass, finding that variations in the latter always exerted an influence upon production. The larger or smaller the soil-mass, the larger or smaller in general was the yield. He also found that an unusually large volume of soil permitted a certain degree of overgrowth, and that complete utilization occurred only after the mass was decreased to a certain point. The relation of the number of plants to the total production of each pot proved less significant, when the amount of soil was equal. For 1, 2, and 3 plants, the total production in round numbers for dry weight was respectively 33, 31 and 31 gm.; for 4 and 6, 39.5 and 38.9 gm., and for 8, 12, 16, and 24 plants respectively 41.8, 41.5, 41.2 and 41.6 gm. From these and similar results was drawn the conclusion that soil-mass or "growth-room" was a factor in production, a view contested by Mayer.

He enunciated the axiom that each plant requires for its existence and development a certain amount of room, qualifying this by saying that this has no practical significance for the aerial parts, since long before these can crowd and hamper each other spatially, competition for gases and light becomes critical. Roots on the other hand do not possess such freedom of development, though in the field there is relatively much room for them by contrast with conditions in pot experiments.

Mayer, 1879—Mayer (1879:249) agreed that room might be regarded as a production factor to the extent that a plant grown in a small pot or in too close proximity to another was under a handicap as to absorption by comparison with one with adequate space for development, water and nutrients being however the critical factors. He also pointed out that by contrast with the other factors light was not at a minimum in field crops and hence was not the decisive factor in crop production. However, in actual practice where the individuals grow too thickly and overshadow each other, it does hold the final decision as to the productivity of a field. He emphasized the importance of utilizing the sun's energy as completely as possible by planting at the optimum density in graduated stands. In connection with the later discussions of the competitive relations between tall and short grasses, it is most interesting to find that he suggested growing a mixture of "upper grass" and "ground grass" for the more complete utilization of the total energy.

Other studies of soil-mass and density—Lemmermann's investigation of maize and barley by means of sand cultures corroborated the conclusions of Hellriegel (1905:1, 279), but the use of water cultures led to the opposite result, namely, that space had no essential influence and could not be regarded as a production factor. In tests of six crops grown in water cultures of varying concentration, Mitscherlich (1916:346) found that in general the production was decidedly higher with a larger soil volume, though in the case of hemp this effect was less pronounced. In his researches into the significance of water and light as production factors, Pfeiffer (1912, 1915, 1918) has reached the conclusion that light operates as such, a reduction of intensity decreasing the yield. Reciprocal shading affects assimilation on the one hand and transpiration on the other, the combined effect being to lower growth. On the contrary, he found that the soil-mass was without significant effect on the yield of oats and carrots, and that only as a matrix for water and solutes was it of importance.

With respect to density, Cserhati (1890:349) determined that in moist soil thick seeding gave a larger return than thin. In general, the individuals were smaller the thicker the seeding, but the compensation of a greater number brought about a larger harvest. Weiser and Zaitschek (1913:49) have likewise shown that with favorable weather conditions denser planting gave the larger production, while in a dry summer the thinner yielded better returns, thus serving as an adjustment to dryness. Wollny (1881), Seelhorst (1899), and others have also found that density affects the height and thickness of the stem, the length of spike and amount of grain, the time of blooming and maturing, the chemical composition, etc. Widely spaced beets contain less sugar but more nitrogen and ash than those sown more closely, while the spacing of potatoes is well known to have a marked effect upon size and yield.

Later Studies

As has been indicated earlier (p. 14), Montgomery (1910, 1912) was the first to study competition in cereals with more or less reference to the process itself, and Kiesselbach (1917-1926) has carried out by far the most comprehensive series of investigations in this field, several of which have already been abstracted (p. 15). Most of the investigations have naturally had practical ends in view and hence have concerned themselves chiefly with density or rate of planting. Corn and the small grains have necessarily received the chief attention, but practically all the annual crops have been passed in review. At Illinois, Hume, Center and Hegnauer have carried on extensive experimentation with respect to the distance between hills and the number of plants in each hill (1908). Montgomery (1910, 1912) has determined the effects of different densities with oats, wheat and corn, and Kiesselbach has taken rate of planting into account in his comprehensive investigation of competition in the same varieties as well as in barley.

Stewart (1919, 1920) has studied the spacing of potato plants and in particular the effect of missing hills upon the yield. Brown (1924), Musgrave (1926), and Livermore (1927) have also carried on similar experiments with the potato. Atkinson and others (1919), Hansen (1920), Putnam (1920), and Vinall (1923) have worked out the methods and rates of seeding for sunflowers as a silage plant. Sieglinger (1923) has made studies of the optimum spacing for milo and kafir, McKeever (1924) and Mooers (1928) for cotton, and Beattie (1927) for peanuts. The effects of adjacent rows and especially the error in yield arising from the dominant plants of the outside rows of plots have been fruitful objects of experiment. Investigations in this field have been made by Kiesselbach (1917 et seq.), Hulbert and Remsberg (1927) and Stringfield (1927) for small grains, by Cole and Hallsted (1926) for kafir and milo, Garber and Odland (1926) for soybeans, and by Musgrave (1926) and Livermore (1927) for potatoes.

A number of researches during the past decade, notably those of Howard (1915, 1916), and of Hole and Singh (1916), have dealt with the effects of deficient aeration, and hence have a bearing upon competition for soil oxygen. All of these have been discussed at considerable length in "Aeration and Air-content" (1921), and do not require further mention here, since none were directly concerned with competition. They do serve, however, as a background for the whole problem of toxic secretions and soil toxins, and their possible rôle in competition. In spite of the excellent work done by Bedford and Pickering (1914) in this field, the existence, nature and rôle of supposed toxic substances are still subject to grave doubt (Clements, 1921), and much more extensive ecological research in various climates and soils will be necessary to a solution.

The entire course and outcome of competition furnish positive proof that crops do have definite and more or less permanent effects on soil factors. Such reactions may persist for a year or even much longer, producing a sort of "belated" competition of much interest and importance. They have been intensively studied by Hartwell and his associates for a decade (1918, 1919, 1927), and more recently by Garner, Lunn and Brown (1925), and by Robertson and Kezer (1927).

METHODS OF EXPERIMENT

The methods employed for the experimental study of competition are based primarily upon those developed from 1900 to 1905 and described in "Research Methods in Ecology." These have been supplemented by additional ones devised in connection with the study of ecesis ("Experimental Vegetation"), and in particular by the phytometer method (Clements and Goldsmith, 1924). They differ in most important respects from the earlier experiments of Sachs, Wollny, Hellriegel and others in that these were carried out in pots as rule, with the consequence that shoot relations were usually better than in the field and root relations regularly worse. Instrumentation had barely been thought of at this time and phytometers were wholly unknown. This was equally true of the quadrat method, with the possibility of employing unit cultures under the widest range of conditions.

Equally significant has been the utilization of a complete series of cultures from wholly natural conditions in grassland through the partial control afforded by field plots to the completer control of garden and greenhouse. The latter has rendered possible the use of laboratory methods for the measurement of stomatal behavior, transpiration, root pressure, conduction, etc. The sequence between the extremes has been continuous, and it has thus been possible throughout to analyze prairie and plot results in terms of the laboratory and to check determinations in the latter by referring them to outdoor cultures under more natural conditions. Of the first importance have been the steps taken to secure replication of experiment and culture and the confirmation of results. Not only have cultures been installed in duplicate or triplicate, but many have also been repeated the following season. The same species has been employed in a number of combinations, and frequently in more than one community, and the most important ones through field, plot, and greenhouse. Moreover, each life-form has been represented by a number of species, and the cultures and series for both species and life-forms have been consistently repeated, with a corresponding increase in comparative values.

The detailed technique for the various types of culture is described in the chapters concerned, and it is necessary here merely to call attention to some of the new devices for analyzing the course of competition or segregating the factors. Important among these are level and insert phytometers, artificial shoots or leaves for regulating the degree of shade, cylinders for preventing root competition, manipulation of shoots and leaves to increase light, various types of underplanting, factor sequences, etc.

2. TRANSPLANT CULTURES IN SUBCLIMAX PRAIRIE

Subclimax or low prairie—This community occupies the broad lowland valleys of the larger streams in the true prairie association. It is a part of the extensive subclimax prairie found to the eastward within the climatic region of the deciduous forest and represents its persistence in areas where the holard or water-content of the soil is high. It is characterized by tall-grasses, often 6′–8′ high, of which *Andropogon furcatus, A. nutans, Panicum virgatum* and *Elymus canadensis* are the chief. With these occur regularly two mid-grasses[1], *Andropogon scoparius* and *Bouteloua racemosa,* dominant in both subclimax and true prairie, while such dominants of the latter as *Stipa spartea, Sporobolus asper, Agropyrum glaucum* and *Koeleria cristata* appear in varying abundance, especially in the ecotone. *Spartina cynosuroides* is usually present as an extensive pure consocies, which constitutes the final stage of the hydrosere in the low prairie. Where annual mowing is the rule, *Poa pratensis* invades in force and assumes the rôle of a dominant, as a result of the high holard and the increased light intensity (Clements, 1920).

The area of low prairie employed in the present studies has served for experimental work in vegetation since 1918. It is situated in the flood-plain of Salt Creek just north of Lincoln, Nebraska, and is about 60′ lower than the high-prairie station. The soil is a fertile dark silt-loam of the Wabash series, fine in texture because of the preponderance of silt and clay. This is correlated with a high water capacity, which is about 70% for the surface foot (Hilgard method), and by high hygroscopic coefficients, which are 10%–12% at the several levels down to 4′. There is no trace of acidity with the Truog test, and the volatile matter and nitrogen-content are more favorable to plant-growth than in the high prairie (cf. "Experimental Vegetation," p. 42). The portion of the flood-plain along the base of the hills is being gradually built up, and the subsoil, though constantly moist, is never water-logged except at great depth. This permits the growth of such coarse herbs as *Vernonia baldwini, Helianthus rigidus, Solidago rigidus, Liatris scariosa, Kuhnia glutinosa,* and *Petalostemon candidus.* However, these are all suppressed in some degree by the overshading produced by the tall-grasses, and the societies formed by most of them reach the best expression on the high prairie. The low light intensity also explains the absence of short-grasses, though these appear here and there, where close grazing keeps down the tall-grasses and correspondingly increases the light values. The characteristic pioneers of the subsere in lowlands are *Ambrosia trifida, Iva xanthifolia* and *Amarantus retroflexus,* especially in bare areas due to flooding.

[1] The distinction between tall-grass and short-grass is not sufficiently exact, and the term mid-grass has been used for some time to include *Stipa, Sporobolus, Agropyrum,* etc., which are intermediate between *Andropogon* and its allies, and *Bouteloua* and *Bulbilis.*

Methods—An extensive experience with the germination of grass seeds in nature made it desirable to devise new methods for competition cultures in the prairie (Clements and Weaver, 1924). These were based upon the greenhouse method of sowing in flats and later transplanting in the field, the essential difference consisting in the transfer of the entire community. The flats employed were 14″×14″ square and 4″ deep, with removable bottoms. These were filled with a fertile silt-loam, well-screened and settled into the flats by repeated watering for several weeks before sowing. This also permitted the germination of the seeds already in the soil and the eradication of the seedlings. The actual seeding of the cultures was done on May 1, 1924; much care was taken to plant the seeds of the various species at the optimum depth, the flats receiving attention twice each day. By May 16–17 the seedlings had developed an even stand 1″–4″ high, and the cultures were then transported to high and low prairie respectively. Here a trench was cut in the native sod sufficiently wide and deep to receive the soil-mass; the grass on the high prairie had been mown the preceding fall and the low prairie had been burned in the spring. The bottoms of the flats were next removed and the culture held together by the mass of roots was pressed firmly into place. Sections of sod were then cut and fitted across the trench in order to separate the cultures by a distance of 4″–6″. The surrounding vegetation was disturbed just as little as possible, and all the surplus soil and debris were removed.

Immediately after transplanting, a strap-iron frame 14″ square and divided by strong cords into 100 inch-squares was employed to define the competition quadrat, the frame being fastened at the level of the soil. This outlined the cultures permanently and made it possible to count the number of individuals at various periods, as well as to secure an accurate record of the course of competition in terms of growth, tillering, dominance, suppression and mortality. In order to insure typical conditions, only the central 100 square inches of the total 196 was counted. The technique of counting or charting in such seedling groups entails even greater care and accuracy than in charting meter quadrats. There were often as many as 10 seedlings per square inch, and it sometimes required as much as two hours to record a single quadrat. The record itself was made in such a manner that the number of individuals of each species, and whether they were dominant or suppressed, could be followed for each square inch of the culture (plate 1).

The cultures were watered from time to time after being transplanted, until they were well established. In fact, during the drought of June 1925 the quadrats of 1924 were suffering so much that enough water was applied to them on both high and low prairie to equal a 2″ rainfall. Watering again became necessary in 1926, when the prairie vegetation actually began to wilt, a phenomenon of rare occurrence. In the competitive groups as well as in the controls, the seedlings of all other species were removed as they appeared. The unavoidable trampling about the quadrats retarded the growth of the native prairie in some degree and reduced the competition

Competition culture of *Elymus* and *Panicum* in low prairie.

A. Seedlings in inch units, June 6, 1924.
B. The same culture on June 23.
C. Growth in late summer, August 8.

from the latter, so that the plants grew better about the margin than in the center of the culture.

Kinds of cultures—The series of competition cultures was identical for the subclimax and true prairie with respect to the life-forms concerned. This was based upon the plan of duplicating the process of natural competition, but in such fashion as to permit an effective analysis. The mixed composition of the prairie rendered it desirable to group competitors in accordance with their rôle in the community, leaving the competition between individuals of a single species for control cultures. As a consequence, dominant was made to compete with dominant, with subdominant, or with ruderal, and subdominant was pitted against subdominant. The general sequence was as follows: (1) dominant against dominant, e.g., tall-grasses, *Andropogon furcatus* with *Panicum virgatum,* on low prairie, and mid-grasses, *Andropogon scoparius* with *Stipa spartea,* on high prairie; (2) dominant tall- or mid-grasses against short-grasses, e.g., *Agropyrum glaucum* with *Bulbilis dactyloides;* (3) dominant with subdominant, e.g., *Andropogon nutans* with *Oenothera biennis;* (5) subdominant with subdominant, e.g., *Liatris punctata* with *Kuhnia glutinosa;* (6) dominant with ruderal, e.g., *Andropogon scoparius* with *Amarantus retroflexus.*

The experiments were continued through four seasons, until the close of 1927. In 1925, a second series of cultures essentially identical in character were installed in high and low prairie to permit the study of differences in behavior due to annual variations in climatic factors. Duplicate or triplicate plots of each species or group were grown in prairie or under cultivation to determine the rate of growth without competition and for the study of root habits and interrelations. In such duplicate quadrats the various species were sown more or less densely as pure stands. These are referred to hereafter as control cultures.

Readings of holard and light intensity were made at frequent intervals both in and around the various quadrats, and continuous records were obtained of temperature, humidity, and evaporation, and often of wind movement as well. The hygroscopic coefficients and hence the approximate non-available water or echard had already been determined for the different soils and stations. Bisects were employed to exhibit the root and shoot relations; together with the factor readings they revealed the course of events in detail with respect to dominance, suppression and disappearance. The record was completed by means of photographs taken at the time of quadrating.

Root relations of dominants—The functions of the root play such an important part in competition as to warrant a detailed consideration of root development in the various dominants. In each case the seed was sown at the same time as that in the competition quadrats. The plants employed for root excavation were grown in cultivated soil free from weeds, and in most cases the growth of the roots was compared with that of individuals grown in control quadrats in the prairie.

Elymus canadensis underwent practically the same development on both high and low prairie, though the seedlings were somewhat larger in the latter. By August 2, after three months' growth, the plants were about 6″ tall and bore 1–3 tillers each. The fine well-branched and widely spreading roots had reached a working level of 18″ and a maximum depth of about 30″. The mature root systems are not so deep as those of the other dominants, but the roots are more finely branched than in any other, with the possible exception of *Andropogon nutans.* Although the laterals on the spreading roots seldom exceed 2″ in length, they are rebranched to the third and fourth degree (Weaver, 1919).

Panicum virgatum had reached an average height of 21″ on July 31, each plant possessing about 3 tillers, 3–4 rhizomes a half inch long and approximately 14 roots. The working level was 19″ and the maximum depth 3.4′, while the spread at the 1′ level was 14″. None of the branches had roots more than 2″ long and no branches occurred at a depth below 2.4′. In the low prairie the subsoil was wet, especially below 31″, and the root system was less extensive. None went deeper than a foot and many of the new ones ended at about 7″. Unlike the vertical roots of *Spartina,* those of *Panicum* spread 5″–6″ in all directions when but half a foot deep. By midsummer of the second season the roots had attained a working level of about 2′, many penetrating more deeply. The roots were very coarse and lacked the profuse network of branches characteristic of upland grasses (fig. 1).

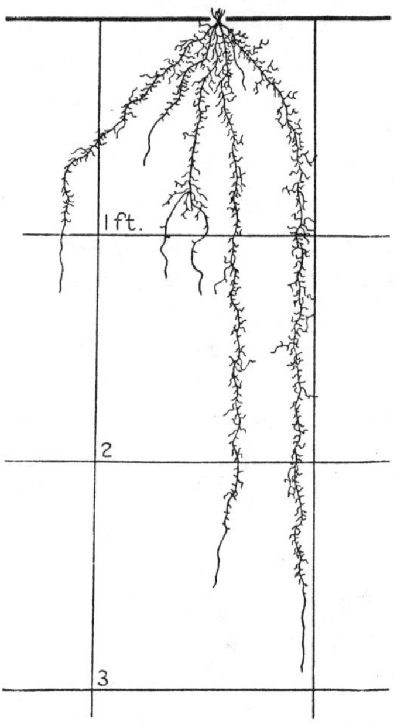

FIG. 1.—Root system of *Panicum virgatum* near end of first summer

Andropogon furcatus in cultivated soil had developed primary roots varying from 14″ to 40″ in depth by July 30, the general working depth being 20″. Many of the roots spread considerably, often 8″–11″ laterally at a depth of a foot. Seedlings in the low prairie possessed a less vigorous root system, chiefly as a response to the higher water-content, but partially also because of lower light values. During the second season the roots reached a working level of 2′ or more. The mature roots are coarse and although better fitted for absorption than those of *Spartina,* are poorly branched by comparison with upland species.

Andropogon nutans had reached an average height of a foot in the cultivated area by August first. As a rule the plants exhibited 3–4 tillers,

2 rhizomes half an inch long and about 19 roots. The latter spread widely in all directions to the extent of about a foot at the 12″ level, the maximum distance being 16″. This spreading habit is significant in the competition of seedlings for water in the surface soil. The general working depth was 19″, but many roots extended to 21″ and some even to 34″. On the high prairie at this time, seedlings 8″ tall and with 2–3 tillers enjoyed a similar working level and maximum penetration. The mature root system reached a depth of 4′–5′, but it is not as extensive as for the other dominants. However, it is more finely and abundantly branched, and succeeds better than its associates in dry habitats, as experiments have shown (Clements and Weaver, 1924).

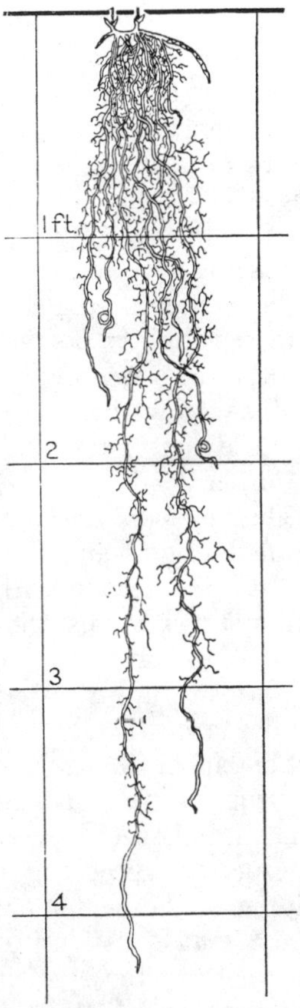

Fig. 2.—Root system of *Spartina cynosuroides* near end of first summer.

Spartina cynosuroides was excavated on July 29, when the plants had an average height of 21″, but only a rare tiller. The roots grew vertically downward, with practically no lateral spread near the surface (fig. 2). The primary roots had reached a depth of 3.5′–4.4′, while the secondary ones were 16″–24″ long, the working level falling at about 2′. Average plants exhibited about 12 roots and 3 short rhizomes up to 4.5″ long. On the low prairie at this time, the fairly open stand in the control quadrat had a height of 17″ and the deepest roots were found at 26″–36″. The older plants had very coarse tough roots, less well supplied with absorbing laterals than those dominants that also grow on the high prairie (Weaver, 1920).

DOMINANT VERSUS DOMINANT

The competition cultures of this group comprised the five tall-grass dominants of the subclimax prairie, including *Spartina,* which is often to be regarded as the final consocies of the preceding successional stage. The particular combinations of dominants were as follows: (1) *Elymus canadensis* vs. *Panicum virgatum;* (2) *E. canadensis* vs. *Andropogon nutans;* (3) *Andropogon furcatus* vs. *Spartina cynosuroides;* (4) *A. furcatus* vs. *Panicum virgatum;* (5) *A. furcatus* vs. *Sporobolus asper.* In 1925 a second culture was started of the first and fourth combinations, for comparison with the initial one.

Elymus canadensis vs. Panicum virgatum

First culture, season of 1924—On June 4, a month after sowing and two weeks after transplanting, the quadrat contained 442 plants, an average of 4.5 per square inch. Of these, 264 or 60% belonged to *Elymus* and 178 or 40% to *Panicum;* the former appeared slightly in the lead, its individuals being 2″–3″ tall with 3 leaves, while those of the latter were 1″–2.5″ tall with 2–4 leaves. By June 27, *Elymus* was represented by 242 plants, a decrease of 8%, and *Panicum* by 367, an increase of 106%, thus reversing the percentage at the first listing. The one averaged 4.5″ tall, with some 7″; it was slightly poorer than the control, but was producing tillers freely. The other also was of the same height, as well as smaller than the control, but it had few tillers, while many new seedlings had appeared. The sod was quite dense; few plants had died, but the lower leaves of both species were often dead (plate 1).

On August 7 the number of plants was found to have decreased 12%, *Elymus* losing 51 and *Panicum* 22; the first with 191 individuals constituted 36% and the second with 345 plants 64% of the stand. However, as to dominant individuals, the one led with 156 or 57%, the other exhibiting but 118 or 43%, since the seedlings that appeared late were all suppressed. The former were 8″–9″, the latter about 10″ tall. The soil was covered to a depth of 1.5″ by a layer of dead leaves and plants of both species. By September 8, the number of plants had dropped to 189, less than 2 per square inch, of which 90 or 48% were *Elymus* and 99 or 52% were *Panicum.* Of the dominants, 52 belonged to the one and 38 to the other, the respective percentages being 58 and 42. The grasses were mostly dead or dying in the center of the quadrat, but were green around the margin, where *Elymus* was most abundant. The latter was also taller as a rule and seemed clearly to be getting the upper hand. The sod was dense, competition severe, and little or no tillering had occurred since August 4; a single *Panicum* came into flower in the autumn.

The end of the first year's competition between two grasses approximately equal in number found *Elymus* somewhat in the lead. Its more rapid growth in height and its ability to grow in lower water-contents apparently gave it a decisive advantage, which was enhanced by its finer and more wide-spreading root system.

Season of 1925—*Elymus* started growth much earlier in the spring, producing abundant shoots 4″–5″ long with 3 green leaves by April 7, when *Panicum* had not yet appeared. On May 28 there were 58 plants of the former, of which but 16 were suppressed, while all of the 18 individuals of the latter were much reduced. The spring drouth had been especially detrimental to *Panicum;* the culms were not over 2″ tall and were greatly attenuated, usually with but one leaf. *Elymus* endured the drouth much better, largely because of its more effective absorbing system, but partly also on account of its earlier start. Its average height was 12″ with as many as 5 leaves per plant.

The approximate chresard of the quadrat was determined on May 26 and again on June 3, following the very dry spring. The values were less at all levels than in the surrounding prairie, especially from 0″–24″, the layer in which *Elymus* carried on absorption.

TABLE 1—*Chresard relations of Elymus-Panicum quadrat*

Depth	Quadrat		Prairie
	May 28	June 3	June 3
	%	%	%
0 –0.5′	2.8	6.3	12.1
0.5–1	10.9	11.7	17.1
1 –2	15.1	15.0	25.7
2 –3	18.4	19.8	22.3

By July 15, *Elymus* had produced spikes at a height of 20″–25″, as a consequence of good growth. The quadrat was still fairly open and the shade cast by this grass on August first at a distance of 3″ above the soil was weak, the light intensity being 63% of full sunlight. *Panicum* had still further decreased in number by August 17, only 6 plants persisting. All of these were much suppressed, without tillers and usually with but 2–4 leaves to a stalk, the tallest only 8″ high. *Elymus* was very dense, except for one or two thinner areas; it had tillered so abundantly and spread so much that the individuals could not be distinguished. The average height of the flower stalks was 20″, the width of the stem 3 mm. The leaf layer on the soil was a half-inch thick and the lower leaves in it were beginning to decay.

Season of 1926—On April 15, *Elymus* was already 6″–8″ tall with broad leaves, while *Panicum* had barely reached 4″. By June 5, the general level of the former was 18″; it was the tallest and best developed of all the grasses, owing in part to its early start. *Panicum* occurred but rarely, and only on the periphery, with a single exception. It was half as tall as *Elymus* and formed not more than 1% of the stand. The light intensity at this time was 5%–7% at the surface of the soil, but the lower leaves of the rye-grass were still green, indicating its tolerance of shade. The density of the stand was such as to hinder water-loss from the soil surface to a marked degree. During two hot days from June 5–7, the evaporation in the open from moist soil in cans 2.5″ across was 35 cc., while from similar containers under the mulch in the quadrat it was only .1 cc. The loss from a white porous-cup atmometer at a height of 8″ within the quadrat was but 42 cc. as compared with 98 cc. in the bare area. At this time the light intensity at the same level was 33%.

On June 30, the leaves of both species were rolled as a consequence of excessive transpiration. The temperature 4″ above the soil surface was 104° F., at 1″ above the surface but under the leaf mulch 105° F., and at an inch below the surface, 83° F. The chresard at the several depths down to 2′ was 3%, 7.4%, and 8.9% respectively. These conditions rendered it necessary to water the quadrat to insure survival. By August 2 this had raised the water-content 5% by comparison with the unwatered prairie, though little change occurred at greater depths. The amounts available were 8.1%, 8.0%, 8.2%, 8.8% and 12.9% at the respective depths to 4′. After watering, the well-lighted leaves of dominant individuals gave a photosynthate rate 60% higher than that of the suppressed.

By August 2, *Elymus* had produced spikes profusely at a height of 24″–32″, some of them 5″ long. The basal leaves on the more vigorous culms were largely dead to a height of 6″, and light penetrated the culture fairly readily, so that *Panicum* was also thriving in some degree Shade was less important as a factor than the water-content, since practically all the plants were dominants. The light intensities were 68% 23% and 20% at the respective heights of 24″, 15″ and 6″. By September first the few stalks of *Panicum* were flowering at a height of 21″–29″, though it still formed but 1% of the stand, while *Elymus* exhibited 24 spikes on stems 26″–33″ tall.

Second culture, season of 1925—This duplicate quadrat was started a year later to permit determining the effect of annual variations of climate on the outcome of competition. Unfortunately, *Panicum* germinated so poorly that on June 9 there were but 27 seedlings of this in contrast to 263 for *Elymus,* and these were clearly suppressed. On July 16 there were 244 plants of *Elymus* and 26 of *Panicum.* The latter were hardly visible as they were little more than an inch in height. The average size of *Elymu* was 5″, though it was denser and taller around the edges, the maximum being 9″. The shade was not dense, and it seemed clear that wate was the decisive factor in the competition. However, by August 19, when there were 212 and 17 individuals respectively, *Elymus* averaged 7″ in heigh and formed a dense stand. The light was correspondingly reduced an played its part in the suppression of *Panicum,* which averaged but 3″.

Season of 1926—On April 15, *Elymus* was 3″–6″ tall and formed dense thrifty stand, while *Panicum* was sparse and but 2″ high. By th first of June the quadrat was completely dominated by *Elymus* with a average height of 13″ and a maximum of 18″. *Panicum* was represente by a few suppressed plants with slender stems and but 2–3 small leaves there were also a few reduced plants of the rye-grass. The more rapi growth of the latter and the greater sensibility to winter-killing of th former were the deciding features of the process. *Panicum* had entirel disappeared by August 2 in consequence of the drouth; the growth c *Elymus* was rather thin with a general level of 16″ and 10 flower stalk with a height of 21″–29″.

Comparative behavior and factors—The decisive success of *Elymus* i competition with *Panicum* was due primarily to its earlier start and mo rapid growth, and in some degree to its better germination. These qualiti enabled it to overtop its competitor, as well as to take more comple possession of the soil. There is also some evidence that it is more tolera of shade, but further studies are needed to confirm this. It was also le sensitive to winter-killing, a fact in harmony with its boreal origin which in sharp contrast to the tropical derivation of *Panicum.*

The behavior of these two grasses in competition cultures furnish the explanation of the differences in their behavior in nature. *Elymus* much more abundant and important in the northern portions of the subclimax prairie and along the valleys that extend westward into the subclimates with less rainfall. On the other hand, *Panicum* is more important to the southeast, where the climate is more humid as well as warmer. In the local area, *Panicum* usually grows in dense but scattered clumps in the moister spots, and is much less aggressive than *Elymus* in taking possession of roadsides and other disturbed places. It appears probable that the vigor of the embryo and the food storage in the seed may play a part in this.

Elymus canadensis vs. Andropogon nutans

Season of 1924—On June 4 the quadrat contained 482 individuals, of which 258 belonged to *Elymus* and 224 to *Andropogon*. The former was beginning to tiller and appeared the more vigorous. By July 8 the total number was 580, but *Andropogon* had increased to 415 and *Elymus* had dropped to 165. The latter was 4″–10″ high and was tillering well; the former was 1″–8″ tall, with few tillers, but both were thriving. A month later *Elymus* had fallen to 157 and *Andropogon* had risen to 445 out of a total of 602 plants; 270 or 61% of the latter and 49 or 31% of the former were suppressed. The one was slightly taller, averaging 9″, while the other was but 7″, and the leaves were narrower. The culture was very dense and many plants of the rye-grass died after reaching 6″–8″, but death in the goldstem usually occurred below this height. There was little tillering, but many individuals had produced new shoots after dying to the base.

On September 4 the total number had been reduced to 285; 65% of *Elymus* had died, leaving 55 plants, while the mortality in *Andropogon* was 48%, giving 230 survivors. Of the latter 145 individuals were suppressed, of the former but 11; the one averaged barely 7″, the other about 12″ in height. Many of the dominants had died, except for a single shoot or leaf from the base.

Season of 1925—On April 7, *Elymus* was 4″–5″ tall with 3 leaves per plant, but *Andropogon* had not yet appeared. By May 28 there were 40 plants and 69 stalks of the former, with but 7 suppressed, while for the latter, 29 were suppressed out of 33. The effect of drouth upon *Andropogon* was so marked that this species could be seen only on close inspection, except for a single dominant individual with 10 stalks. Most of the plants were less than 2″ tall, in contrast with *Elymus*, with an average of 12″ and a maximum of 15″.

On June 3, the chresard in the quadrat and in the surrounding prairie was as indicated in the following table:

Table 2—*Chresard relations of the Elymus-Andropogon quadrat*

Depth	Quadrat	Prairie	Excess in prairie
	%	%	%
0 –0.5′	3.3	12.1	8.8
0.5–1	10.6	17.1	6.5
1 –2	14.2	25.7	11.5
2 –3	20.9	22.3	1.4

The much greater demand made by the culture upon the water-content of the upper 2 feet is strikingly indicated by the excess in the prairie. Most of this demand was exerted by the *Elymus* dominants, as a consequence of which *Andropogon* was steadily dwindling. Light evidently played but a minor part in this, since the midsummer intensity at 3″ above the soil and under *Elymus* was 52% of full sunlight. On July 15 the stand was 14″ high and consisted mostly of rye-grass, which had headed at a height of 20″–23″.

On August 17 but 8 plants of *Andropogon* remained; these averaged 6″ in height and were in poor condition. On the contrary, *Elymus* was very thrifty and had tillered so abundantly that the individuals could not be distinguished. Most of the quadrat was covered with a dense sod and the leaf mulch was one-half inch thick. The light intensity beneath the rye-grass was but 2%–5%, which was too low for growth. The average height of the flower stalks was 18″, the maximum 24″; the leaves were broad and the spikes robust. *Elymus* alone was visible in the quadrat at this time.

Season of 1926—On April 15, *Elymus* was 6″–8″ high with broad leaves, and by June 5 it had made a very vigorous growth, with a general level of 18″. *Andropogon* was sparse and greatly attenuated, none being over 10″ high. It consisted chiefly of single stalks, which constituted not more than 1% of the stand, and had assumed an entirely subordinate rôle. At this time the light value just above the soil surface was less than 10%, but above the mulch it was 25%. The evaporation at the level of 5″–7″ was 45 cc. for a 2-day period, as compared with 98 cc. in the denuded area.

On August 2, *Elymus* was flowering abundantly at a height of 18″–24″ and most of the spikes were of normal size. The stand was quite open and light was still less of a factor, since the lower leaves were dead to a height of 6″–8″, owing to the drouth. The plants of *Andropogon* were scattered and small, not exceeding 10″ and usually consisting of but a single stalk. This condition persisted into September at a time when *Elymus* was developing many tillers from the base. At the end of the third season the latter was in practically complete possession, only 1% of the quadrat consisting of suppressed *Andropogon*.

Comparative behavior and factors—The course of the competition between *Elymus* and *Andropogon* and its outcome were essentially similar to the results from the *Elymus-Panicum* cultures. Like *Panicum, Andropogon* is of southern derivation; hence, it begins growth later in the spring and ripens late in the autumn, with the consequence that it is more susceptible to early freezing and probably to winter-killing as well. It is more robust than *Panicum* and has a more efficient root system, though hardly equal to *Elymus* in this respect, especially in the upper levels. The decisive advantage of the latter was derived from its early start and rapid growth, which enabled it to obtain the lion's share of the chresard and also to overshadow its competitor.

In its climax relations, *Andropogon nutans* further resembles *Panicum virgatum* in being more important southward, from Kansas to Texas. I

Sporobolus and *Andropogon* are both genera of southern derivation and in consequence there was little difference in their response to winter conditions. Winter-killing was most severe in the case of suppressed plants, which were almost twice as numerous in the latter. As to the rainfall relations, there is a difference of 5″–10″ beween the subclimax and true prairie. The competition results make it plain why *Sporobolus asper* is a dominant of the true prairie, and conversely its climatic relations explain in part why it gained the upper hand in low prairie over the chief dominant of the latter, as a consequence of deficient rainfall.

ANDROPOGON FURCATUS VS. PANICUM VIRGATUM

Season of 1924—On June 4 there was a total of 414 plants, making an average of 4 to the square inch. At this stage of development it was practically impossible to distinguish the two species. By July 9 the number was 491, of which 204 were *Andropogon* and 287 *Panicum.* The latter were 7″–9″ high, fully as well developed as the controls and like them had begun to tiller. The best individuals of the bluestem were 5″–6″, but many were much shorter; by comparison with the control they exhibited a certain amount of suppression.

By August 7 the total number of plants had been reduced to 455, of which 206 were *Andropogon* and 249 *Panicum;* 128 or 62% of the former and 141 or 57% of the latter were suppressed. The average height of the dominant individuals of *Panicum* was 11″; that of the suppressed plants and of those of *Andropogon* was about 9″. The leaves of the former had a better color, but few tillers had developed on either species. On September 4 the total was found to be reduced to 335, 167 of *Andropogon* and 168 *Panicum;* since July 8 the latter had met a loss of 119 or 41% by contrast with 37 or 18% for the former. In spite of this, the panic-grass was in the better condition, as only 98 or 58% were suppressed as compared with 131 or 78% of *Andropogon.* The dominants of *Panicum* averaged 16″, while those of the former did not exceed 10″–12″. The sod in the quadrat was dense, and there were many dead and dying plants. In this quadrat the largest number of plants occurred on July 8, the average being 5 per square inch; at the final count on September 4, this had been reduced to a little more than 3, evenly divided between the two species.

Season of 1925—No growth had appeared by April 7, but on June 16 there were 48 *Andropogon* and 8 *Panicum,* a total of 56 by contrast with 335 at the close of the previous season. The plants of the former were 3″ tall with 2–5 tillers, those of the latter about 5″ with 2 tillers. Winter-killing had been so severe that most of the old plants did not even sprout; neither species was in good condition, but *Panicum* had been harmed more than *Andropogon.* The quadrat was very open and there appeared to be little competition above ground.

On August 18 there were 14 plants of *Panicum,* 7″–9″ tall and in as good condition as those of *Andropogon.* The former was much better around the edge of the culture; the plants were taller, with more leaves and thicker culms, and had developed two panicles. *Andropogon* averaged 10″ in height, and had tillered so much that the individuals could not be discerned. This ability to tiller profusely contributed largely to the success of this species in competition.

Season of 1926—By June 5 the general level of the stand was about 12″, with the tallest plants of *Panicum* slightly higher. Nearly half the surface was bare, the ratio of individuals in the remainder being 9:1 in favor of *Andropogon*. The latter had tillered profusely, but the panic-grass, though apparently vigorous, had but one or two stalks per plant. The appearance of the leaves near the soil in light of 5%–10% indicated that the bluestem functioned better in low values. On the other hand, the photosynthate values for the leaves of dominant individuals of *Panicum* were less than twice as great as for those of suppressed ones. On August 2 both grasses were thriving, and there had been no change in dominance. The general level of *Andropogon* was 13″, that of *Panicum* slightly greater, the maximum reaching 21″. September showed no appreciable change in the grasses, except that both were slightly taller. Owing to the drouth of summer, neither had produced flower stalks.

Comparative behavior and factors—The course of the competition between these two species of the subclimax prairie brings into relief the delicate adjustment that usually exists between the dominants of an association. Furthermore, it indicates how the balance is thrown in one direction or the other by seasonal factors, and thus furnishes the approach to experiment in the broader field of edaphic and climatic relations. *Panicum* germinated better and grew more rapidly the first year; at the end of the season the dominant individuals were twice as many as for *Andropogon* and they were a third taller. While winter-killing reduced the total from 335 plants in September to 56 in June, it was much more severe upon *Panicum*, which comprised but 8 out of this number. The toll taken by the next winter was insignificant, as would be expected from the fact that the plants were older and hence in better condition to withstand both winter temperatures and water-losses. Even with its greatly reduced number, *Panicum* held its own in vigor and height-growth, but was much less prolific in tillers.

Thus, the success of *Andropogon* in dominating the culture was due in the first place to its greater ability in resisting winter-killing, which gave it the advantage of numbers the second season, and its capacity for producing tillers, which increased this number greatly in the second and third summers. However, winter-killing was the decisive factor in overcoming the lead that superior germination and growth had given *Panicum* the first year. The evidence as to the relative tolerance of these two dominants was not conclusive, though it suggests a slight advantage for *Andropogon*. Similarly, the root systems are about equally efficient, though there is some indication that *Panicum* grows best in a somewhat higher holard, and this is confirmed by the fact that it is more abundant in the more humid part of the subclimax prairie.

ANDROPOGON FURCATUS VS. SPARTINA CYNOSUROIDES

First culture, season of 1924—On June 4 there were 289 plants, of which 226 belonged to *Andropogon* and 63 to *Spartina*. The former averaged 1″–2″ high with 5 leaves, the latter 1″–4″ with 2–4 leaves. By June 27 the total was 308, 236 *Andropogon* and 72 *Spartina*. The one averaged 4″ in height, was beginning to tiller and was fully as good as the control. The other averaged 3.5″, the maximum being 10″ or 4″ less than the controls, which were otherwise more thrifty as well. On August 7 the number was found to be only slightly lower, namely 301, of which 235 were bluestem and 66 cord-grass; 111 or 47% of the former and 39 or 59% of the latter were suppressed. The average height of the first was 11″, of the second 12.5″.

By September 8 there were 259 plants, 218 of them *Andropogon* and 41 *Spartina*. The mortality in the month preceding had been 7% in the one and 38% in the other; on the other hand 106 or 49%, and 11 or 27% respectively were suppressed. The average height for *Andropogon* was 14″, one less than the check; the dominant individuals of *Spartina* ranged from 15″ to 29″, while the controls averaged 30″. Although the plants of the former resembled the control very closely, those of the latter were not only much shorter, but the stems and leaves were also much narrower. *Andropogon* had tillered fairly well, *Spartina* but little; the sod was dense and the former seemed to be in control.

Season of 1925—By April 7 only a few isolated plants of *Spartina* had appeared, and on June 16 there were but 4 left of the 41 counted the preceding September. These averaged 7″ tall and bore only 2–4 leaves each. The quadrat was now clearly dominated by *Andropogon*, which had produced so many tillers that it was impossible to distinguish the individual plants. On July 1 the chresard in the quadrat was as follows: 1.2% at 0″–6″, 11.1% at 6″–12″, and 14.2% at 1′–2′. These values were but 1–2% lower than in the adjoining prairie.

On July 15 there was a good growth of grasses to an average height of 9″, but the stand was rather thin on one-half the quadrat. On August 17 there were 6 plants of *Spartina;* these averaged about 20″, but one very thrifty individual at the edge was 31″ tall. *Andropogon* averaged 10″ in height and bore abundant tillers, with the consequence that the shade was very dense and correspondingly unfavorable to the lower leaves of *Spartina*.

Season of 1926—The spring was very late and cold, and neither species had resumed growth on April 15. By June 3 only three plants of *Spartina* had developed, each about 18″ tall; drouth seemed the clear explanation of its failure. The stand of *Andropogon* was open but good, the average being 10″. The soil was protected from drying-out by a loose mulch of dead leaves about 1″ thick. The light value just above this mulch was 10%, and was largely responsible for the increasing suppression. The relative rate at which photosynthate was produced by dominant and suppressed plants was 3:1, the respective values being 0.3082 gm. and 0.1100 gm. per 100 sq. cm.

The 3 individuals of *Spartina* grew to 25″–29″ in height by August 2, but they produced little effect upon the quadrat as they had only 3–4 stalks each. On account of drouth the culture was watered from time to time and *Andropogon* made a fairly dense and uniform growth in consequence, to an average height of 12″.

Second culture, season of 1925—In the spring these species were again sown in a competition culture, which was seeded more thinly than the previous one. On June 13 there were 167 seedlings of *Andropogon* and 14 of *Spartina*. The former were 3″ tall with 3 live and 1–2 dead leaves; they were thrifty and had produced 1–2 tillers, some of them 2″ long. The latter averaged 3″ tall with 3 leaves per plant. On July 16 there were 147 of the bluestem and 8 of the cord-grass. The first averaged 6″ tall, ranging from 3″–13″, and bore 2–5 tillers per plant. The second was 4″–8″ tall, with 1–2 dead leaves per plant and possessed no tillers. The rapid deterioration of *Spartina* during the month was due to its poor absorption and to its intolerance of shade.

By August 19 the number of *Andropogon* was 150 and of *Spartina,* 12, the increase arising from dormant seeds that germinated after rain in July. The former had tillered considerably and formed a dense stand 8″ high; the leaves were broad, the shade heavy and a thin layer of dead leaves covered the soil. *Spartina* was best developed along the edge of the quadrat, where it was 8″–10″ high, but more or less attenuated and with few leaves.

On June 27 the light intensity on the ground in the quadrat was 10%–14%; on August 1 it averaged 13%.

Season of 1926—On April 15 the few individuals of *Spartina* present were 2″–3″ high. By June first, *Andropogon* entirely dominated the quadrat with an average height of 9″, forming a dense and fairly uniform stand. *Spartina* was still represented by a few individuals, to be found only on close inspection. On August second it could no longer be discovered, while *Andropogon* had made an excellent growth throughout the quadrat.

Comparative behavior and factors—From the outset *Spartina cynosuroides* was handicapped by its very poor germination, which was but 30% and 9% of that of *Andropogon* in the two cultures. This placed it at a critical disadvantage in the competition and doubtless rendered the outcome more decisive than if they had started on more nearly equal terms. On the other hand, it appears probable that the poor germination was as natural a response as the relatively less effective absorption, to soils drier than those to which it is accustomed and that similar results are to be expected in nature. Like other plants that grow readily in water or wet soils, the relation of germination to water and air differs materially from that of mesophytes. This applies also to the water relation of seedling and adult, the high holard producing a root system with fewer branches and hence less capable of absorption, especially in drier soils. The hydrophytic character of *Spartina,* its tall growth and its habit of forming dense pure consocies indicate its intolerance of deep shade and support the experimental results to the same effect.

On climatic grounds, *Andropogon* might well be expected to suffer more from winter-killing than *Spartina,* but this effect is usually more a matter of winter drouth than of actual cold. In consequence, *Spartina* in dry soil or in a dry season would be more susceptible, and this is indicated by the mortality. The loss was 31 plants during the first summer and 37 in the following winter and spring. A definite comparison with *Andropogon* is impossible because of the abundant tillers of the latter.

The experimental study of *Spartina* in competition with *Andropogon* confirms the conclusions drawn from its successional behavior in the field. It is not to be regarded as a dominant of the subclimax or low prairie in spite of its frequence and abundance within the general area of the latter. It grows typically in the edge of sluggish streams or ponds, and in water-logged or wet soil, rarely occurring in moist soils except in dry seasons. As already stated, it is properly the final consocies of the hydrosere leading to subclimax or true prairie.

TALL-GRASS VERSUS SHORT-GRASS

The competition in this series is essentially that of dominant against dominant also, but one of the competitors is a short-grass, which does not belong climatically to either the subclimax or the true prairie, though it occurs in both. It is properly a relict of a former dry phase during which the mixed prairie moved eastward in consequence of a pronounced climatic change. Such relict communities have sometimes been greatly extended as a result of overgrazing. The cultures utilized *Andropogon furcatus* as a tall-grass dominant of the subclimax prairie and *Agropyrum glaucum,* a mid-grass dominant of the true prairie in competition with a short-grass, *Bouteloua gracilis.* Owing to the scarcity of seed, this species was not available for the duplicate quadrat installed the second spring and hence its close relative, *B. hirsuta,* was substituted.

Root relations—*Agropyrum glaucum* possesses a fairly fine root system that develops rapidly; plants only 3 months old had begun to tiller and produce rhizomes. The working level was at 2′, but some of the roots extended more deeply. They do not spread widely, penetrating the soil more or less vertically and reaching depths of 5′–7′ in mature plants. In comparison with *Andropogon furcatus* they are better adapted to dry soil, though they are not as well fitted for absorption as the two species of *Bouteloua,* since they do not occupy the soil so thoroughly.

Bouteloua gracilis and *hirsuta* develop similar root systems, which are very fine, minutely branched and ramify throughout the soil. They likewise develop very rapidly, plants 1.5 months old bearing roots 10″ long. Up to this time the seedling has a single well-branched primary root as a rule, but with the production of tillers the secondary root system appears and soon extends into the moist soil below the surface. Until this is accomplished, drouth has a critical effect. By late summer the wide-spreading roots extend well into the second foot and some end in the third.

Andropogon furcatus vs. Bouteloua gracilis

Season of 1924—On June 4 there were present 195 seedlings of *Andropogon* and 54 of *Bouteloua,* ranging from 0.5″–2″ high. By July 9 the former had increased to 272, while the latter had lost 5, leaving a total of 49. *Andropogon* averaged 5″ high, was thriving and exhibited many new plants 1″–2″ in height. *Bouteloua* was 4″–5″ tall and, though not abundant, was vigorous, tillering and forming dense small mats. By August 8 the bluestem had decreased to the extent of 15 individuals and grama had increased by 5 plants. Of the total of 257, 103 or 40% of the former and 18 or 33% of the latter were suppressed; the respective heights of the dominants were 9″ and 7″. On September 4, *Andropogon* was found to be reduced 19% or to 209 plants of which 36 were suppressed; *Bouteloua* had lost 6, leaving 48 with but 4 suppressed. The maximum heights were respectively 14″ and 7″. *Bouteloua* was thriving generally, but was best around the edge of the quadrat, while *Andropogon* was better toward the interior, perhaps owing to the competition of the other tall-grasses on the outside. At this time it was impossible to foresee the outcome, in spite of the fivefold greater germination of *Andropogon.*

Season of 1925—Neither of the species had appeared by April 7, but on June 16 there were 21 clumps of *Andropogon* and 32 of *Bouteloua.* Six of the former ranged in diameter from ¼″ to 1.5″, while 17 of the latter fell between ¼″ and 2.5″, all the other plants of both being smaller. The mats of *Bouteloua* were both larger and more numerous, making a stand of varying density throughout the quadrat. *Andropogon* was in poor condition, being mostly suppressed except in open ground or near the margin; the best plants in the quadrat were but 2″–3″ tall. There appeared to be little competition above ground, except in one corner. On July 15, *Bouteloua* was 6″–10″ tall, but no spikes had appeared; little *Andropogon* remained and this was evidently suffering from drouth.

On August 17 *Andropogon* was represented by 7 plants, bearing 1–4 stalks and averaging 7″ in height. There were also 7 mats of *Bouteloua,* reaching a maximum of 3″ in width and a height of 5″. The stand was open and the effect of shading slight. The roots of *Andropogon* at this time resembled those of yearling plants, while those of *Bouteloua* were fairly mature. By August 25 the latter was dominant over most of the quadrat and by October 5 it had seeded abundantly.

Season of 1926—*Bouteloua* started growth before *Andropogon* and on April 15 the leaves were 6″ long. By June 6 its average height was 8″, with a maximum of 12″, and it formed a uniform and fairly dense stand. *Andropogon* was abundant in only one corner of the quadrat; it had practically disappeared except near the margin and did not constitute 5% of the cover. Competition between the two species had nearly ceased, since they occupied different areas. On August 2 both were flourishing, at a height of 8″, and neither was invading the area of the other. The light intensity beneath *Bouteloua* was 11% at the 2″ level. By September first this species had produced an abundance of flower stalks 12″–16″

tall and extending well above the general level. *Andropogon* was beginning to invade the grama area, the best bunch being 11″ tall. This was a consequence of late summer rains, indicating that the drouth had been chiefly responsible for its failure to overtop the grama and to obtain the upper hand by reducing the light available.

Andropogon furcatus vs. Bouteloua hirsuta

Season of 1925—On June 9 the quadrat contained 206 seedlings of *Andropogon* and 163 of *Bouteloua.* The former averaged 4″ tall with 2–4 leaves; the latter was somewhat shorter and hence slightly shaded; it had 3–4 leaves per plant and was beginning to tiller. Both species looked thrifty, but the soil had washed away from some plants, leaving them attached by a single root. Light readings taken on June 27 in various parts of the quadrat at the ground level gave a range of values from 21% to 33%.

By July 16 there were 181 individuals of *Andropogon* and 119 of *Bouteloua;* both were thriving and bore 1–5 tillers per plant. The bluestem had an average height of 4″ and a maximum of 9″, while the grama was but 2″ and 3″ respectively. Many of the leaf tips of the former had dried, but the latter showed practically no dead tips or leaves, thus indicating its greater resistance to drouth. The stand was quite dense, the light value at the ground level being 11% on August first, which was to the disadvantage of *Bouteloua.* The chresard in the upper 6″ of the soil was 6%, but at greater depths it was 11%–15%, thus accounting for the excellent growth of *Andropogon.*

On August 20, *Andropogon* comprised 159 and *Bouteloua* 123 plants; the mortality since June 9 had been the same for both, namely, 23%–24%. The quadrat was in excellent condition, presenting a uniform mixture of tall- and short-grass. *Bouteloua* averaged 4″–5″ with some 6″–8″ tall and the flower stalks around the margin reached 8″–14″. Both species had produced tillers, but not so abundantly as when growing alone. Many individuals of each were suppressed, but there was no sharp line between these and the dominant ones. By October 5 the grama had produced a large amount of seed.

Season of 1926—*Bouteloua* had started rapid growth before April 15 and by June first formed a dense layer over the quadrat 3″–5″ in height, while *Andropogon* constituted an upper layer 9″–12″ tall. The latter was more abundant and thrifty around the edge and *Bouteloua* in center. In this area occurred a number of suppressed plants of bluestem and the grama in turn was reduced under the best marginal growth of its competitor. However, the dry weather favored the short-grass decidedly and it controlled most of the quadrat except at the border. The shortage of water rendered it impossible for *Andropogon* to grow sufficiently to overtop and shade out the grama, though the evidence indicated that the competition was intense.

On August 2 the two species dominated their respective areas in much the same degree, but the effect of *Andropogon* was clearly revealed in the few flower stalks produced by *Bouteloua* in the quadrat in comparison with adjacent areas in which it was the dominant. At this time the outcome was still undecided; however, the light available for *Bouteloua* under the densest mass of *Andropogon* was about 18%, suggesting the ultimate triumph of the latter. A dense leaf mulch 2″ thick was also a contributing factor in this.

Comparative behavior and factors—The two species of *Bouteloua* are so nearly identical in life-form and requirements that they can be treated as one in discussing the competition with *Andropogon*. *Andropogon* possessed a distinct advantage in its far better germination, but this was more than offset by its greater susceptibility to winter-killing and in smaller degree by the earlier start of *Bouteloua* in the second and third years. Both its dwarf life-form and its original home on high foothills and plateaus serve to explain why this lost but a third of its number the first winter in contrast to a loss of tenfold the survivors for *Andropogon*. With respect to water requirements, the more shallow fibrous root system appropriated a larger relative share of the soil-water during dry years or drouth periods, and the sod habit reduced the loss by transpiration. This provides the obvious explanation of the dominance of grama over much of the quadrats through two more or less critical drouth seasons. On the other hand, when *Andropogon* was able to obtain more water for growth, as around the edge of the quadrat or from late rains, it began to forge ahead as a consequence of its great stature and resulting ability to overshade its short-grass competitor.

As stated earlier, neither species of *Bouteloua* belongs climatically in the subclimax prairie, of which *Andropogon* is the chief dominant. However, short-grass and tall-grass are frequently associated as a consequence of movements due to climatic shifts. Thus, the moist-phase migration that carried the three species of *Andropogon* northward and westward explains why they are found today along the foothill edge of the mixed prairie, just as the wedge driven by the dry phase into the deciduous forest of the East brought the short-grasses into contact with the tall-grass subclimax. In the latter case it seems certain that the short-grasses would long ago have disappeared except on warm dry slopes as a result of overshading by their tall competitors, had it not been for the decisive effect of grazing. This has carried the short-grass relicts far and wide through both the subclimax and coastal prairies and given these the character of a mixed community of tall-grasses and short-grasses. These however are not mixed prairie in the climatic and climax sense, but represent the outcome of equilibrium between the climax and grazing disturbance.

Agropyrum glaucum vs. Bouteloua gracilis

Season of 1924—On June 4 the number of *Agropyrum* seedlings was 37, of *Bouteloua* 297, the wide difference arising from the defective germination of the former. The one was 1″–3″ tall, the other 0.5″–2″; both were thriving. On July 9, *Agropyrum* had increased to 41 and *Bouteloua* had

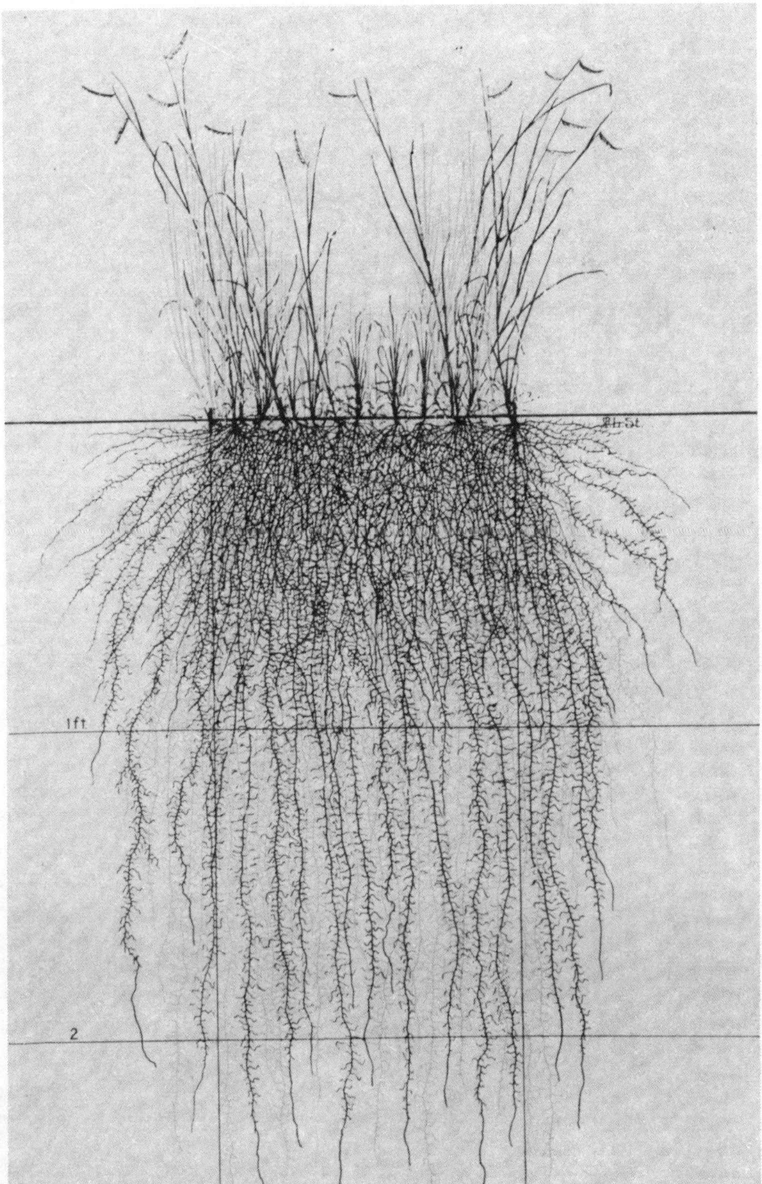

Bisect of culture of *Agropyrum* and *Bouteloua gracilis* in August of 1st year.

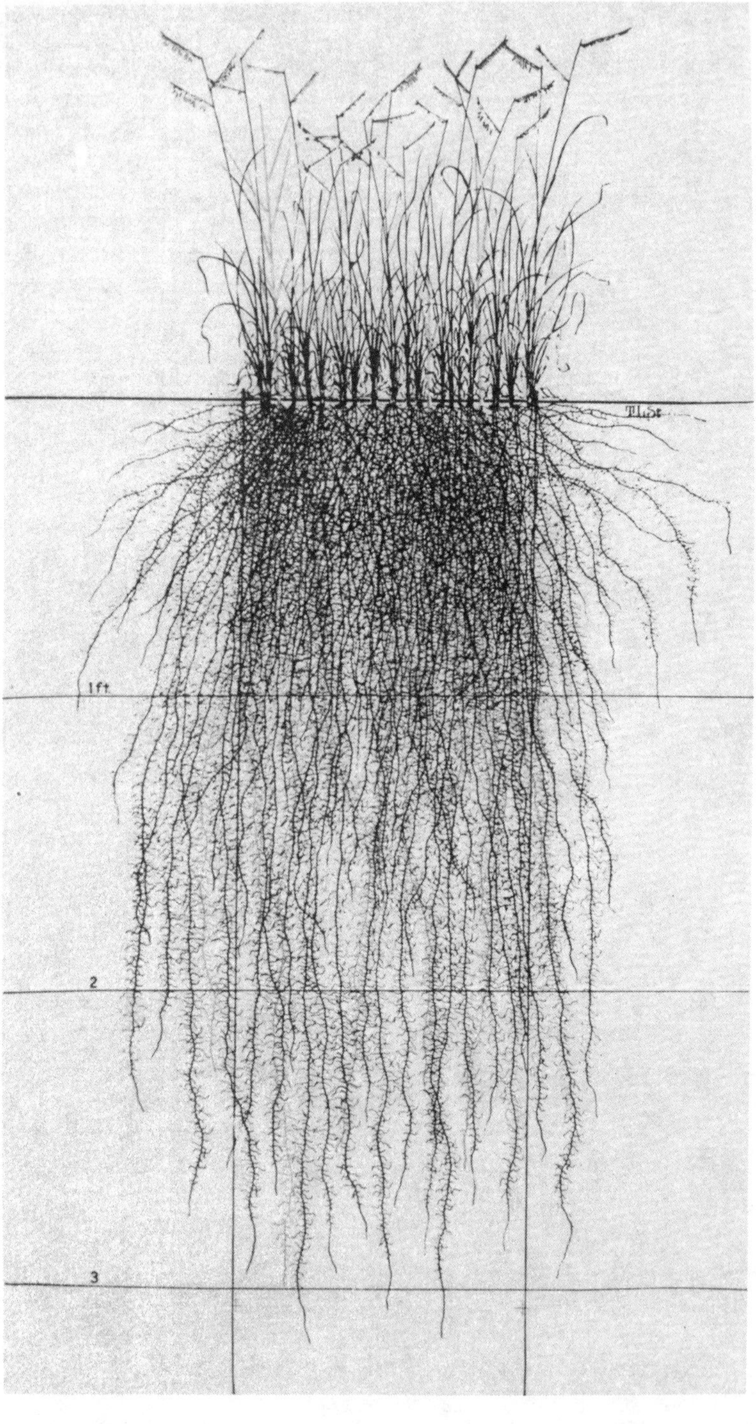

Bisect of culture of *Agropyrum* and *Bouteloua gracilis* in August of 2nd year.

decreased to 260. The latter was thriving; it had tillered heavily, as many as 7 on some plants, and was forming dense mats with an average height of 5″. Like its control, *Agropyrum* was sparse, 5″–7″ tall, and some plants bore 2 tillers. By August 7, *Agropyrum* had lost 13 plants, leaving only 28, while *Bouteloua* had lost 26, leaving 234. The tallest of the former were 11″, but few were more than 4″, owing to the death of the tops. The few tillers and many deaths indicated clearly that it was losing ground. On the contrary, the grama formed a dense cover 5″–6″ high; it had tillered profusely, as many as 14 occurring on some of the larger individuals. A third of the wheat-grass was suppressed, but none of the grama. On September 10 but 10 plants of *Agropyrum* were present, the tallest being 12″ high. *Bouteloua* had lost 28 and now numbered 206, of which 73 were suppressed. The stand was 8″–9″ high and so dense that it was difficult to distinguish the individuals (plates 2 and 3).

Season of 1925—Of the 10 plants of *Agropyrum* at the close of the preceding season, only two were alive on June 16. These were 9″ tall and very slender, with but 2 leaves each. *Bouteloua* formed a continuous dense mat over practically the entire quadrat; it averaged 5″ tall and was very thrifty. On May 26 the chresard was 7.5% in the first 6″ of soil and 13%–20% at greater depths. These values were 3%–6% higher than in the adjacent grassland and were sufficient for good growth.

On July 15 the quadrat appeared to be composed solely of *Bouteloua;* it formed a dense stand with shoots 9″ tall, though no spikes had appeared as yet. By August 17 the last individual of *Agropyrum* had vanished. The grama now averaged 8″ in height, with flower stalks reaching up to 12″. The water-content on August 18 ranged from 16% in the upper 6″ to 10%–12% at greater depths. By October 5, *Bouteloua* had seeded abundantly.

Season of 1926—On April 16, *Bouteloua* was green at the base and its leaves were 6″ long; a good stand 5″–7″ high covered practically the entire quadrat. The light intensity under the leaf-mulch and an inch above the soil surface was 6.7%. No *Agropyrum* was found at this time or later in the season; its utter disappearance constituted an exception to the usual behavior of the unsuccessful competitor in this series of cultures. On June 30 the temperature of the air at 8″ above the soil was 101° F.; an inch above and under the mulch it was 104° F., an inch below the surface, 90° F. and at 8″, 81° F. The chresard at the various depths to 3′ was 2.6%, 9.5%, 16.3% and 17.9% respectively. By August 2 the grama had reached a height of 12″ and a few flower stalks had appeared. On September first the growth was remarkable and the flowering profuse, such as is typical of pure stands.

AGROPYRUM GLAUCUM VS. BOUTELOUA HIRSUTA

This culture was in essence a duplicate of the preceding, but was installed in the spring of 1925. An especial effort was made to secure a much thicker stand of *Agropyrum* in order to afford an adequate test of its ability to compete with *Bouteloua.*

Season of 1925—On June 9 there were 357 seedlings of *Agropyrum* and 169 of *Bouteloua,* making a total of 526 with twice as many of the one as the other. The first averaged 3.5″ in height, the second was somewhat shorter and more or less shaded, but was beginning to tiller. The plants of *Agropyrum* were slender with 2–3 leaves and the tips had dried for 1″–2″. On July 16 there were 254 plants of wheat-grass and 162 of grama. The former was in good condition, averaging 8″ with a maximum of 13″; the stalks were usually single, but occasionally as many as 3–4 per plant. The grama was mostly 2″–3″ high and producing tillers up to as many as 8 to an individual.

By August 20 there were only 179 plants of *Agropyrum* to 152 of *Bouteloua.* The mortality since June 9 was 50% for the former and but 10% for the latter. The quadrat presented a good mixture of mid and short grasses. The wheat-grass was very irregular in height, ranging from 2″–3″ to 9″–10″; it had not tillered much, the best plants having but 2–3 tillers and many none at all. The grama was distributed rather uniformly over the plot, averaging 3″–4″ in the center, but flowering abundantly at 10″–14″ around the margin. A third of the quadrat was bare, except for a cover of suppressed plants and a thin layer of dead leaves. By October 5, *Bouteloua* had produced seeds in quantity.

Season of 1926—On April 15, *Agropyrum* was already 2″–3″ tall and *Bouteloua* was turning green at the base. By June first the latter formed a dense thrifty stand over the area to a height of 3″–4″; the former was scattered and less flourishing. The plants were 7″–12″ tall, but cast little shade. It had suffered seriously from winter-killing and was greatly reduced in number. By August 2, *Bouteloua* formed a dense mat 4″–8″ thick over the entire quadrat and bore an abundance of flower stalks 10″–14″ high. The scattered stalks of *Agropyrum* could be seen only by close observation. The best plants were only a foot high with short narrow leaves, all the basal ones being dead.

Comparative behavior and factors—Low germination was a decisive feature in the failure of *Agropyrum* in competition with *Bouteloua.* However, this does not explain the outcome in the second culture in which its seedlings were twice as numerous as for the latter. The primary cause of its handicap must be sought in the water relations, since its greater stature practically excluded light as a factor. In both the true and mixed prairies in which it is a dominant, *Agropyrum* grows at the lower levels as the most mesophytic member of the association. As a mid-grass its demand for water is materially greater than that of a short-grass and it is correspondingly handicapped during dry periods. It is at this very time that the root system of *Bouteloua* with its numerous branches gives the short-grass a further advantage in absorption. This is reflected in the fact that grama produced tillers early and in great abundance, while in the wheat-grass they were regularly few or wholly lacking, with a consequent reduction in its ability to overshade.

Since *Agropyrum* is a typically boreal genus of grasses, it was somewhat unexpected to discover the heavy toll taken by winter. This was obviously not a matter of the direct effect of low winter temperatures, which should affect the southern *Bouteloua* even more seriously. It was due to the action of winter drouth upon plants that had been weakened by a water deficit through most of the season, and furthermore had been unable to make and store an adequate supply of food for growth in the spring.

The course and outcome of this competition throws much light upon the relations of *Agropyrum* and *Bouteloua* in the mixed prairie, where they are regular associates. Toward the northeast where the rainfall-evaporation ratio is more favorable, the two grow together on nearly equal terms, often associated with another short-grass, *Bulbilis dactyloides*. A similar relation obtains over much of the Great Plains during wet years, but in the drier years and in the southern part of the association the short-grasses always have a potential and often an actual advantage. This has everywhere been greatly augmented by overgrazing in this most extensive and important of grazing associations, with the consequence that the turf of short-grass represents its most familiar form today (p. 144).

DOMINANT VERSUS SUBDOMINANT

The third series of competition cultures in the low prairie dealt with the relation between the dominant grasses and the subdominant forbs. Herbaceous species with conspicuous flowers are characteristic of all grassland associations where the rainfall is ample. As it diminishes toward the West for example, such subdominants decrease in number and in abundance and may finally disappear in large measure. This has long been recognized as a matter of competition for water especially ("Plant Indicators," p. 125; "Plant Succession and Indicators," p. 256), and hence furnishes an additional objective for the experimental attack.

The tall-grass employed in each culture was *Panicum virgatum;* this was sown with *Oenothera biennis* in one quadrat, with *Liatris punctata* in another, and *Solidago rigida* in a third. A duplicate of the first quadrat was started in 1925.

Root relations—The root system of *Panicum virgatum* has already been described (p. 40). The roots of *Oenothera biennis* develop rapidly and spread widely. By July 30 of the first year, plants in the control quadrat were 22″ tall, somewhat larger than in the culture and the longest taproots penetrated to 35″. The laterals spread widely and penetrate deeply and a single plant occupies a larger volume of soil than a seedling *Panicum* of the same age, but not so thoroughly (fig. 3).

Liatris punctata rapidly develops a deep taproot that is all out of proportion to its shoot. By August 2, plants but 5″ tall and with 2 leaves possessed taproots 33″–38″ long, which were very poorly equipped with short unbranched laterals. A year later the taproots had extended to a depth of nearly 4′; numerous long branched laterals arose from the thickened portion and passed obliquely outward and downward, greatly extending the absorbing area (fig. 4).

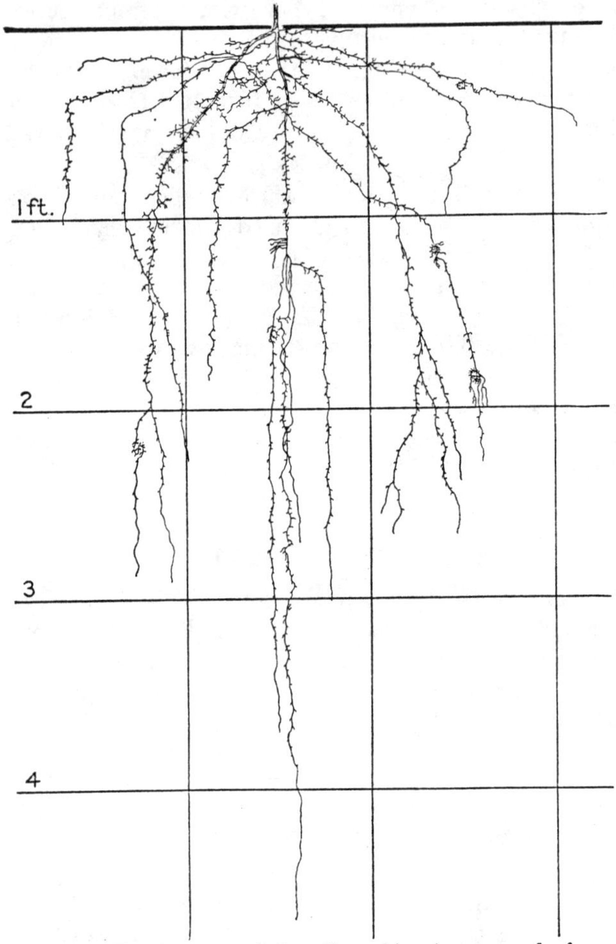

FIG. 3—Root system of *Oenothera biennis* near end of first summer.

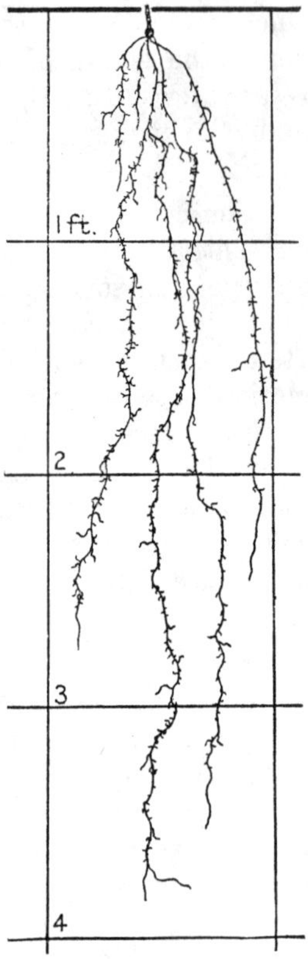

FIG. 4—Root system of *Liatris scariosa* near end of first summer.

Solidago rigida possesses roots that are fibrous, very numerous and wide-spreading, and penetrate deeply. In these respects they resemble those of many grasses, but they are much less well-branched.

PANICUM VIRGATUM VS. OENOTHERA BIENNIS

First culture, season of 1924—On June 4 there were 264 seedlings of *Panicum* and 48 of *Oenothera;* the one was 1″–2″ tall with 4 leaves, the other slightly taller and mostly with 4–5 leaves. By June 23 the grass had increased to 319 plants or 21%, but no change occurred in the number of the forb.[1] The height of the latter was but 2.5″ by contrast with 4″ in the

[1] The term applied to herbs other than grasses and sedges; cf. "Experimental Vegetation," p. 21.

control, and the lowest leaves were mostly yellow. Many plants of *Panicum* were but 2″–2.5″ tall and distinctly poorer than those of the control; the leaves were yellowish, the two lower usually being dead.

On August 7 the grasses were found to have decreased to 274 or 14%, without any change in *Oenothera*. Of this number 130 or 48% were suppressed as against only 7 plants of the latter. The forb had made an excellent growth to a fairly uniform height of 9″, casting a dense shade and dominating the area. Its tolerance was indicated by the absence of dead leaves; three flower stalks 9″–24″ had appeared. The light intensity at a height of 3″ under the leaves was 18%. Dominant plants of *Panicum* had reached a level of 11″, forming a thin layer above the leaves of *Oenothera*. Many of the plants were weak and dying, and on all the lower leaves were nearly dead, the stems yellowish and the upper leaves greatly attenuated. No tillers were produced, owing to the unfavorable conditions for growth (plate 4).

On September 9 the 48 original plants of *Oenothera* were still present, though 10 of them were suppressed. One was in full bloom at 23″ and another fruiting at 30″. The broad leaves cast a dense shade and the grass had fared poorly. Of the 206 plants remaining, 60% or 122 were suppressed. The maximum height of the few tall individuals was 15″–18″; the plants in general were attenuated and yellowish below because of the deep shade. The growth of *Oenothera* indicated that it was well supplied with water at all times, while *Panicum* suffered chiefly from the resultant overshading.

Season of 1925—On April 7, *Oenothera* had reached a height of 2.5″ but *Panicum* had not yet begun growth. By May 28 there were 29 of the former with only 5 suppressed, and 47 of the latter, all of which were suppressed. The forb averaged 15″ in height and the plants were 5″–9″ wide; the basal leaves were yellowing from drouth. The grass was in very poor condition, the tallest but 8″ high; the plants had not tillered and were much attenuated, with but 2–3 leaves per plant. The mortality since the preceding September had been 159 or 77%. The loose debris on the soil was 3″ thick, composed of the stems and leaves of both species; the light value under this on June 5 was but 0.7%.

By June 22 the height and spread of *Oenothera* had increased materially. Light readings made on June 20 just above the leaves of *Panicum* but under the former gave a value of 20%. On June 27 the light intensity at the same level was 12%, while on the ground it was but 2%. By July 15, *Oenothera* averaged 30″ in height and the scattered *Panicum* but 12″. The shade was much less dense than earlier, since many of the basal leaves of the forb had died. On August first the light intensity was 40% at the general level of the grass leaves.

On August 17 there were 24 plants of *Oenothera*, of which 8 were suppressed, while there were 32 of *Panicum*, all very much suppressed. The dominant individuals of the first averaged 40″, while the tallest were 44″, and all were in blossom. The suppressed plants were about 15″ tall, with reduced flower clusters. The grass averaged 9″ in height; there were almost no tillers and the leaves were sparse and pale. On August 25 fruits were developing in the forb, but no flower stalks had appeared on the grass. The chresard in the surface 6″ was 7% on August 28 and varied from 9%–11% at greater depths. These values were about 3% below those for similar levels in the adjoining prairie.

Season of 1926—*Oenothera* did not appear in the spring, since the mature plants had died in the fall, and no seedlings were present in spite of the abundant seed produced the previous year. The old stems were broken at the surface of the soil and removed. *Panicum* formed a scattered stand as the result of its suppression for two seasons, but the disappearance of its successful competitor gave it a new opportunity. The plants averaged 9″ in height on June 5 and were very thrifty, being in the best condition since the culture was started. For the next two months the grass gained considerable ground by means of tillers and by August 2 formed an open cover over the quadrat. At this time the plants were 9″–12″ tall; by September first the height had increased 3″ and tillers were abundant.

Second culture, season of 1925—In duplicating this experiment the plan was to afford *Panicum* a better opportunity by reducing the number of *Oenothera,* but germination was so defective that the actual ratio of seedlings was only 3:1 instead of 5:1, as in the original quadrat. On June 9 the number of *Panicum* was 63, of *Oenothera* 18; the total number of plants was so small that the initial competition was slight. The grass average 2″ tall and usually had 2–3 short leaves. The forb was very thrifty, with a spread of 1.5″–3″ and 4–5 leaves. On July 16 there were 66 plants of *Panicum* and 19 of *Oenothera;* the former averaged 6″ in height, with a maximum of 12″. Most of the individuals were without tillers, but some bore 1–2 each; the uppermost leaves extended above the plants of *Oenothera.* The latter was very uniform in height, averaging 6″ and with 8–12 leaves; the width was usually 5–6 inches, with 9 as the maximum. It was in excellent condition and cast a dense shade; the light intensity on August first was 16% under the leaves, while near the soil it was but 4%.

By August 20, *Panicum* had lost but three plants and *Oenothera* none. About a third of the individuals of the former were suppressed and did not extend above the level of the forb. The dominant ones averaged about 10″ tall and occasionally had a vigorous tiller. The lower leaves were dead in the shade, and the suppressed plants seemed doomed. A single plant of *Oenothera* was suppressed, though the individuals were so crowded that the leaves were more or less vertical. The average height was 7″ and the number of leaves 8. Neither species grew as tall as in 1924, owing to drouth, but the course of competition was similar, *Panicum* faring somewhat better.

Season of 1926—By April 15 the rosettes of *Oenothera* had a spread of 4″–5″, but *Panicum* had not yet started growth. On June first the former ranged from 4.5″–11″ in height, forming an even stand; the spread of the leafy tops was 2″–6″. The grass was sparse and mostly but 5″–8″ high, being overtopped by its competitor, which was rapidly gaining the ascendency in consequence. By August 2 there were 9 dominant plants of *Oenothera* more than 13″ tall and 12 partly suppressed ones. The stems were bare to a height of 6″–9″ and the remaining leaves were small on account of the severe drouth. The shade cast by them was light and probably more beneficial to the grass than otherwise. The entire stand was open and the growth of *Panicum* was good, attaining a height of 10″–19″. The light at its upper level under the forb was 28%, which was much greater than usual.

Comparative behavior and factors—The two cultures of *Panicum virgatum* and *Oenothera biennis* serve to bring out clearly several facts in the competition between grass and forb. Opposed to a number of seedlings several times greater, the more rapid growth, broad leaves and spreading rosettes of *Oenothera* gave it a marked advantage. These features enabled it to overtop and overshadow the grass so decisively that many or all of the individuals were suppressed. The larger demands of the forb for soil-water were undoubtedly a factor in this as well. On the other hand, enough of the grasses persisted during two years of unfavorable competition to enable them to take possession of the quadrat after the short-lived *Oenothera* had died out.

It is especially significant that an abundant crop of seed did not permit *Oenothera* to maintain its hold, confirming the view that annual, biennial or short-lived perennial forbs can not make headway against perennial grasses after the initial invasion permitted by more or less complete denudation. In the subclimax, true and mixed prairies, annuals and biennials are practically absent in the typical climax cover, being found only in successional areas, especially those of secondary origin. They occur abundantly in the desert plains grassland, but chiefly in the winter rainy season when the grasses are dormant. The single rainy season of the bunch-grass prairies of California also favors annuals, but these are abundant in the climax cover as a rule only when the rainfall is well above normal.

Panicum virgatum vs. Liatris punctata

Season of 1924—On June 4 there were 350 seedlings of *Panicum* and 1 of *Liatris;* the grass was 0.5″–1″ tall with 1–3 leaves, the forb bore only its cotyledons and the first leaf, which was 2″–4″ tall. By June 23 the number of *Panicum* had increased to 430, while *Liatris* had dropped to 61. The plants of the former were 1″–3″ tall with a maximum of 7″; the leaves were yellowish, the two lower often being dead. *Liatris* was 3″–5″ tall and had 2 leaves for the most part, as in the control. By August 8 but one *Panicum* had disappeared in contrast to 9 *Liatris,* leaving 429 and 52 respectively; the suppressed individuals of the former numbered 336 or 78% and of the latter 23 or 44%. The grass formed a dense stand at a general level of 16″, while the forb averaged 5″ and attained a maximum of 9″. Competition was so severe that few of the grasses, even of the largest, had tillered; many of the smaller individuals were dying and nearly all had dead leaves at the base. The leaves were narrower than those of the control and slightly yellowish. Even the dominants of *Liatris* were poorer than the control, the leaves being fewer and narrower.

By September 4 the plants of *Panicum* were reduced to 376, of which 263 or 70% were suppressed. *Liatris* had lost but one plant, leaving 51, of which 20 or 40% were suppressed. The average height of the grass was 18″, with a maximum of 2′. The forb gave almost no competition and as a whole the dominant individuals of *Panicum* were well-tillered and thriving. Although quite tolerant, *Liatris* played an insignificant rôle in the culture; the tallest were but 6″–7″ high and many possessed only 1–2 leaves. As a result, the grass had general control of the quadrat at the end of the first season, due chiefly to its more rapid growth.

Season of 1925—Neither species had appeared by April 7. On June 13, *Liatris* was represented by only 18 of the 51 plants present in September Each bore 1–3 leaves 10″–11″ long; these extended as high as the grass level, but were not sufficiently numerous to cast an appreciable shade On the other hand, they were not injuriously shaded by the grass. But 3 plants had developed erect stems 7″ tall and with leaves 2″ long. *Panicum* was very dense through the quadrat, having produced so many tillers that the individuals could not be recognized; the average height was 8″ the light intensity on the ground under the debris was 2% on June 5 while on June 20 at a level of 2″ it was 6%. On July 15 there was a fine growth of *Panicum* to a height of 15″; the stems of *Liatris* were scattered and ranged from 12″–15″.

On August 18 there were 21 plants of *Liatris*, much suppressed; only 3 of these had developed flower stalks and they were along the edge. These varied from 13″–18″ in height and were beginning to bloom profusely The suppressed individuals bore 1–2 leaves, 9″–12″ long and 2–8 mm. wide their ability to endure the dense shade of the grass is probably explained by the erect position of the grass-like leaves. *Panicum* formed a dense cove 16″ high, best developed in the outer half of the quadrat. The central por tion was covered with a mulch of dead grass leaves and the plants wer more suppressed. While the culture was clearly dominated by the grass the density was too great for the best growth.

Season of 1926—By June 5, *Panicum* formed a fairly dense and uniforr sod to an average height of 10″. The old stems and dried leaves persiste to the same level and with the 1″ mulch of dead leaves greatly reduce the shade at the soil surface. The light value at a level of 3″ was 7.5% and at 8″ it was 29%. The vertical leaves of *Liatris* gave a fairly goo starch test throughout and a photosynthate value of 0.190 gm. per 10 sq. cm., indicating their high tolerance. There were 17 plants with 1– leaves but no stems, and 6 with stems and numerous leaves, mostly well exposed to light near the top of the grass level. Evaporation from the so under the mulch was 13.8 cc. in comparison with 35 cc. in the open.

On August 2 the general level of *Panicum* was 11″–13″, although som plants reached a height of 17″. There were many narrow-leaved suppresse individuals, with dead leaves reaching a height of 2″–3″. The light valu in the densest cover was 11% at a height of 5″. The 6 stems of *Liatris* had reached heights of 10″–13″, the other plants being still in the grass-like stage. By September first the forb had attained a stature of 9″, but the general level of the grass had not changed.

Comparative behavior and factors—In both germination and rate of growth *Liatris* proved to be much inferior to *Panicum*. Its slow growth was due in large measure to its poorly branched root system, during the first year especially, and to the necessity for storing food in its thickened taproot. It survived the deep shade produced by *Panicum* only by virtue of its exceptional tolerance, which appears to be associated with the long narrow erect leaves. The first season *Liatris* lost 30 out of 81 plants, largely on account of ineffective absorption, and 33 more disappeared during the winter, probably as an outcome of the water relation also. It began the second season with 18 plants, which increased to 21 later in the summer and to 23 in 1926. None of these disappeared during the two summers of overshading by the dominant *Panicum*, while 3 and 6 respectively developed stems to a maximum of 18″ and bloomed profusely.

Liatris punctata forms a serotinal society, open and often sparse in character, from the true prairie through mixed prairie to the desert plains. It is best developed in the mixed prairie, where the stand is less tall and dense, and the light relations distinctly more favorable. It thrives under a rainfall which is inadequate for *Panicum*, indicating that its lack of success in competition with the latter was more a consequence of slow ecesis than of the actual inadequacy of the adult plants. This is corroborated by the fact that the species of *Liatris* more typical of subclimax and true prairie, namely, *scariosa* and *pycnostachya*, are much taller and more floriferous.

Panicum virgatum vs. Solidago rigida

Season of 1924—The enumeration of June 4 yielded 413 seedlings, of which 305 belonged to *Panicum* and 108 to *Solidago*. The former averaged 2″ high, the latter 0.5″, with 3–4 small leaves. On June 27 the total number was 560, the grass having increased to 464, while the forb had dropped to 96. The average height of the former was 6″, the tallest reaching 10″; the plants were thriving, being nearly as good as the check, though slightly smaller. No tillers had appeared as yet. The best plants of *Solidago* were 2″–3″ tall, but many were suppressed and much smaller. The leaves were more or less erect, with an average of 4–5 per plant, the dimensions of the largest being 2″×1″.

By August 8, *Panicum* had further increased to 495, of which 221 were suppressed, while *Solidago* had dropped to 91. The former was very thrifty, growing in a very dense stand with few dead leaves; the average height was 13″. It had produced few tillers, probably because of its density. *Solidago* was clearly suppressed, averaging but 2.5″ high; it was much smaller in the center than on the margin, where a few plants reached 7″. The leaf-blades in the one case had an average size of 1.5″×0.75″, in the other of 3.5″×1.25″.

By September 8 the total number of individuals had fallen to 515, of which 433 were *Panicum* and 82 *Solidago;* suppression had operated on about half of each species, namely, 288 of the former and 40 of the latter. The grass had made a good growth throughout, averaging 16″ high, with some attaining 22″. The best plants bore 1–3 tillers but many had none. *Solidago* was also flourishing, but none were more than 4″–5″ tall, with 6 leaves or less. While those in the middle of the culture were smaller, they were also in good condition. The outcome at this time was still indecisive, since *Solidago* had survived much shading, as well as competition for water, and was holding its own at the end of the season. In view of the large number of plants in the quadrat, which was a total of 515, the mortality had been low, namely, 13% for the grass and 24% for the forb.

Season of 1925—On April 7, *Panicum* had not yet appeared, though the plants of *Solidago* ranged from 4″–7″ in height. By May 28 there were 184 individuals of *Panicum,* all of them suppressed, and 47 of *Solidago,* of which 31 were suppressed. The latter averaged 9″ in height; the dominant plants extended well above the layer of dead grass. They bore 3–5 leaves about 4″x1″ in size. Most of the suppressed ones were much attenuated, and hardly rose above the layer of old grass, which gave a dead appearance to most of the culture. All of the grass culms were greatly attenuated, and practically all the leaves were in the 8″ layer of dead matter. The few that had projected above this were badly frozen. Winter-killing had been marked, 249 of *Panicum* and 35 of *Solidago* succumbing.

The cover had reached an extreme of density and the reduced light intensity was an exceedingly important factor. On June 5 this was but 0.6% on the ground under the layer of dead leaves, while above this but under the leaves of the goldenrod it ranged from 17% to 28%. By June 20 the light had become further reduced; below the grass layer the values were less than 1% and above it but 7%. The stem relations at this time are shown graphically in plate 5; several flower stalks had developed in the case of *Solidago.*

On July 15, *Solidago* was thriving at an average height of 14″, and several inflorescences had reached 18″–24″. The plants of *Panicum* were scattered; they averaged about 13″ tall. Owing to the death and drying of the plants, the light intensity had increased by August first, when it averaged 20% at a level of 4″ above the soil.

By August 18, *Solidago* numbered 31 individuals, of which 12 were suppressed; the smallest of these were 4″ tall with 4 small leaves about 1.5″ long. The dominants had 4–5 leaves each, with blades 6″x2.5″; there were 4 flower stalks 27″–37″ high and with many buds. *Panicum* was in fairly good condition only on the margin of the quadrat, where it received more light. Some of these plants were 16″ tall with 4–6 living leaves, but most of the culms were only 8″ or less in height, with 2 narrow attenuate leaves often with dead tips. The center of the quadrat contained only an occasional grass. On October 5, *Solidago* bore 6 flower stalks 23″–38″ tall, but *Panicum* gave no evidence of blooming.

Competition culture of *Panicum* and *Solidago* in low prairie.

A. Growth on June 3, 1925.
B. Growth on June 10, 1926.

Season of 1926—On April 15, *Solidago* was 2″–3″ tall with 3–6 leaves; some individuals had persisted above ground through the winter, partly as a consequence of the dense layer of debris. By June 5 the goldenrod was about 15″ in height, the thrifty plants constituting a uniformly dense stand. *Panicum* was very much suppressed and was thinly scattered in the quadrat. The plants were but 6″–9″ tall, with a single stem and 2–3 leaves; dead leaves were abundant to a height of 3″. The light intensity approached the limit of tolerance for this species; it was 2% at the 3″ level, 8.8% at the 7″, and 21% just below the upper leaves of *Solidago*. Starch tests demonstrated that the lower leaves of the latter were making little or no starch, while the upper ones contained it in abundance. Evaporation at a height of 5″–7″ was 57 cc. in the center and 65 cc. near the edge, by contrast with 98 cc. in the open. On June 30 the upper leaves of the forb gave a photosynthate value of 0.265 gm. and those at the 8″–10″ level in the center of the quadrat one of 0.222 gm.

By August 2 most of the plants of *Solidago* were 23″–29″ tall; the stems were leafy throughout and the lower leaves drooped over the grasses. The latter were scattered thinly over the quadrat and were especially sparse in the densest shade. A leaf mulch 3″ deep was an important factor in hindering their growth, and none exceeded 10″ in height. On the whole, the culture had become more open, the light values at the level of 3″, 9″ and 17″ being 8%, 10% and 12% respectively. By September 1 some plants of *Solidago* had grown to 3′; 27 plants bore large thyrses, some of which were already blooming. The stems were still leafy to the base and the presence of new offshoots with basal leaves also increased the effective competition that *Panicum* had to meet. In spite of this, it maintained its place and reached an average of 10″ and a maximum of 18″ in the best lighted area, where it even tillered.

Comparative behavior and factors—During the first season *Panicum* possessed a distinct advantage over *Solidago* by virtue of its better germination and more rapid growth. Moreover, it did not reach the maximum number of individuals, 495, until August 8, while the forb was at its maximum of 108 on June 4. The mortality for the summer was nearly twice as great for the latter, and the survivors were but 4″–5″ in height in comparison with 16″–22″ for the grass. Nevertheless, *Solidago* was in good condition at the end of the season and possessed the advantage of a larger supply of stored food. This gave it a much quicker start the following spring and soon led to a reversal of conditions, both as to shading and absorption. Moreover, the grass suffered more severely from winter-killing, partly because of its more southern derivation and partly because the crowded plants had not matured sufficiently. By the beginning of summer all the grass individuals were suppressed and they remained in this condition through the two seasons. At the end of the three-year study, *Solidago* was nearly thrice as tall as *Panicum*, and the final disappearance of the latter seemed certain.

Solidago rigida constitutes a serotinal society in the true prairie; it forms a corresponding socies in the subclimax grassland, but persists in the mixed prairie only in well-watered valleys. It makes exceptionally vigorous clusters, the dense basal mat often persisting through mild winters. It has the advantage of *Panicum virgatum* in the storage capacity of its underground parts, its vigor of growth and great leafiness. As a typical subdominant of the true prairie, it withstands drouth better than its grass competitor, while its northern derivation and more northerly range explain in part its greater resistance to winter-killing. This species is usually less abundant in the subclimax prairie, owing to the competiton of *Andropogon* especially, but where the tall-grasses are handicapped by pasturing or other disturbance along roadsides, it may attain actual dominance.

SUBDOMINANT VERSUS SUBDOMINANT

The fourth series of competition cultures involved the relation between two or more subdominant forbs, and was intended to permit analysis of the subdominance exerted by such species within the limits set by the dominance of the grasses. For the most part the species employed are common to both the subclimax and true prairie, though their abundance is usually somewhat greater in the latter. The combinations were as follows: (1) *Petalostemon candidus* vs. *Brauneria pallida;* (2) *Kuhnia glutinosa* vs. *Liatris punctata;* (3) *Oenothera biennis* vs. *Liatris scariosa;* (4) *Helianthus rigidus* vs. *Amorpha canescens* and *Kuhnia glutinosa.*

Root relations—The root system of *Helianthus rigidus* develops rapidly, penetrating deeply and at the same time forming an effective network in the surface soil. It spreads by means of rhizomes, making dense groups which are a feature of its keen competition.

Kuhnia glutinosa develops a deep taproot at an early period. By July 30 of the first summer when the shoots were 11″ tall, the taproots had reached a depth of 2′–2.5′. The thickened upper portion of the root, the manner of branching and the variation in habit are characteristic, while the great depth of penetration and the increased branching during the second summer were striking features. The roots of mature plants reach a level of 16′–17′.

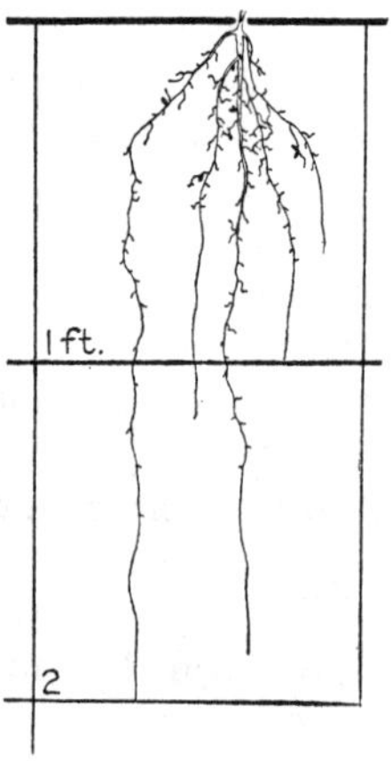

FIG. 5—Root system of *Petalostemon candidus* near end of first summer.

Unlike the taproot of *Liatris punctata,* that of *L. scariosa* is fibrous. Seedlings soon form a corm at the base, from which roots extend into the soil to a depth of 2.5′–4′ by July 31. This has taken place while the plantlets are as yet but 4″–5″ tall with only 4 or 5 leaves. The spread in the surface layer is not great and the short laterals rarely exceed an inch in length.

Petalostemon candidus was 8.5″ tall in the control quadrat by August first, and the taproots extended downward for 14″–22″. In comparison wih most species the roots are meager, the laterals being long but poorly branched and with prominent nodules. The roots of adult plants reach a level of 5′ or more (fig. 5).

Brauneria pallida possesses a taproot even more poorly provided with laterals. It grows with much the same rapidity as that of *Petalostemon* and attains practically the same level.

Petalostemon candidus vs. Brauneria pallida

Season of 1924—On June 4 there were 100 plants in the quadrat, of which 66 belonged to *Petalostemon* and 34 to *Brauneria;* the former was about 1.5″ tall, the latter showed leaves 2″ long and both were thriving. By July 9 there were 120 plants, or 79 and 41 respectively. The one was 3″–10 high and the other bore 4 leaves as a rule. On August 8 the total was 113 individuals, of which 80 were *Petalostemon* and 33 *Brauneria;* 20 of the former and 9 of the latter were suppressed. The best plants of *Petalostemon* were 19″–22″ tall, thrifty and well-branched, and 6 were in bloom. The largest of *Brauneria* were 4″ high with 4 leaves per plant. There was apparently little competition as yet and the ground was not fully shaded.

By September 8 the number had fallen to 101, with 76 of *Petalostemon* and 25 of *Brauneria,* 17 and 7 being suppressed respectively. Of the first, 16 were 15″–21″ in height; they were flourishing and producing seeds in normal spikes; the best plants of the second averaged 4.5″ with 5 leaves. *Petalostemon* was clearly in the ascendant, owing to its more rapid growth. The mortality during the season was but 5% in contrast to 39% for *Brauneria;* it occurred during the drouth of late summer and affected severely the species with the poorer root system.

Season of 1925—On April 27, *Brauneria* was growing well and *Petalostemon* was making a start. By May 28 there were 39 plants of the latter of which 13 were suppressed and 23 of the former, with 17 suppressed. *Brauneria* averaged 4″–6″ tall, *Petalostemon* 6″; the latter was fairly thrifty, with 3–6 branches per plant. At this time the soil was hard and dry, with several large cracks running through the quadrat. On July 15 there was a fine stand of *Petalostemon* 12″ high, with flowers appearing at a height of 16″–18″. *Brauneria* was but 6″–8″ tall, in spite of the fact that the shade was not dense. On August first the light intensity at the top of these plants was 77%.

By August 17, *Petalostemon* was reduced to 30 individuals of which 6 were suppressed and *Brauneria* to 13, all suppressed; the mortality during the summer had been 23% and 57% respectively. *Petalostemon* was clearly dominant, averaging a foot tall, and fruits were ripening on 5 stalks. The shade was still moderate, but *Brauneria* averaged only 4″–5″ high and the largest plants had but 4 leaves.

Season of 1926—On April 15, *Brauneria* occurred as small rosettes with 3–5 short leaves. By June 5, *Petalostemon* made a fairly open stand 12″–20″ high; the 30 individuals were mostly single stems with few leaves on the lower 6″. *Brauneria* numbered 16, scattered through the quadrat; they ranged from plantlets 3″ tall with a single leaf to vigorous plants 9″ tall with 5–7 leaves. The soil was nearly bare and lost water readily.

In consequence, the evaporation from the soil in the quadrat was nearly as great as in the open, being 31.8 and 35 cc. respectively.

By August second, 6 plants of *Petalostemon* were fruiting, but the spikes were small. Both species were suffering from drouth, the leaves showing this effect by dropping off or rolling up. By September first 4 individuals of *Petalostemon* had died. The remaining plants were in fair condition, but they were defoliated to the height of 9″ and many of the stems were eaten by insects. *Brauneria* persisted, but was not thriving, the height remaining stationary.

Comparative behavior and factors—*Petalostemon candidus* and *Brauneria pallida* are both subdominants of the subclimax and true prairies, forming a mixed society in the estival aspect, often with *Erigeron strigosus*, *Psoralea tenuiflora* and others. They extend into the mixed prairie, but much restricted in area and reduced in abundance, especially in the case of *Brauneria*. This difference in behavior is reflected by the response to competition. *Petalostemon* grew more rapidly to a height several times that of *Brauneria* and produced flowers and fruits the first season. It consequently had an overwhelming advantage with respect to light, and this was rendered even more decisive by its greater ability to withstand drouth. This resistance is due in part to its better root system, and in part to its small leaflets. Because of its more northern range, *Brauneria* made an earlier start in the spring, and was distinctly less affected by winter-killing, but these advantages were unable to offset the handicap arising from water and light relations during the summer.

Kuhnia glutinosa vs. Liatris punctata

Season of 1924—On June 4 there were 12 plants of *Kuhnia* 0.5″–1.5″ tall and 6 of *Liatris* 2″–3″ high and with but a single leaf. By July 9 the one had increased to 40 individuals averaging 5″–7″ and the other to 20, which were 3″–5″ high and with about 6 leaves. *Kuhnia* was branching freely at its base. On August 8 there were 41 of *Kuhnia* averaging 11″ and with a maximum of 13″, and 23 of *Liatris*, which were smaller than the check plants. The former dominated the quadrat but cast only a slight shade, so that only one plant and four of its competitor were suppressed. By September 8, *Kuhnia* numbered 42, many reaching a height of 16″, and *Liatris* 17″, the tallest being but 6″ high. Nearly all of the former were thriving and well-branched, and 13 were blooming or about to bloom. The latter possessed but 2–3 leaves and practically all the individuals were suppressed. The shade was moderate and the soil surface hardened. The mortality for the season was nil for *Kuhnia* to 26% for *Liatris*.

Season of 1925—By April 7 both species had begun to grow and on June 16 there were 31 plants of *Kuhnia* to 8 of *Liatris*, by contrast with 42 and 17 the preceding autumn. The former was clearly dominant and was very thrifty; it ranged from 4″–8″ in height with 3–12 stalks per plant, throwing a dense shade throughout the quadrat. The latter was much suppressed, with a range of 2″–7″ in stature, and but a single flowering shoot had developed. By July 15 there was a uniform stand of *Kuhnia* 13″ tall; the shade was so dense that the lower leaves were dead to a level of 4″–5″. *Liatris* had become even more suppressed, the light intensity on August first at 5″ above the soil being but 9%.

By August 17 the 43 plants of *Kuhnia* had produced 170 stems, some plants consisting of as many as nine. The average height was 14″, but the dominants reached 19″ and formed a fairly definite upper layer. The stems of these were about 4 mm. in diameter by contrast with 1 mm. for the suppressed individuals. The subdominant plants were nearly as well developed, with a stature of 14″. The stems had dropped their leaves to a height of 5″ and the latter had become yellow and functionless to 6″–7″, while a few reduced individuals had died. All of the dominants and some of the others bore buds and were about to bloom profusely. *Liatris* was represented by 13 plants, only one bearing a flower stalk, which was 5″ tall. The others had but 2–5 narrow leaves and were often too delicate to stand upright. By October 5, *Kuhnia* had produced fruits in abundance.

Season of 1926—On April 13 a few leaves were appearing from the crowns of *Liatris*. By June 5, *Kuhnia* constituted a fairly dense stand at a level of 8″, while of the 7 plants of *Liatris*, 5 were very much suppressed. These lacked stems or the latter were very short. The light value at 3″ was but 6% and the lower leaves of *Kuhnia* were turning yellow. Starch tests on dominant individuals showed an abundance in the 16 upper leaves, a fair amount in the next three, a little in the next one which was still green, and none at all in the last pair of yellow leaves. In the case of a suppressed plant, starch was abundant in the 7 upper leaves, fairly so in the next two and scanty in the lowest four.

The air temperature just above the *Kuhnia* crowns on June 30 was 100° F.; at 7″ below it was 98.5° F., while it was 84° and 75° at 1″ and 6″ below the soil surface respectively. The light intensity just below the upper leaves was 22%, but at 3″–5″ where the leaves were yellow it fell to 4%. A test of suppressed plants with green leaves yielded no starch, and the photosynthate value was very low, namely, 0.053 gm. The chresards at the various depths down to 2′ were 4.0%, 9.6%, 10.6% and 10.4% respectively.

By August 2 the vigorous *Kuhnia* had reached a height of 16″ and flower buds were abundant. The lower leaves were dead or yellowed to an average of 7″, where the light value was 9%. The 7 plants of *Liatris* still persisted, and the tallest was about to bloom at 17″. On August 27 leaves exposed to full sunlight made photosynthate at the rate of 0.1805 gm. per unit area, while *Kuhnia* gave practically the same result, 0.1752 gm. By September first all the individuals of the latter were blossoming profusely; none were suppressed. The exceptional tolerance of *Liatris* enabled it to persist under the dense shade, but only a single plant came into bloom.

Comparative behavior and factors—*Kuhnia glutinosa* and *Liatris punctata* occur as subdominants from the subclimax and true prairie through much of the mixed prairie, though the serotinal society constituted by them is most striking in the latter. *Liatris* ranges more widely to the north and south, while *Kuhnia* is found much further eastward in the typical form *eupatorioides*. The two genera are closely related phylogenetically, but differ markedly in form and growth habit of the plant-body. *Kuhnia* grows rapidly and vigorously, producing many branches and numerous broad leaves; *Liatris* on the contrary grows very slowly, the stems are few and erect, the leaves narrow and upright. As a consequence, light was the decisive factor in the competition between the two, though drouth played a part also. *Kuhnia* suffered practically no mortality during the three years, while *Liatris* dropped from the original number of 23 plants to 17, 13 and 7 during the successive summers. The effect upon reproduction was equally marked, all the individuals of the former producing flowers in abundance, while but one of *Liatris* bore an inflorescence.

Oenothera biennis vs. Liatris scariosa

Season of 1924—On June 3 the number of plants was 57 for *Oenothera* and 172 for *Liatris;* the former had rosettes with a spread of 3″, while the latter bore only cotyledons or a single leaf in addition. By June 23, *Liatris* had decreased to 71, making a loss of 59% by contrast with none for *Oenothera;* its lack of vigor was clearly indicated by the greatly attenuated and moribund plants throughout the quadrat. On the other hand, the plants of *Oenothera* were 5″–6″ tall; they were thrifty and beginning to put forth flower stalks. On July 25–26 the light values under the cover of this species were 13% by the chemical method.

On August 8, *Oenothera* still numbered 56, of which but 3 were suppressed. The general level was 10″ and the stand so thick that the shade was very dense. Four plants at the margin were 16″–26″ tall and bore flower stalks. Of the 71 plants of *Liatris* on June 23 but 35 remained; all of these were badly suppressed, not over 2″–3″ tall and some were mere remnants.

By September 9, *Oenothera* had decreased to 51, of which 14 were suppressed. *Liatris* now numbered but 17, a loss of 50%, all of which were greatly reduced; most of these had but a single narrow leaf and were not over 4″ in height. The general level of *Oenothera* was 9″; the leaves were 1″–2″ broad and spread in such a way as to produce a dense shade. There were 4 flower stalks, 20″–30″ tall.

Season of 1925—On April 7, *Liatris* had not yet appeared, but the quadrat was covered with a dense carpet of *Oenothera.* The rosettes of the latter reached a diameter of 5″ and a height of 2″–3″. The chresard on May 26 was 11%, 12%, 16% and 18% at 0.5′, 1′, 2′, and 3′ respectively. These were 3%–4% higher than for the same depths in the undisturbed prairie, except that the surface level was 6% higher. On May 28 *Oenothera* numbered 37 of which 11 were suppressed and *Liatris* 7, all suppressed. The toll taken by winter comprised 14 of the former and 10 of the latter. The average height of *Oenothera* was 15″, the maximum 18″, and the spread was 4″–6″. The basal leaves were dead, perhaps as a result of drouth. For the most part, *Liatris* possessed but a single leaf, and the tallest plant was but 6.5″. On June 5 the light intensity just above the layer of debris averaged 4%.

By July 15, *Liatris* had practically disappeared, while *Oenothera* was 2.5′–3′ tall; the stems were large and woody and the leaves were dead to a level of 10″–12″. On August 17 the stand consisted of 30 plants in the quadrat and 15 more on the margin; of these but 5 were suppressed, averaging 2′ or less in height. The dominants were 3.5″–4.3″ high and the maximum diameter was 1 cm. All the plants had been blossoming for some time. The leaves had fallen to a height of a foot and formed a mulch a half-inch thick over the quadrat. *Liatris* had completely disappeared.

Comparative behavior and factors—*Liatris scariosa* is a serotinal subdominant of the subclimax prairie, as well as of the eastern portion of the true prairie. By contrast, *Oenothera biennis* is a subruderal, present in the climax areas only as a result of more or less disturbance, and usually abundant only in roadsides and waste places. Both species range widely and differ little in their response to winter conditions. However, the much more rapid growth of *Oenothera* and its rosette habit were so decisive that more than half of the plants of *Liatris* succumbed during the first month of competition. This high mortality during a brief period in early summer indicates that light rather than water was the limiting factor, which is in accord with the early spread and later stature of *Oenothera*.

Helianthus rigidus vs. Amorpha canescens and Kuhnia glutinosa

First culture, season of 1924—On June 4 the total number of plants was 34, of which 22 were *Helianthus* and 12 *Amorpha*, and by July 9 the number had increased to 63, 28 *Helianthus*, 19 *Amorpha* and 16 *Kuhnia*. The first was much the tallest, the others being respectively 2″ and 5″–8″ tall. On August 11 the total had fallen to 54, made up of 25, 13 and 16, with 6, 4 and 10 suppressed respectively. *Helianthus* dominated the quadrat, the vigorous plants averaging 15″ high; *Kuhnia* was 8″ or less and *Amorpha* not over 4″. The shade was not dense, and the soil was dry and cracked.

On September 8 there were 51 plants, 24 of *Helianthus*, 13 *Amorpha* and 14 *Kuhnia*, with the number of suppressed individuals the same as on the previous date. The dominance of the sunflower had become augmented, 12 plants averaging 18″ in height with some stems 6 mm. in diameter. Some also bore axillary branches with flower buds. *Amorpha* was growing slowly, and though quite normal, did not exceed 3.5″ in stature. *Kuhnia* was in poor condition; the plants were attenuated, none were over 7″, and no flower buds had appeared.

Season of 1925—On April 7, rosettes of *Helianthus* were abundant; they bore 3–4 leaves and had a spread 0.5″–1.5″. By May 28 there were 80 sunflowers, or over thrice as many as on the preceding September. Of this number 15 were suppressed, while the 10 plants of *Amorpha* and the 15 of *Kuhnia* were all suppressed. The sunflower was clearly dominant at a height of 12″; the plants were fairly uniform in size, with usually 6 pairs of leaves per plant. The best plants of *Kuhnia* were but 5.5″ tall, those of *Amorpha* 6″ and very delicate. Light was very important at this time, readings on June 5 at a height of 4″ yielding values of 4% and 5%. This was confirmed by the suppressed plants, which were very slender with many of the lower leaves yellow or dead. Drouth also played a part, the poorer root systems of *Kuhnia* and *Amorpha* being a handicap in the competition with the much more vigorous *Helianthus*.

On June 27 the light intensity near the soil varied from 1%–2%; at the 6″ level where *Amorpha* and *Kuhnia* were growing beneath the sunflower, it averaged 12%. The chresard on July 29 was 8% in both the 6″ levels of the first foot and about 10% in the deeper soil. This was slightly more than in the surrounding prairie, except below the 2′ level.

By July 15 there was a dense stand of *Helianthus*, while *Amorpha* and *Kuhnia* were very much suppressed; the respective numbers on August 17 were 72, 8 and 14. The sunflower averaged 20″ high; the stems were leafy and flower buds were appearing. The stalks projected over the edge of the quadrat and at midday cast a shadow to a distance of 14″. The rhizomes had pushed out of the quadrat as far as 7″ on two sides. The shade was dense, the light value being but 11% at a height of 5″ on August first, and the effect of this was seen in the amount of starch present. Starch tests made on a sunny morning showed that the sunflower leaves possessed but a small amount below 6″ and that there was a steady increase with height. Leaves of *Kuhnia* contained no starch below 6″, a moderate amount between 6″ and 8″, and an abundance above this. The tallest plants of this species were 11″, the average about 7″; they were in poor condition with weak reclining stems, small pale leaves and no flower buds. *Amorpha* had attained a height of only 4″ with slender trailing stems. The leaves were 7 or less per plant and the leaflets very tiny, but the plants were exhibiting a remarkable ability to persist under such severe conditions. On October 5 the sunflowers had one or two heads each at heights of 30″–33″ and were ripening seed.

Season of 1926—By June 5, *Helianthus* had developed a uniform and fairly dense cover at a height of 13″. It had extended its area nearly a square meter as a consequence of the movement of its rhizomes, but was now cut back to the limits of the quadrat. *Kuhnia* was represented by only 10 plants; these were 3″–8.5″ tall and were in fair condition since the shade was not dense at this time. *Amorpha* comprised but 6 individuals, 2″–5″ tall. Most of these were very delicate, since at this lower level the shade was much denser, being but 12% at 3″. Tests of the sunflower indicated that starch was abundant in the upper pair of leaves but scanty in all the others, a result probably due in part to drouth. In a plant of *Kuhnia* growing beneath the dominant sunflowers, starch was scarce in the upper portion and entirely absent in the lower leaves. On the other hand, *Amorpha* exhibited a considerable amount in all its leaves, in spite of the fact that it was the most shaded. The evaporation at a height of 5″–7″ in the center of the quadrat was 63 cc. in contrast to 98 cc. on the outside.

On August second the 50 sunflowers averaged 18″ high. The stems bore only remnants of leaves to 7″–10″, and the bare soil below was not densely shaded. The leaves were narrower than those of the preceding year and the plants were obviously suffering from drouth. Of the 12 plants of *Kuhnia* 5″–8″ tall, none were thriving because of combined drouth and shade. The 8 individuals of *Amorpha* were greatly dwarfed with tiny leaves.

By September first the tallest sunflowers were 30″ high, but only two blossomed owing to the drouth. The stems of *Kuhnia* sprawled over the ground to a length of 14″, but none stretched up more than 5″. *Amorpha* was in a similar condition, none of the plants rising above 3″.

Second culture, season of 1925—The preceding quadrat was duplicated in 1925, with the substitution of *Liatris punctata* for *Amorpha canescens.* Moreover, fewer seeds of *Helianthus* were planted in order to afford the other species a better start. By June 6 there were 8 plants of sunflower, 26 of *Liatris* and 18 of *Kuhnia;* the first was about an inch tall, but one bore as many as 8 small leaves. *Liatris* was in fair condition only; the tallest was 4″ and the average number of leaves three. *Kuhnia* averaged 3″ in height with 10 leaves per plant, and was thriving.

On July 16 there was but one sunflower, with 23 plants of *Liatris* and 19 *Kuhnia.* The latter was dominant, averaging 8″ high with a maximum of 18″; it was very leafy, but many of the lower leaves were turning yellow and dying. Most of the individuals of *Liatris* were in poor condition, the tallest being 4.5″ with 4 leaves. On August 20 there was a single *Helianthus,* 24 *Liatris* and 20 *Kuhnia.* The latter was still dominant at a height of 12″–14″ and was preparing to bloom abundantly. *Liatris* was in essentially the same stage as it was a month previous. The one individual of *Helianthus* formed a large rosette with 14 leaves. The quadrat was very open and all the plants were growing normally. By October 5, *Kuhnia* had produced seeds in fair abundance.

Season of 1926—By April 15, *Helianthus* alone had begun to grow; it was represented by the single rosette. On June first there were 55 stalks of *Kuhnia,* but only 11 plants of *Liatris;* the stems of the former had almost trebled in number by budding from the base of the original stems. This species formed a dense stand 6″ high, which shaded the *Liatris,* the dominant individuals of the one being in sharp contrast to those of the other. *Helianthus* had disappeared from the enclosed area.

By August second the 11 plants of *Liatris* were still persisting, but they were not more than 6″ tall and were without flower stalks. *Kuhnia* averaged a foot in height and formed a fairly open stand. The lower leaves were dead and the quadrat consequently fairly well lighted. In the absence of *Helianthus,* it had easily outgrown *Liatris* and had taken possession of the area.

Comparative behavior and factors—In these cultures the outcome of the competition between three subdominants was determined by the vigor, rate of growth and the relative height, but it was also profoundly influenced by the number of individuals. In the first culture *Helianthus* had become dominant by the middle of the first season, its stature at this time being twice that of *Kuhnia* and four times that of *Amorpha*. The shade cast by the dominant was so deep that the greater vigor of *Kuhnia* was of little advantage in the secondary competition with *Amorpha,* and in starch-making the latter was the more successful because of its greater tolerance.

The slow growth of *Helianthus* in the second culture and its rapid disappearance are apparently to be ascribed to the small number of individuals which was less than one-sixth of the total. After it had vanished, the superior vigor and height of *Kuhnia* became decisive, and it readily suppressed the slow-growing *Liatris.*

The number and position of buds play an important rôle in the final outcome of competition, as these three species of composites demonstrate. They are alike in belonging to the serotinal aspect of the prairie, but differ much in abundance and density. *Helianthus rigidus* sends up new shoots from its rhizomes, making a compact family that defies invasion, but these groups are usually well scattered. *Kuhnia glutinosa* is more abundant, but the buds arise from a crown producing a much branched plant. In *Liatris punctata* the buds arise in much the same manner, but they are fewer except in sparse areas, and the stems and narrow leaves are consequently less effective in competition.

DOMINANT VERSUS RUDERAL

Ruderals are the weeds of waste places, fallow fields and roadsides and in the prairie region are chiefly annuals introduced from Europe They occur in the climax stage only under conditions of disturbance, bu their vigor and rate of growth are such that a measure of their ability to compete successfully with grass dominants is important in connectio with the stability of climax and the course of succession. The followin cultures were based upon *Andropogon furcatus* as the tall-grass dominan in competition with *Amarantus retroflexus* and with *Ambrosia trifida.*

Andropogon furcatus vs. Amarantus retroflexus

First culture, season of 1924—On June 3 there were 210 seedlings c *Andropogon* and 53 of *Amarantus.* The grass was 1″–1.5″ tall with 3– leaves, the weed ¼″–1″ tall, some having a spread at the top of 3″. B July 8, the grass had increased to 386 and the weed had decreased to 4 Most of the latter were producing flower clusters at 5″–6″ with none abov nine, and they were yellowish in color, apparently because of a deficie nitrogen supply. The grass ranged from 4″–8″ tall, and differed from th control only in being more slender; it was also yellowish. At this tim *Amarantus* already appeared to be suffering severely from the competitio

By August 8, *Andropogon* had lost 54 plants, leaving a total of 33 while but 3 of *Amarantus* had disappeared; of the grass 189 or 57% an of the ruderal 22 or 48% were suppressed. A single individual of the latt was over 7″ high, but all had fruited, some at a height of 3″. Nearly a had shed their leaves and were now of little effect in the culture. Th grasses formed a layer at an average height of 8″; all showed the effect of considerable shading and many had died.

By September 10, *Andropogon* had dropped to 244 plants, of which 99 or 40% were suppressed. The individuals of *Amarantus* were all dead or nearly so, although many of the dead stems were standing. The general level of the dominant grasses was 10″; tillering was poor on the whole, though fairly good near the margins.

Season of 1925—Neither species had appeared by April 7, but on June 16 there were 70 plants of *Andropogon* averaging 3″ tall and for the nost part with 1–5 tillers. The effect of winter had been marked, since 174 plants had died, making a mortality of 71%. No seedlings of *Amar-antus* were seen, in spite of the seed production of the preceding year, and none appeared during the summer. On August 17 there were 82 single stalks or clumps of grass, some of the latter with as many 15 culms. The stand averaged 8″ in height, being rather open and occupying less than half the area of the quadrat. The plants were thrifty and well-tillered.

Second culture, season of 1925—The poor growth of *Amarantus* in 1924 seemed to be due in part to the lack of adequate nitrates in the soil, and hence this was remedied in the duplicate by watering the culture from ime to time with Knop's nutrient solution. By June 9 there were 65 eedlings of the ruderal and 275 of the grass; the former was thrifty and ″–2.5″ in height with about 7 leaves per plant, the latter was 2″–3″ tall nd somewhat shaded by *Amarantus*.

By July 16, *Amarantus* had decreased to 44 and *Andropogon* to 268, he loss being threefold greater in the former. The former averaged 6″ nd the tallest were 10.5″ high; the maximum spread was 5″ and they haded the grass somewhat. They were thrifty but yellowish and 10 of hem were in bloom, but the growth at this time was little if at all better han in 1924. The grass was very thrifty, averaging 6″ with a maximum f 11.5″ and with 1–4 tillers per plant. The light intensity on August rst was 5% at the soil surface and 31% at a height of 3″.

On August 20, *Amarantus* numbered 34, of which 18 were suppressed, nd *Andropogon* 246; the mortality during the season had been 48% for he former and 11% for the latter. The ruderals averaged 12″ in height, ith a range of 3″–19″. Their color was good and practically all had loomed, while seed production was copious on the taller plants. The grass veraged 8″ tall and had tillered abundantly. The leaves were somewhat arrower than normal, but as a whole this species showed little effect from ne competition of the weed. The much better growth of *Amarantus* uring this season was due chiefly to the addition of the nutrient solution, ut some effect is to be ascribed to the decreased number of the grass and lso to the higher water-content by comparison with 1924.

By June first, 1926, *Andropogon* formed a dense uniform stand to a eight of 8″. On account of the drouth it grew rather poorly, but was in omewhat better condition than in quadrats where it had a competitor. n spite of the great seed production of the preceding season, no *Amarantus* appeared, suggesting the impossibility of the germination of the seeds of ruderals in a dense stand of grass.

Andropogon furcatus vs. Ambrosia trifida

Season of 1924—Of the total of 461 plants in the quadrat on June third, 51 belonged to *Ambrosia* and 410 to *Andropogon.* The former were 6″–8″ in height, with a top spread of 2″–5″ and usually with 6 leaves; the latter were 2″–3″ tall with 2–3 leaves. On June 27 there were 566 plants, 63 of the ruderal and 503 of the grass. The former was 11″ high for the most part and bore an average of 8 leaves. The shade was continuous in the quadrat and as a result the grass was pale green in color, the blades were narrow and there were no tillers. The average height was 3″, while the check was 4.5″ tall.

During the next period the mortality was slight; the total on August 8 was 561, of which 62 were *Ambrosia* and 499 *Andropogon.* Of the former 17 and of the latter 313 were suppressed. The dominant *Ambrosia* ranged from 20″–36″ tall, and 12 plants bore well-formed spikes about ready to bloom. The suppressed individuals were about 8″–10″ high, green in color and thriving. The dominants among the grasses averaged 9″ in height, but tillers were few except along the edges. The plants were attenuated, the leaves narrow and many of the lower ones dead. The grasses were noticeably yellow in the middle of the quadrat.

By September 8 the total had been reduced to 526, owing to the death of 35 grasses; 16 of the ragweed and 247 of the grass were suppressed. The average height of *Ambrosia* was 39″, though a few were fruiting at a foot. The stems had shed most of their lower leaves, which had fallen on the grass beneath. *Andropogon* averaged 10″ tall except for those on the margin, which were much taller. The effect of the taller and wide-spreading competitor was evident in the slender culms, narrow leaves and sparse tillers, but in spite of this, the mortality for the summer was only 8% This was probably due to the fact that water relations were more favorable under the ragweed than in the adjacent prairie.

Season of 1925—As early as April 7 a number of seedlings of *Ambrosia* had unfolded their cotyledons, but *Andropogon* had not yet made its appearance. By August 17 no *Ambrosia* was to be found, and the absence of stems indicated that this had succumbed in the seedling stage. *Andropogon* comprised 16 clumps, ranging from 2–6 stalks per plant; these averaged 8″ and were broad-leaved and vigorous. The striking reduction in the number of grasses was the evident outcome of overshading the previous summer and a consequent heavy mortality during the winter.

Comparative behavior and factors—*Ambrosia trifida* proved to be a much more effective competitor of *Andropogon furcatus* than did *Amarantus retroflexus.* This was evidently the result of its more rapid growth, broader leaves and greater stature. *Amarantus* was at no time a handicap to the grass, while *Ambrosia* suppressed it more or less completely during its season of growth. The most significant outcome, however, was that neither annual was able to germinate and make a place for itself as a consequence of the seed produced the preceding year. This accords with the almost complete absence of annuals in the climax matrix of the several prairie associations, and also serves to explain why they are abundant in xerophytic grassland only in seasons of high rainfall.

GENERAL SUMMARY

In analyzing the behavior of the paired competitors and seeking the reasons for their success or failure, it is necessary to take into account the features of each species, the course of the process itself, and the factors concerned. To employ Warming's figure, it is the "weapons" of the competing species that are of the first importance, though the effectiveness of these is greatly influenced by the rainfall of summer and the severity of winter. The actual course of competition, the period of time involved, the final fate of the competitors, and the significance for the development or persistence of the community will depend upon the relative advantages afforded by structure and function, and the degree of control exerted by habitat or season.

Structural features concerned—The so-called weapons comprise all the structures of the plant capable of making an effective demand upon a limited supply, directly or indirectly. These demands must be expressed in terms of function, and preliminary attempts to determine these by quantities have been made throughout the present investigation, particularly by means of phytometers. However, in all cases a fair understanding is possible of the functions involved by virtue of the measurement of the factors concerned, the known correlation between structure and function, and the detailed behavior of the competitors through the course of competition. More important still is the fact that the measurement of reactions in terms of chresard, reduced light intensity and nutrients is itself a determination of functional activity.

While every structure has a necessary bearing upon the outcome of competition, some produce a much more direct and important effect than others. The leaf bears a relation to both light and water, the root to water and nutrients; the rôle of the stem is chiefly secondary or indirect in that it determines the position of the leaves with respect to energy absorption and water loss. Growth is naturally of paramount importance, since it not only determines the total demand, but the time and rate at which it is made. It is expressed in germination, development, tillering or branching, size and stature, reproduction, and the maturing or "ripening" of the plant body in preparation for winter. The structural features that result from growth are summed up in the life-form, such as annual or perennial forb, sod or bunch grass, shrub or tree, all of which exhibit secondary characters of importance in the various forms of roots, stems and leaves. Finally, there are other less obvious but equally significant qualities, such as vigor and hardiness or resistance to winter-killing. Number, which is often the most decisive advantage of all, is the primary outcome of growth, in terms of reproduction and germination.

It is evident that practically all the advantages or weapons of competing species are epitomized in two words, amount and rate. Greater storage in seed or rootstock, more rapid and complete germination, earlier start, more rapid growth of roots and shoots, taller and more branching stems, deeper and more spreading roots, more tillers, larger leaves and more numerous flowers are all of the essence of success. Apart from the quantity relation, form and minute structure are chiefly important in connection with the habitat and seasonal factors rather than in the process of competition itself.

Competition between tall-grass dominants—In the competition between *Elymus canadensis* and *Panicum virgatum,* the advantages of the former were decisive. Germination was better, growth more rapid and the height greater; the root system was more efficient and tillering more abundant while it was less subject to winter-killing and made an earlier start in the spring. The last two features are a consequence of its boreal origin. *Panicum,* on the other hand, is more southern and was further hampered by its habit of growing normally in a higher holard. As a consequence water was rather more effective than light, and particularly so when its relation to winter-killing is taken into account. *Andropogon nutans* is more vigorous than *Panicum* and somewhat more xeroid; it gave better final germination than *Elymus,* the seedling mortality was lower and it tillered more abundantly. However, the winter loss was much heavier in correspondence with its southern affinities, and its start in spring likewise later. It prefers somewhat moister soil and hence this lowers the effectiveness of its absorbing system. As with *Panicum,* both water and light were intimately concerned in the outcome between *Andropogon* and *Elymus* the success of the latter being due almost equally to both factors. However, in the subclimax prairie where the rainfall is abundant and winter temperatures not too low, the vigor and stature of *Andropogon* have their full effect, and this species is usually more important than either of the others.

The ability of *Andropogon furcatus* as a competitor has been tested against two of its tall-grass associates, *Panicum virgatum* and *Spartina cynosuroides,* and a mid-grass of the true prairie, *Sporobolus asper.* It excels the first in the production of tillers, in being more tolerant and more xeroid, and also more resistant to winter-killing, but it germinated less well and the seedlings grew more slowly. *Spartina* was inferior in germination, absorption and tillering to a decisive degree and also in resistance to cold. These were largely results of its hydroid nature and are connected with its position as a seral dominant at the end of the hydrosere, thus readily explaining why it yields regularly to the various Andropogons. Both water and light were about equally critical in the case of *Panicum* since it requires more water than *Andropogon,* has a poorer root system and is usually overtopped by this competitor. The greater success of *Sporobolus asper* by comparison with *Andropogon* was due to its more rap growth, better absorption and tillering, which more than offset the potential height and better germination of the bluestem. Because of its better adjustment to the climate of the true prairie, its water requirement was lower, with the result that water was first decisive and then suppression by shading in consequence of more rapid growth. This culture was conclusive as to the difficulty experienced by tall-grasses in invading the true prairie of mid-grasses.

Competition between tall and short grasses—In these cultures, first *Bouteloua gracilis* and then *B. hirsuta* was tested against a tall-grass, *Andropogon furcatus*, and also against a mid-grass, *Agropyrum glaucum*. The taller species possessed a decided advantage in height, which was felt especially during wetter seasons, but the recurring drouth intervals tended to favor the more xeric short-grasses. The water requirements of the latter were lower, they produced tillers more abundantly, and they were hardier and also began growth earlier in the spring. The sod habit was usually an advantage likewise. In the case of *Andropogon* and *Bouteloua*, the light factor favored the former when the water-content was good, while the short-grass prospered when the holard was too low for normal growth of the bluestem. The usual result is a delicately balanced structure with the tall-grass forming the upper, the short-grass the lower layer. Between *Agropyrum* and *Bouteloua*, the ability of the latter to form tillers in much greater abundance and to grow with much less water during drouth intervals gave a clear-cut advantage to this species, in spite of its much lower stature.

Competition between dominant and subdominant—*Panicum virgatum* in these cultures was pitted against three species of forbs of very different life-form, namely, *Oenothera biennis, Liatris punctata,* and *Solidago rigida*. In spite of much better germination, the grass was handicapped by the lack of tillers, by winter-killing, and by slower growth than in the case of *Oenothera* and *Solidago*. The rosette form of *Oenothera* with the later tall stems and broad leaves gave this species a marked superiority for the first two years, both as a result of shading and absorption, but like the annuals, it disappeared completely at the end of the biennial period. *Panicum* and *Solidago* competed on nearly equal terms during the first summer, but the boreal nature of the latter gave it the lead as a consequence of resistance to winter-killing and of a much earlier start the following spring. The resulting rapid growth enabled the golden-rod to suppress the grass completely after this time. *Panicum* was successful only in competition with the very slow-growing *Liatris*, confronted by the early necessity for storage in its tuber. Water was the critical factor, since the tolerance of the forb enabled it to persist in spite of its handicaps.

Competition between subdominants—With the exception of *Oenothera*, the species employed were all perennial forms of the true prairie, found to some extent also in the subclimax. In the culture of *Petalostemon candidus* and *Brauneria pallida*, the former germinated more readily and growth was more rapid, the absorbing system giving a further lead. The latter endured winter better and made an earlier start in the spring, but this did not compensate for reduced absorption and light intensity during the summer. In the competition of *Liatris* with the much more vigorous and rapidly growing *Kuhnia* and *Oenothera,* both species, *punctata* and *scariosa,* were at a great disadvantage, especially with respect to light. The one persisted greatly suppressed, the other vanished completely. In the case of slow-growing forbs, as with *Agropyrum,* the abundance of *Liatris* in high or low prairie according to the species indicates that it may take more than four years for thorough ecesis, but once established, these species are very persistent.

In the triple competition between *Helianthus rigidus, Amorpha canescens* and *Kuhnia glutinosa,* the first had the advantage of vigor, rapid growth and height, and of forming a dense community. As a consequence, it quickly became dominant and suppressed both competitors to a degree approaching extinction. When the individuals were fewer, however, they were unable to make headway against *Kuhnia,* which became the dominant, *Liatris* assuming its usual subordinate position in cultures. In both cultures, stature was decisive after establishment, and light was in consequence the paramount factor.

Competition between dominant and ruderal—*Andropogon furcatus* was matched against two ruderals, respectively *Amarantus retroflexus* and *Ambrosia trifida.* The grass germinated about four times as well as either competitor, and possessed the further advantage of potential height and a better root-system. For a time the competition beween *Andropogon* and *Amarantus* was keen for both water and light, but lack of water and nutrients was chiefly responsible for the marked suppression of the latter. The rapid growth of *Ambrosia* enabled it to take the lead and hold it during the first summer, chiefly in consequence of the reaction of the tall spreading stems and broad leaves upon light. Unlike *Amarantus,* some of the seeds produced germinated the second spring, but the seedlings were unable to compete with the better established grass and the final fate of the two weeds was the same.

3. TRANSPLANT CULTURES IN TRUE PRAIRIE

True or high prairie—This is the characteristic climatic community of the rolling hills of the Missouri Valley within the general limits of 25″–33″ of rainfall. It differs from the included areas of subclimax or low prairie in the valleys by virtue of consisting of mid-grasses rather than tall-grasses. The dominants are *Stipa spartea, Sporobolus asper, Andropogon scoparius, Koeleria cristata, Bouteloua racemosa,* and *Agropyrum glaucum,* with considerable *Stipa comata* in the ecotone with the mixed prairie. The major and typical dominants are *Stipa spartea* and *Sporobolus asper,* since they are more abundant in this association than in any other, while the remaining dominants are equally important in the subclimax or in the mixed prairie.

Both tall-grasses and short-grasses are found in the true prairie, but they are properly relicts due to former climatic changes, or are the result of disturbance in some degree. Since *Andropogon furcatus* and *nutans* occur in abundance in the valleys, they have found their way up the mesocline slopes of the north and east. As sod-formers they have taken advantage of mowing and grazing to spread gradually at the expense of the bunch-forming mid-grasses into those situations with relatively higher holard. *Poa pratensis* has similarly moved up the swales from the low prairie, though it apparently makes no headway against the native mid-grasses in the absence of mowing or pasturing. The three short-grasses, *Bouteloua gracilis, B. hirsuta* and *Bulbilis dactyloides,* are all relicts of a recent dry phase of the climatic cycle, as is shown by their persistence on dry ridges and xerocline slopes of the south and west. These too have been favored by grazing, and *Bulbilis* in particular has extended its range and increased its abundance as a result of pasturing, owing to its migration by stolons.

Station and cultures—The high-prairie station is located on a rather flat hill-top about 60 feet above the general level of the flood-plain of Salt Creek and about ¼ mile from the low-prairie station. The soil is Carrington silt-loam, a fertile glacial drift with a maximum water capacity of about 60% (Hilgard method). This soil is fine in texture and is sufficiently supplied with calcium to lack acidity (cf. "Experimental Vegetation," p. 11). Determinations through many years indicate that the subsoil is usually moist to great depths, but during drouth periods of great intensity the chresard may be reduced to the point at which daytime wilting occurs. Even this is of very infrequent occurrence, since the native species of the true prairie possess deep root systems, some of them penetrating to 20′ or more.

The general organization of the cultures was much the same as for the low prairie, six series being represented. The additional series was devoted to competition between short-grasses, and a larger number of cultures was employed to test the course of competition between dominants and ruderals. The species utilized were the same for the most part, though the combinations differed in a number of instances.

DOMINANT VERSUS DOMINANT

The competition cultures of this series employed five of the six dominants of the true prairie in competition with a tall-grass or with each other, and in addition two tall-grasses in competition to test the influence of the high-prairie habitat. The particular combinations were the following: *Stipa spartea* vs. *Andropogon nutans;* (2) *Sporobolus asper* vs. *Andropogon nutans;* (3) *Andropogon scoparius* vs. *Agropyrum glaucum;* (4) *Agropyrum glaucum* vs. *Bouteloua racemosa;* (5) *Andropogon nutans* vs. *Elymus canadensis.* Finally, it proved possible to follow the course of competition in a pure culture of *Andropogon nutans,* which resulted from the failure of *Verbascum thapsus* to grow in a quadrat of dominant versus ruderal.

Root relations—The root systems of *Andropogon nutans, Agropyrum glaucum,* and *Elymus canadensis* have been described in the previous chapter. *Stipa spartea* possesses a rapidly growing root system which reaches an average depth of 18″ and a maximum of 28″ by midsummer, when the culm is 15″ tall and bears 3–4 tillers (cf. "Experimental Vegetation," p. 30, fig. 13). Older roots, even those of mature plants, scarcely extend more deeply in the prairie, but are exceedingly well-fitted for absorption in the upper levels. Numerous profusely branched smaller roots fill this region, and the larger ones send off many laterals that divide into fine branches in the deeper soil.

Sporobolus asper produces coarse fibrous roots that penetrate the soil to an average maximum of 2′, the greatest depth found being 3.5′. As a compensation, the root system is very dense and thoroughly occupies all of the soil, spreading from the base of the plant almost horizontally to a distance of 12″–20″. In consequence, a soil area of 6–7 square feet may be preempted to a depth of 18″ by the roots of a single bunch. The roots branch profusely, but the laterals are only 0.5″–2″ long, though they may branch again. Below the 18″ level the roots are comparatively few in number.

Andropogon scoparius possesses an exceedingly efficient absorptive system. By late July seedlings 5.5″ tall had developed 3–4 tillers and 8 roots each, reaching a working level at 12″ and a maximum depth of 18″. The roots are fine and extremely well branched, some of the branches attaining lengths of 3″–4″. In the cultivated soil the roots were more numerous and reached slightly greater depths. The extent of the root system under competition is shown in plates 8 and 9; it penetrated to 1.5′ the first season and extended beyond the third foot by the end of the second year.

Bouteloua racemosa develops a root system that is moderately deep, wide spreading and exceedingly well branched, in which respects it resembles its associates of the high prairie. In tilled soil the seedlings were 11″ tall by July 30, when the roots had reached a working level of 18″ and a maximum of about 3′. The plants bore 7–8 tillers, 15 roots and 1–2 rhizomes. Even at a depth of a foot the laterals were sometimes 9″ long, and they spread abundantly through the upper levels of the soil.

Stipa spartea vs. Andropogon nutans

First culture, season of 1924—On June 4 there was a total of 438 plants, 240 of them *Stipa* and 198 *Andropogon;* the former was 3″–5″ tall, the latter 1″–2″. *Stipa* was more abundant in parts of the quadrat and *Andropogon* in other parts. On July 8 the total number was 423, but *Stipa* had decreased by one hundred and *Andropogon* increased by about the same amount, and were respectively 119 and 304. The first ranged from 3″–5″ high and was tillering nicely, but half of the original stand had died. *Andropogon* was 5″–8″ tall where there was little of its competitor, closely resembling the controls, but in the spots where *Stipa* was abundant, it was only 3″–5″.

By August 11, *Stipa* had dropped to 74 and *Andropogon* to 275, making a total of 349; of the former 38% and of the latter 10% had died, and about a third of each was suppressed, namely 22 and 97. The average height of the stand was 8″, some of the *Andropogon* reaching 12″–14″, with *Stipa* somewhat shorter. The former was tillering better and appeared more thrifty. There were many dead leaves at the base of the plants, but the cover was not especially dense.

The number of plants on September 5 was 270, of which 58 were *Stipa* and 212 *Andropogon;* the mortality during the preceding month had been about 23% for each. *Andropogon* made up about 72% of the population in the quadrat on July 8, while at present this had increased to 80%. Of this species 97 or 46% was suppressed, while of *Stipa* but 16%. The bluestem averaged 9″ high with a maximum of 14″; the plants were about 2″ shorter than the controls and the leaves narrower. *Stipa* averaged 10″ in height, some extending to 16″. They likewise were 2″ shorter than the controls and bore but 1–2 tillers instead of 3–4. The leaves were green almost to the tip, while those of its competitor were dry and brown for 1″–3″ at the tip. At this time *Stipa* appeared to be more thrifty than *Andropogon,* in spite of the fact that over one-half of it had died during the summer, by contrast to one-third of the bluestem.

Season of 1925—On April 7, *Stipa* was already in vigorous growth with a stature of 4″–5″ and 2–3 leaves, while *Andropogon* was just beginning to appear. By June 16 there were 34 bunches of *Stipa* consisting mostly of 2 to many stalks and 154 *Andropogon,* all but 3 with single stalks. The former was very thrifty, averaging 15″ and with a maximum of 21″; the bunches were large, one comprising as many as 23 stalks. The bluestem was 3″–5″ tall with 2–4 fairly broad leaves per plant. *Stipa* was dominant over most of the quadrat, but *Andropogon* formed a dense mass where this was less abundant.

On July 15, *Stipa* averaged 18″ tall to 10″ for *Andropogon,* while a month later some plants had stretched up to 25″. Under this dense mass *Andropogon* had died out almost entirely, the survivors averaging 6″ tall, though the light value was 20% or more. At its best the bluestem was 9″ in height, but even in this area *Stipa* was beginning to overtop and shade it. Determinations of the chresard on August 28 gave values of 5.5% in the

6″ level and 9%–10% at the several depths to 2′. These readings were 2.4% lower for the first 6″ than in the prairie, 3.4% lower for the second, but about the same for the second foot. They indicate the heavy demand made by the competing species, especially during the drouth of early spring.

Season of 1926—By April 15 the bunches of *Stipa* were in active growth with leaves 3″ long, but the shoots of *Andropogon* were just beginning to appear. On June 9 the former was fairly dense with a height of 25″ and was fruiting abundantly, while the latter was in poor condition and was suffering from drouth. The respective abilities of the two species to resist drouth was well evidenced on August 4 when the bluestem was drying badly, while the leaves of *Stipa,* though tightly rolled, were green and the plants were thrifty. These bore flower stalks 3′ tall and had developed seeds. On September first *Stipa* was still thriving, the foliage reaching a level of 2′; *Andropogon* was much overshadowed, but 7″ high and in poor condition. The plants were greatly dwarfed and the leaves noticeably narrower.

On June 30, 1927, practically the entire area was dominated by *Stipa,* though some *Andropogon* persisted along the edges of the quadrat.

Second culture, season of 1925—*Stipa* germinated poorly and the seedlings were not so far advanced as those of *Andropogon* on June 9 when the respective numbers were 15 and 355. *Stipa* had been at a disadvantage from the outset. It was not only soon overtopped by *Andropogon,* but the rootlets of the latter occupied the soil so completely as to deprive it of water. The bluestem was thrifty, averaging 4″–5″ and with 2–3 leaves, which had dried back 1″–3″ from the tip in consequence of drouth. On the contrary, the seedlings of *Stipa* had suffered greatly from the drouth, were much dwarfed and the leaves were rolled. On June 27 the light intensity near the soil was but 6%, so that shading was also a decisive factor.

On July 21 the number for *Stipa* was 12 and for *Andropogon* 322; the former was in poor condition, with rolled leaves and but 2″–3″ in height. The bluestem formed a dense stand 7″ high and was clearly dominant. The leaves were broad and tillers so abundant that it was becoming difficult to distinguish the parent plants. On August 20 there were but 8 plants of *Stipa* to 381 of *Andropogon. Stipa* had now become negligible; none of the culms were over 5″ tall and scarcely any tillers had developed.

The chresard for the latter part of the summer is indicated in the following table; the figures in parentheses indicate the amount in the adjoining grassland. The samples of August 18 were taken only a few days after good rains.

Table 3—*Chresard in Stipa-Andropogon culture*

Date	0.0′–0.5′	0.5′–1′	1′–2′	2′–3′
	%	%	%	%
July 6........	2.1	6.9	8.5	
July 29........	2.4 (1.5)	4.0 (5.3)	5.1 (7.4)	(8.2)
Aug. 18........	18.9	16.2	13.9	18.4
Aug. 28........	2.5 (8.4)	7.7 (13.7)	8.3 (9.3)	(7.2)

Season of 1926—On April 10 the new leaves of *Stipa* were 1″–1.5″ long, while those of *Andropogon* were just appearing. On May 31 there were but 106 of the 381 individuals of the latter present the preceding September and more than half or 65 were suppressed. *Stipa* numbered 54, of which all but 3 were suppressed by the dominant bluestem. Since but 8 plants remained at the close of the previous summer, most of these were derived from seeds that had lain dormant since the original planting. *Andropogon* averaged 8″ with a maximum of 14″; it was quite thrifty, but tillers were not abundant on account of the dense stand. *Stipa* was only 3″–5″ tall and most of the leaves were tightly rolled; its poorer root system caused it to endure drouth less well than the bluestem. By the final examination in September all of the *Stipa* had disappeared, and the quadrat was covered with a dense uniform sod of *Andropogon* to a height of 10″

Comparative behavior and factors—The outcome of the competition between a mid-grass and a tall-grass indicates that dominance may often depend upon what seem to be slight and fortuitous differences. As the most characteristic dominant of the true prairie, *Stipa spartea* should possess a decisive advantage over *Andropogon nutans,* a conclusion further supported by its stature and root system. This is confirmed by the course of events in the first culture in which the two species germinated and grew about equally well. Although *Andropogon* appeared to be in the lead during the summer, the season closed with *Stipa* in better condition and this was much enhanced by its early start the following spring. In spite of the fact that the bluestem was present in much larger number, the early start and much more rapid growth of its competitor proved conclusive and the bluestem dwindled steadily throughout the summer. This was primarily a consequence of shading, but the larger demands of *Andropogon* for water, accustomed as it is to a higher holard, played some part.

The results from the second culture afforded a striking contrast. *Stipa* germinated so poorly that at the end of the first month the ratio of seedlings was 1:22 in favor of *Andropogon.* It was handicapped from the outset as to absorption, and this told so strongly that the bluestem soon overtopped and overshaded it. *Stipa* was negligible before the close of the season and, notwithstanding the recruits secured from belated seeds the next spring, it vanished before the end of the summer. Poor germination is usually slow germination also and is often accompanied by slow growth. In a dense stand of seedlings, differences of an inch or two in root length, representing the growth of but one or two days, may easily prove decisive both as to the absorption of water and the consequently rapid growth that results in overtopping. Such a contingency probably arises but rarely if ever between the dominants of different associations, except in the ecotone, but must always be kept in mind in a particular community.

The early germination and growth of *Stipa spartea* under normal conditions throws much light upon its behavior in nature. Not only this species but others of the genus, such as *comata, pennata, setigera,* etc., have disappeared under the impact of overgrazing, except in protected places. This is chiefly a consequence of their early growth in spring at a time when other grasses have started but little or not at all. Such grazing has greatly decreased their later growth and seed production, and placed them at an increasing disadvantage with such competitors as *Andropogon* and other tall-grasses. The bluestems not only start materially later, but their sod-form is more of an assurance against overgrazing than the bunch-form of *Stipa*. These differences are also effective in the case of fire, especially in late winter or early spring, and they serve to explain the gradual extinction of *Stipa spartea* over much of the southern portion of the true prairie and the steady encroachment of the three species of *Andropogon, furcatus, nutans* and *saccharoides*. A similar fate is overtaking *Koeleria cristata* in the same region, though to a less degree, owing to its starting somewhat later.

Sporobolus asper vs. Andropogon nutans

Season of 1924—By June 12 there was present a fairly uniform stand of *Andropogon,* but *Sporobolus* was not thriving, the tips of half the plants being dead. On July 8 the total number of plants was 492, 355 of them *Andropogon* and 137 *Sporobolus*. The dominant individuals of the former averaged 6″–8″, the suppressed but 1″–4″; those of the latter were 6″–8″ high also, but fewer were suppressed. Both species were as well developed as in the control plots and were tillering. The stand was dense and the competition correspondingly keen.

By August 8 there was a total of 476 plants, of which 365 were *Andropogon* and 111 were *Sporobolus;* the former had increased slightly in number while the latter had dropped 19%; 233 or 64% of one and 46 or 41% of the other were suppressed. The general grass level was 10″, but many leaves of *Sporobolus* reached upward to 16″–17″, while the tallest of *Andropogon* were about 12″. The dominant individuals of the first were more thrifty than those of the second; there were many dead leaves and tillering was not abundant in either.

On September 4 the number was 274, of which 198 were *Andropogon* and 76 *Sporobolus*. The mortality during the preceding month had been 46% for one and 31% for the other; 147 or 74% and 31 or 41% respectively were suppressed. Although the tips were much attenuated and often dead, *Sporobolus* was in much better condition at the end of the season; it averaged 16″–22″ tall as against 10″–12″ for its competitor. The stand was dense and there were many dead and dying plants. In spite of the fact that *Andropogon* was much more numerous at the start and the mortality was practically the same for both, the suppression in autumn was nearly twice as great, as a consequence of the much greater stature of *Sporobolus*.

Season of 1925—Growth had not begun in either species by April but on June 13 there were 77 plants of *Andropogon* and 62 of *Sporobolus*. The toll taken by winter had been but 14 for the latter to 121 for the former, a result to be ascribed to the poorer condition of *Andropogon* at the close of 1924. *Sporobolus* was distinctly dominant, averaging 12″ and with a maximum of 22″; it formed thrifty clumps with many shoots and leaves

On the other hand, *Andropogon* was 8″–9″ high with but 2–3 culms per plant and 2–3 leaves per stalk, and in consequence the quadrat appeared to consist entirely of *Sporobolus*.

On July 15 the stand was 16″ in height. Determinations of the chresard at this time gave 3% in the upper 6″ level, 7.7% in the second and 9%–13% at the further depths to 3′. The upper layer was 1% drier than the corresponding one in the prairie, but the latter contained only 6%–7% of chresard at the other levels. The competition in consequence was keenest in the upper layer and affected most the subdominant individuals with shallow roots.

By August 17, *Andropogon* numbered 40 plants, while *Sporobolus* bore about three times as many stalks. The latter was so dense in consequence of abundant tillering that individuals could not be distinguished; it averaged 17″ tall and some leaves were 31″ long. The tallest bluestems were n the best lighted area, where they averaged 12″ high and still carried on effective competition. However, in most of the quadrat it had lost in the struggle and was but 6″ tall, except for the extreme margin. The shade was fairly dense on the ground, but the light value at 3″ on August first was as high as 36%. By October 5, *Sporobolus* had produced abundant seed.

Season of 1926—By June 9, *Sporobolus* averaged 13″ in height and was four times as abundant as *Andropogon*. The latter occurred only on he margin and in the more open spaces, and was clearly losing ground because of the drouth and its inability to endure shade. On August 4 *Sporobolus* was 18″ tall and still flourishing in spite of drouth conditions. *Andropogon* was much affected by the low light intensity, which was but % at 6″; it was drying badly and had many dead leaves. By September rst the dominant had reached a height of 20″ and bore many flower stalks 30″–33″ high. The plants were well tillered, green to the base and hriving. The late summer rains enabled *Andropogon* to make a more igorous growth and thus regain some of the ground it had lost. On une 30, 1927, both species were growing well and the final outcome was not decided, in spite of the lead held by *Sporobolus*.

Comparative behavior and factors—The general course of competition avored *Sporobolus asper* at the expense of *Andropogon nutans*. This was argely the consequence of the more rapid growth of the former, its more xtensive tillering and the ability to form a denser stand. It was also better adapted to obtain water during dry periods and it suffered much less om winter-killing, chiefly as a result of its better condition at the close f autumn.

The somewhat indecisive outcome after four seasons is to be assigned n part to the timeliness of late summer rains for *Andropogon*, but even more to the ecological resemblance of the two species. Both are of southern derivation and occupy much the same range; they are also vigorous rowers and late bloomers, often flowering in late September and October. Their climatic relations are more nearly alike than would be expected, nce *Sporobolus* occurs much more abundantly in the subclimax and coastal prairies than any other of the typical dominants of the true prairie.

Andropogon scoparius vs. Agropyrum glaucum

First culture, season of 1924—The total number of plants on June 12 was 269, of which 201 belonged to *Andropogon* and 68 to *Agropyrum*. The latter averaged 3″ and were vigorous, the former but 1″ and they were less vigorous. On July 8 the number was the same, 8 plants of *Agropyrum* having died and as many seedlings of *Andropogon* having appeared. The first ranged from 3″–9″ high and bore 2–3 tillers per plant; the second was 2″–3″ tall and was tillering abundantly. The quadrat was still open and the competition did not seem severe.

On August 11, *Agropyrum* was represented by 28 and *Andropogon* by 189 plants, making a total of 217. The respective losses were 53% and 10%, and the suppression affected 9 and 64 individuals. Most of the *Agropyrum* was in poor condition and only a few reached a height of 11″–12″. The average for *Andropogon* was 7″, but it was tillering abundantly and was quite thrifty. By September 5 there were 4 *Agropyrum* and 139 *Andropogon* present, making a mortality since July 8 of 93% and 33% respectively. In two months the former had fallen from 22% to about 3% of the total stand. At this time about half of the individuals of each species were suppressed; the two dominant plants of *Agropyrum* were 11″ tall, but not thrifty. *Andropogon* averaged 8″ and was prospering.

Season of 1925—On April 17 both species were beginning to grow and were in good condition. During the ensuing dry period the chresard on May 26 was 6.5%, 11.8%, 12.3% and 13.1% respectively at the several levels. On June 16, *Agropyrum* numbered 5 plants, which were in good condition, averaging 9.5″ tall with 3–5 leaves each. *Andropogon* formed a stand 7″ high and was so dense that the individuals could not be distinguished. The competition between these was severe and most of them were attenuated and bore narrow leaves.

By July 15 there was a good growth of bluestem, 8″ tall, while but 2 plants of *Agropyrum* remained. These were still present on August 18; they were 11″ high and the tips were dead. *Andropogon* was extremely dense to a height of 9″; the shade on the ground was heavy and the many dead leaves formed a layer 3″ thick. By October 5 a few flower stalks had developed on *Andropogon*.

Season of 1926—On April 15, *Agropyrum* alone had made new growth, a few plants reaching a height of 6″, but by June 9 *Andropogon* was 8″–14″ in height and formed a fairly dense and uniform stand. The folded and dried leaves indicated that it was suffering from drouth. There were 3 plants of *Agropyrum* 11″–15″ tall that were apparently thrifty, but only one had produced tillers. By August 4 the dominant bluestem was distinctly brown and the three individuals of *Agropyrum* had the leaves tightly rolled. Late in September the former constituted a dense cover 11″ tall but bore no flower stalks, while the single culms of *Agropyrum* were well developed. These three plants still persisted on June 30, 1927, in a very dense growth of the dominant bluestem.

Second culture, season of 1925—In this culture, care was taken to have the seedlings of *Agropyrum* outnumber those of *Andropogon,* with the result that on June 9 there were 282 of the one and 54 of the other. The former averaged 4″ with a maximum of 5″, and bore 2 leaves per plant. The tips were very slender and were dried back for 1″–3″. *Andropogon* was but an inch tall, with the leaf tips and lowest leaf dry, and obviously had suffered more from the drouth. On July 21 there were 205 plants of *Agropyrum* and 43 of *Andropogon;* the former averaged 6″ and was thrifty, but usually without tillers. The drouth was more severe on the bluestem, which was in poorer condition, with a height of 2″ and 3–4 tillers on the best plants. There were many dead individuals, in the center of the quadrat especially.

By August 20, *Agropyrum* numbered 204 and *Andropogon* 40 plants, the mortality to date being 28% in the one and 26% in the other. *Agropyrum* averaged 6″ tall with a range from 2″–10″, while *Andropogon* was but 2″–3″ high and had not tillered so well as in the other quadrats. The culture at this time was very open and the stand of both species was scattered and poor. The large number of dead plants and basal leaves reflected the severe drouth, in spite of the fact that the quadrat had been repeatedly watered.

Season of 1926—On April 10, *Andropogon* was just beginning to turn green, while *Agropyrum* was 2″–4″ tall. Winter-killing had exerted practically no effect, the numbers of May 31 being 218 for *Agropyrum* and 48 for *Andropogon,* with 83 and 34 respectively being suppressed. The former reached a level of 8″ with a maximum of 12″; it was fairly thrifty in an open stand, but the rolled and dead leaves showed that it was suffering from drouth. On the other hand *Andropogon* was so much suppressed that it could be found only upon close examination. The soil had been disturbed by repeated watering and the plants were attached by only a few roots, which were often plainly visible.

On August 4, *Agropyrum* formed a scattered growth throughout the quadrat, reaching a maximum of 12″. The stand was open and a layer of debris covered the surface of the soil. *Andropogon* persisted only as remnants, which had entirely disappeared by September. At this time *Agropyrum* formed a medium stand with an average height of 10″ and a maximum of 14″. The late rains had produced many new shoots 3″–4″ tall, but no flower stalks had appeared.

Comparative behavior and factors—The outcome of the competition in these two cultures was determined primarily by the initial number of seedlings, though the difference in the character of the first season also played some part. The maximum numbers for the first culture were 209 for *Andropogon* and 68 for *Agropyrum;* for the second the situation was reversed, the maxima being 54 and 282 respectively. In each case the higher number determined the dominant and finally in the second culture the survivor. The advantage of *Andropogon* in the first quadrat was increased by the greater rainfall, since it is less xerophytic than *Agropyrum.* Its bunch habit is the result of abundant tillering and this gave it a further lead in securing the larger share of the water available. Though it was much denser than its competitor, the leaf level was lower and overshading played little part in the outcome.

The drier season experienced by the second culture favored the more xerophytic *Agropyrum,* and the greater stature and higher leaf level of this species further augmented the handicap that small numbers placed upon *Andropogon.* The effect of shading was seen especially in the suppression during the second season; this amounted to 70% in the latter and 37% for the former.

The competition results accord with the abundance and rôle of *Andropogon scoparius* and *Agropyrum glaucum* in the true and mixed prairie. The former is much more abundant in the eastern and especially the southeastern part of the true prairie, while the latter increases in importance to the west. In the mixed prairie *Agropyrum* occurs as a major dominant, except under overgrazing, while *Andropogon* is restricted to the "breaks" of ravines and bluffs, where the cover is open and the demand for water less.

Agropyrum glaucum vs. Bouteloua racemosa

First culture, season of 1924—The total number of plants in the quadrat on June 12 was 125, of which 94 belonged to *Agropyrum* and 31 to *Bouteloua.* The former was quite vigorous at 3″–4″, while the latter reached 1″–1.5″. By July 8 the number had decreased to 115, but this affected *Agropyrum* alone, its mortality being 29%. *Agropyrum* ranged from 2″–9″ in height, with an average of 5″ and of 2 tillers for the largest plants. *Bouteloua* averaged 2″ and was spreading widely, the numbers of tillers reaching as many as 8 per plant.

By August 8, *Agropyrum* had undergone a further loss of 84%, leaving but 11 plants, while *Bouteloua* had decreased about 40% or to 29 individuals. The total number was only 40 and in consequence the quadrat was very open. A few plants of *Agropyrum* had reached 8″–12″ before dying, while the average for *Bouteloua* was 3″. On September 5 there were but 19 plants, one *Agropyrum* with a single stalk 8″ high and 18 *Bouteloua.* The latter were 4″–6″ tall and formed good bunches about an inch in diameter at the base. The high mortality during the month had evidently been in large measure the result of the openness of the quadrat, increasing both evaporation from the soil and water-loss from the plants.

Season of 1925—On April 7, *Bouteloua* had started to grow, but no plants of *Agropyrum* had appeared by June 16 and later observations demonstrated that this species had vanished finally. There were 6 clumps of the former 1″–2″ wide and 6 smaller ones consisting of individuals with 4–9 tillers each. All the clumps were poorly rooted, many of the roots being exposed, and the general appearance was unsatisfactory. By July 15 the sparse stand was blooming at a height of 10″–12″. On August 17 there were 15 bunches ranging from 0.5″–2″ wide; the leaves were about 6″ long and the 7 flower stalks 14″ tall. The leaves were broad and the plants appeared very vigorous. By October 5 there had developed 6 flower stalks 16″–18″ tall.

Season of 1926—By April 15, *Bouteloua* had become green at the base and on June 9 it occurred in 9 distinct areas in bunches an inch to several inches in extent. The plants were thrifty in appearance, but occupied only about a third of the quadrat. On August 4 the leaves were rolled on many of the plants and there was a single dwarf flower stalk. However, the late summer rains produced a marked effect, the plants making a striking development and covering two-thirds of the area. They blossomed profusely, the numerous flower stalks attaining heights of 2′–2.5′ and giving the look of a normal prairie sod.

Second culture, season of 1925—This was intended to yield a larger number of seedlings of *Agropyrum* than of *Bouteloua,* with the consequence that there were 315 of the former and 40 of the latter on June 13. The average height was 3.5″ and the plants were fairly thrifty. Both had 2–4 leaves per plant, but *Agropyrum* was without tillers, while they were just appearing on some plants of *Bouteloua.* The stand of the latter was poor owing to low germination; a few plants had died and many of the lower leaves were dead and the tips of others dry for 1″–3″.

On July 21 the number was 137 for *Agropyrum* and 26 for *Bouteloua.* The former averaged 4″, with a maximum of 8″ and had not tillered; half of the plants bore dead leaves or leaf tips. *Bouteloua* was 1″–3″ tall and possessed 2–5 tillers to each plant. Dead leaves were few, but there were many dead plants in the quadrat. By August 20 the number of *Agropyrum* was 144 and of *Bouteloua* 28; the mortality for the season was 54% and 30% respectively, which was lower than for 1924. *Agropyrum* averaged about 5″ tall and had tillered somewhat, but was in only fair condition. *Bouteloua* was somewhat more thrifty, but the plants were very small; they averaged 3″ in height and tillering had produced bunches 0.5″–1″ wide.

Season of 1926—By April 15, *Agropyrum* had reached a height of 4″, while *Bouteloua* was barely turning green at the base. On May 31 the former appeared throughout the quadrat, while the latter was hardly to be seen without close inspection. There were 22 individuals and 8 small bunches; these attained a height of but 3″–4″ and were of secondary importance.

On August 4 both species were in poor condition as a consequence of the drouth, the leaves being rolled and wilted, but on September first, after good rains, *Bouteloua* was growing vigorously and rapidly increasing its area as a result of many new tillers. It was clearly dominating *Agropyrum* in those portions where it was most abundant. The two species were about equally abundant, *Agropyrum* being about a foot high at its best, and neither gave any signs of blossoming. On June 30, 1927, there was a rank growth of both species.

Comparative behavior and factors—The first culture was inconclusive, owing to the high mortality the first season, which seems to have been due to the relatively small number of plants. This greatly increased both evaporation and transpiration and serves to explain the death-rate, reducing *Agropyrum* from 94 to 1 and *Bouteloua* from 48 to 18. The stand itself was so open that little of this effect can be ascribed to competition.

At the outset *Agropyrum* was several times as abundant in the second culture as *Bouteloua,* and in spite of a mortality of 54% as against 30%, it retained the advantage through the first and most of the second season. However, it was handicapped by its inability to form tillers readily during the dry summers, in contrast to the behavior of *Bouteloua.* The two species exhibited little direct competition with each other, and the final success of *Bouteloua* in holding its own with *Agropyrum* was apparently due to its better response to late summer rains in 1926. It suffered more in the preceding drouth, but its ability to produce tillers quickly when the ground was moist enabled it to overcome this.

The experimental results are in harmony with the fact that *Bouteloua racemosa* ranges more widely than *Agropyrum,* spreading further into the subclimax prairie on the east and into the desert plains of the Southwest. Over much of this area, however, their requirements are much the same and their relative abundance is determined largely by grazing, which is less harmful to the sod-forming *Agropyrum.*

ANDROPOGON NUTANS VS. ELYMUS CANADENSIS

First culture, season of 1924—On June 4 the total number of plants in the quadrat was 484, of which 239 belonged to *Andropogon* and 245 to *Elymus.* The latter was in better condition, averaging 3″ high to 1 for the former. By June 27 the total was 629 plants, 423 *Andropogon* and 206 *Elymus.* They were of about the same height, 3″–4″, but the latter had more tillers and seemed more thrifty.

By August 11, *Elymus* had suffered a mortality of 60%, leaving 83 plants and *Andropogon* had lost 34%, leaving 279. The total number of survivors was 362, of which 63 and 202 respectively were suppressed. The average height was 4″–5″, some blades extending upward to 9″–12″. The taller plants bore broad leaves and were thriving, but the shorter ones were sickly in appearance with narrow dry or dying leaves. There was practically no tillering and half of the quadrat seemed dead. The most thrifty plants belonged to *Andropogon* at this time.

On September 9, but 70 plants were left, 56 *Andropogon* and 14 *Elymus;* the respective losses during the month were 80% and 83%, and the suppression 43% and 10%. The cover appeared to be dead; few of the survivors reached a height of 4″, except at the edges where they were 8″–9″.

Season of 1925—By April 7, *Elymus* was 3″–4″ in height, while *Andropogon* had not yet resumed growth. On June 16 the respective numbers were 11 and 5, and the losses from winter-killing 3 and 51. *Andropogon* was in poor condition, averaging 4″ high and with 2 tillers per plant; *Elymus* was 8″–9″ tall, with two vigorous tillers each. It was more thrifty, reaching a stature of 12″ on the margins. The quadrat was bare for the most part and was but little shaded. This condition was reflected in the fact that on June 17 atmometers at 6″–9″ in the quadrat lost 24 cc. in comparison with a loss of 26 cc. in a bare area outside. The evaporation from a moist soil surface at the soil level in the quadrat was 12.8 cc. in contrast to 16.1 cc. from a bare area.

On July 15, *Elymus* was flowering sparsely at 18″; the general cover was thin and little *Andropogon* was in evidence. A month later the latter numbered 15 and the former 10 plants. So much of the quadrat was bare and the insolation consequently so great that the soil dried rapidly and the plants grew poorly. The tallest individuals of both species were 10″; *Elymus* was dominant, the flower stalks reaching a height of 18″. *Andropogon* failed to produce flowers and was clearly lagging at the close of the season.

Season of 1926—By April 15, *Elymus* had developed a few broad leaves that were 6″ long. On June 9 the quadrat was very open and the shade slight. The culms of *Elymus* were scattered through half of the area, but were mostly on the margin. They were about 14″ high, consisting usually of 1–2 stalks but occasionally of 5–7. *Andropogon* occurred for the most part where there was no *Elymus;* it was 10″ tall and fairly thrifty.

On August 4 the scattered individuals of *Andropogon* bore many dead leaves and were suffering severely from drouth. The chresard at the several depths to 2′ was 3.5%, 7.4% and 11.8% respectively. Eight stalks of *Elymus* had reached a height of nearly 2′ and bore small spikes. The lower leaves were dead to a level of 8″–13″. During September this species produced a number of tillers as a result of good rains, and *Andropogon* also appeared more thrifty. The latter still persisted in late June of 1927, but the quadrat was clearly dominated by *Elymus.*

Second culture, season of 1925—This competition group was duplicated in 1925, in the hope of obtaining a stand with fewer seedlings and correspondingly less severe competition. The maximum number of plants was 431 on June 9, in contrast to a maximum of 629 in the first culture. This comprised 239 individuals of *Andropogon* and 109 of *Elymus;* the latter averaged 5″, with a maximum of 7″. It was somewhat taller and bore broader leaves than its competitor, and consequently possessed a distinct advantage over it. There were many dead leaves through the center of the quadrat, but the stand was better along the edge.

On August 20 there were 217 plants of *Andropogon* and 88 of *Elymus.* The mortality during the season had been respectively 57 or 21% and 79 or 50% by contrast with 80 and 83% for the first culture during 1924. The average heights were 4″ and 6″; both species had produced tillers and were in good condition. The greater stature and broader leaves of *Elymus* gave it the advantage in overshading its competitor.

The stand was less dense than for the original culture in 1924 and shade proved less critical in the outcome. The following table permits a comparison of the course in the two quadrats:

TABLE 4—*Comparative numbers in cultures*

Genus	Number at the start		Number at the close		% mortality for summer		Condition at close	
	1924	*1925*	*1924*	*1925*	*1924*	*1925*	*1924*	*1925*
Andropogon .	423	274	58	217	80	21	poor	good
Elymus	206	157	14	88	83	50	poor	good
Total	629	431	72	305	81.5	35.5		

This comparison reveals the fact that the initial density is a primary consideration in competition, a conclusion also borne out by the poor condition of the 1924 quadrat in 1925. This indicates that the effects of severe competition may be more or less permanent.

Season of 1926—By April 10, *Elymus* was 4″–5″ high, the leaves living through the winter, while *Andropogon* was just beginning growth at the base. By May 31 the average height of the former was 11″, of the latter but 4″–5″. The one was very thrifty, forming a dense stand, while the plants of the latter were suppressed, and many had died. On August 4, *Elymus* was in poor condition as a result of drouth, the leaves being rolled and often dead below 6″; however, 8 of the tallest were producing spikes. *Andropogon* was suffering even more severely, the leaves being tightly rolled at a level of 6″–8″. By September first *Elymus* was about 10″ and *Andropogon* 6″ tall, so that the latter was considerably shaded. No plants of the latter produced flowers and it was evident that *Elymus* had a decisive lead. This was confirmed by the situation on June 30, 1927, when *Elymus* dominated the quadrat.

Comparative behavior and factors—In both cultures the initial advantage lay with *Andropogon,* which began the season with about twice as many seedlings as *Elymus.* The mortality during the summer was slightly higher for *Elymus,* but this was more than offset by winter-killing, which took a toll of 51 from *Andropogon* as against 3 for its competitor in the first culture and was almost equally decisive in the second. *Andropogon* was unable to produce seed in either quadrat and by the end of the second or third year had definitely yielded the dominance to *Elymus.*

In view of the above, the greater abundance of *Andropogon nutans* in the subclimax prairie is probably to be ascribed to its sturdy rhizomes and the consequent sod-forming habit, which enables it to hold ground once occupied. The wider range of *Elymus* to the north and westward is reflected in the better response to winter-killing and the further advantage it derives from an earlier start in the spring.

Andropogon nutans

Season of 1924—This quadrat was seeded with *Andropogon nutans* and *Verbascum thapsus,* but the latter germinated very poorly and the few seedlings were removed to permit a study of the course of competition between individuals of the grass. By June 12 a uniform and vigorous stand of *Andropogon* had developed, but no count was made at this time. On July 8 there were 940 seedlings, which gave an average of practically 10 to the square inch. These averaged 6″–8″ tall, were beginning to tiller, and were in as good condition as the check plants. By August 11 the number had dropped to 633, of which 431 were suppressed; this gave a mortality of 33%, indicating the severity of the competition. This was further shown by the attenuated condition, the small amount of tillering and the mass of dead leaves at the base of the stand. The average height of the dominants was 9″; the cover was dense and the shade correspondingly deep.

By September 12 the number had become further reduced to 462, of which 290 were suppressed, the mortality being 27% or nearly as much as for the preceding month. The height and density remained practically the same as at the last examination, but tillers were abundant. This was the evident result of the decreased density, namely, from 10 to 5 per square inch.

Season of 1925—Growth had not started by April 7, but the quadrat was covered with a thick mat of grass on June 16. The plants had produced so many tillers and were so crowded that it was impossible to count them without injury. The average height was 6.5″ and there were 2–4 leaves per plant. *Poa pratensis* had invaded the area to some extent, but was pulled out. A mulch of dead grass leaves 0.5″–2″ thick covered the ground between the culms and was effective in reducing evaporation, as the following test shows. A soil-can filled with moist soil and placed under the mulch lost but 5.3 cc. in comparison with a control in a bare area, which gave a loss of 16.1 c.c.

On July 15 the growth had attained a fairly uniform height of 10″ and this condition still persisted on August 18. The cover was too dense for optimum development, but was fairly typical of the usual stand in the prairie.

Season of 1926—By June 9 the grass formed a dense cover over the entire quadrat to a level of 8″–10″. The soil was well protected by a layer of dead leaves 2″–4″ deep. On August 4 the plants were suffering from drouth, about a third of the leaves being dead to the base. By September there was a dense uniform growth to a height of 13″, which closely resembled that in the prairie proper. However, the plants were too crowded and the season too dry to permit the formation of flower stalks. The essential process in the course of competition was the elimination of the weaker plants through shading and drying of the soil on the part of the dominant individuals, but it was impossible to follow up the different plants after tillering became pronounced.

TALL-GRASS VERSUS SHORT-GRASS

In the following cultures mid-grasses or tall-grasses were grown in competition with short-grasses, simulating the condition that prevails in both mixed and coastal prairie. The combinations were as follows: (1) *Agropyrum glaucum* vs. *Bulbilis dactyloides;* (2) *Stipa viridula* vs. *Bouteloua gracilis;* (3) *Andropogon nutans* vs. *Bouteloua gracilis.*

Root relations—*Stipa viridula* is not native to southeastern Nebraska, but is a dominant of the mixed prairie, especially northward, and extends into the western portion of the true prairie. It is typical of broad shallow swales where the water-content is at the maximum for the mixed prairie.

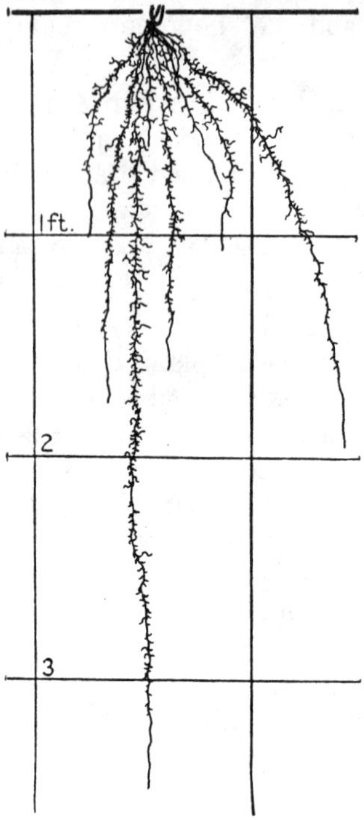

FIG. 6—Root system of *Bulbilis dactyloides* near end of first summer.

The general development of the root system is similar to that of *Stipa spartea*. The mature roots spread widely in the surface soil to depths of 12″–18″, with profuse and delicate branches, and may penetrate as deeply as 11′.

The root system of *Bulbilis dactyloides* closely resembles that of *Bouteloua gracilis* and *hirsuta* with respect to fineness and degree of branching. Its development is rapid. On July 31, when the seedlings in the prairie were only 3″ high and with 5–6 tillers, the roots already formed a working level to 16″ and had attained a maximum of 3′. The surface spread was not great, but this was compensated in some degree by the delicate and profuse branches (fig. 6).

AGROPYRUM GLAUCUM VS. BULBILIS DACTYLOIDES

First culture, season of 1924—On June 12 there were 218 plants in the quadrat, of which 130 belonged to *Agropyrum* and 88 to *Bulbilis*. Those of the former were 3″–4″ tall and very vigorous, of the latter about an inch in height and with many tillers. The total number on June 27 was 206; of the wheat-grasses 25 had died and 23 new seedlings of *Bulbilis* had appeared. The stand was open and there was little shade. Both species were tillering freely and seemed to be thrifty; *Agropyrum* averaged 5″–8″ high, while *Bulbilis* was already forming little mats.

By August 11 the total had been reduced to 139, comprising 46 of the wheat-grass and 93 of the buffalo-grass, the respective losses being 46% and 8%, and the suppression 18 and 40. The majority of the plants of the former were 3″–6″ tall, about a fourth reaching 10″–12″; they had tillered but little and many were drying. While the buffalo-grass was only 2″–3″ high, it was thriving and had formed mats 0.5″–1″ wide.

On September 12, but 93 plants remained, 20 of them *Agropyrum* and 73 *Bulbilis;* the one had lost 57% and the other 21% during the preceding month, and 13 and 27 plants respectively were suppressed. The few scattered individuals of *Agropyrum* that remained were 11″–12″ tall, the best bearing 7 leaves. *Bulbilis* had formed many mats an inch in diameter and short runners had appeared.

Season of 1925—By April 7, *Agropyrum* was well started and *Bulbilis* was beginning to turn green. On June 13 there were 20 plants of the former and 36 clumps of the latter. The one averaged 7″ high with about

5 leaves per plant and had no tillers; it was scattered through most of the quadrat and cast little shade. The other formed clumps an inch or less in diameter and to a height of 2″; runners extended out for 3″–6″ and even 10″, and were often rooted. Both pistillate and staminate flowers were present in the quadrat and a few fruits were ripening. Both species were thriving, but *Bulbilis* covered most of the area.

On July 6, the chresard at the usual levels to 2′ was as follows: 5.9%, 12.7% and 13.5%. There was a difference of but 1%–2% between the quadrat and the prairie outside. On August 17 but 12 plants of *Agropyrum* remained; these consisted of only 1–2 culms, ranging from 5″ with 1 leaf to 10″ with 9 leaves, and were of little consequence in the stand. The buffalo-grass completely covered the quadrat with a dense mat and extended beyond to a distance of 8″

Season of 1926—Both grasses had started growth by April 15, and on June 9 *Bulbilis* formed a closed mat over practically the entire area. It was 2″–5″ tall with well-developed stolons, many rooting in the bare area outside, and a few staminate flowers were blooming. *Agropyrum* was scattered thinly throughout and was not thriving; there were 28 plants, either single stems or with 1 tiller, and none were more than 10″ tall.

On August 4 the short-grass was still thriving in spite of the drouth and rooted stolons 13″ long were present. The scattered culms of *Agropyrum* were also fairly thrifty. The chresard at this time at the usual depths was 3.7%, 7.9% and 8.9%. There was no perceptible change in the cover during the fall and on June 30 of the following year a few plants of wheat-grass still maintained themselves in the short-grass mat.

Second culture, season of 1925—This quadrat was duplicated in 1925 in order to test the effect of a much larger number of seedlings of *Agropyrum*. On June 13 there were 318 plants of the latter to 13 of *Bulbilis*. The former were thrifty, averaging 3.5″ and with 3 leaves, but bearing few tillers. The plants of *Bulbilis* were very small and sparse. By July 21 there were 241 *Agropyrum* and 9 *Bulbilis;* the former averaged 7″ and reached a maximum of 12″, a few plants having 1–2 tillers. The buffalo-grass was in good condition, the tallest being about 2″ high and the tillers usually 2–3 per plant.

On August 20 there were 249 plants of *Agropyrum* and 11 of *Bulbilis*. The wheat-grass was in very good condition with an average of 7″ and a maximum of 12.5″. There were as many as 3–4 tillers per plant and these often extended several inches from the parent, so that it was rather difficult to distinguish individuals. *Bulbilis* was much suppressed at a height of 2″ or less and most of the plants had not formed tillers. This was primarily due to the reduced light intensity, since the chresard on August 18 was 20%, 17.5% and 10% in the respective layers.

Season of 1926—On May 31 *Bulbilis* numbered 10 plants, but none formed good clumps owing to the spring drouth. *Agropyrum* constituted only a scattered growth on August 4, when the plants were about a foot tall and fairly thrifty. *Bulbilis* occurred as many small clumps distributed throughout the quadrat, with leaves 2″–5″ tall but no flower stalks. By September first the scattered plants of wheat-grass ranged from 6″–12″ in stature. Good clumps of *Bulbilis* were found only sparingly, but they had tillered abundantly since the rains of late summer. The quadrat was still open and the final outcome could not be foretold, a conditior that still prevailed on June 30, 1927.

Comparative behavior and factors—The general course of competitior in the two cultures was unfavorable to *Agropyrum*. This was particularly true of the first one, in which it was reduced from 130 to 20 the first season leaving *Bulbilis* with thrice as many survivors, in spite of a much smalle initial number. The drouth of the second year prevented it from forming tillers, a result to which the dense sod of *Bulbilis* also contributed. However, *Agropyrum* survived the third year and was still present in th quadrat during the summer of 1927. In the second culture the seedin was so overwhelmingly in favor of *Agropyrum* that it held the dominanc from the outset, but the greater ability of *Bulbilis* in the production o tillers proved a distinct asset and it seemed probable that it would persis as a layer.

In spite of its ability to make tillers and stolons and to form a dens sod, as well as its greater resistance to drouth, *Bulbilis* was unable to com pletely replace *Agropyrum*. This is in harmony with their relation in th mixed prairie, over extensive areas of which *Bulbilis* forms a layer o short-grass under the wheat-grass. Grazing has probably played a par in this at some time, enabling the buffalo-grass to spread rapidly whe *Agropyrum* was cropped off, but they seem to be in fair equilibrium a present. The effect of pasturing on these two grasses has been studie by means of an exclosure in connection with the general inquiry int competition in nature (p. 144).

Stipa viridula vs. Bouteloua gracilis

Season of 1925—On June 13 after the seedlings were well up, th number was 250 for *Stipa* and 41 for *Bouteloua*. The former averaged 2 and appeared to be dying out as a result of drouth; none had more tha 1 tiller or 3 leaves. *Bouteloua* was 1.5″ tall with 4–5 tillers per plan and was thriving. By July 15 there were 96 plants of *Stipa* and 43 *Bouteloua;* the first averaged 4″ with a maximum of 8, but there were fe thrifty individuals. Most were without tillers or had but one or two at th best and bore but a single live leaf, many of them being on the poi of dying. On the other hand, the plants of the short-grass were thrift averaging 3″ high with a maximum of 5″. The number of tillers range from 6–10, and the plants were little affected by the drouth.

On August 20, *Stipa* numbered 68 and *Bouteloua* 43 plants; the respe tive losses had been 73% and 0%, and *Bouteloua* was clearly in th ascendant. It had tillered freely, forming clumps an inch wide. Th general level of the blades was 5″ and of the flower stalks 8″–12″. *Sti* was but 3″–4″ tall and in poor condition, the tips of the leaves being dea

Many of the plants were without tillers, while some had 2–4 each. About a half of the quadrat was bare, and light was not a critical factor, the intensity at the level of the sod being 38%. By October the grama had seeded abundantly at 12″–15″.

Season of 1926—By April 10, *Bouteloua* was turning green at the base, while *Stipa* was already 1″–2″ tall. On May 21 there were 48 individuals or small bunches of the former and 21 of the latter. The short-grass had grown to a stature of 5″, but was bunched and scattered. *Stipa* was very sparse and on August 4 was represented by only a few spears, being of slight importance in the quadrat. The drouth was so severe that the short-grass was dwarfed and bore flower stalks but 5″ high. By September first no trace of *Stipa* could be discovered in the dense growth of *Bouteloua,* which had produced abundant flower stalks to a height of 13″.

Comparative behavior and factors—In spite of its much better germination, *Stipa* was handicapped from the beginning by its greater water requirement, especially in view of the drouth more or less prevalent through both seasons. Its number was reduced nearly two-thirds in the first month and more than three-fourths by the end of the season. It entered the winter in poor condition and scarcely a third of the plants survived. While it started earlier in the spring, the number was too small and conditions too severe to enable it to make headway during drouth that greatly dwarfed the short-grass.

The complete disappearance of *Stipa viridula* in competition with *Bouteloua gracilis* in the face of drouth seasons is explained by their relation in nature. While both are dominants of the mixed prairie, *Stipa,* in its typical form, is confined to swales and meadows with high holard and in the variety *robusta* to disturbed areas with little competition. While it has the advantage of height and early growth, it possesses much less vigor than grama and is greatly inferior to it in the production of tillers. The two practically never occupy the same ground in nature, since they stand at two extremes of water relation in the association to which they belong. This is evidently in accord with the results of their competition, though the outcome would probably have been less decisive in normal years.

Andropogon nutans vs. Bouteloua gracilis

Season of 1924—On June 12 the total number of plants in the quadrat was 516, of which 310 belonged to *Andropogon* and 206 to *Bouteloua.* Neither was thrifty and the average stature was only 2″. On June 25 there were 503 plants, 401 of them *Andropogon* and 102 *Bouteloua.* Both were now thriving, the one being about 3″, the other 2″ tall.

By August 11 the total number had been reduced to 479, comprising 392 plants of bluestem and 87 of grama; the respective mortalities had been 2% and 15%. About half of each species was suppressed, amounting to 193 for the tall-grass and 47 for the short-grass. The average height of the former was 10″ and it was thriving, though many of the leaf-tips were dead. The average for *Bouteloua* was 5″, but many were much shorter and all showed signs of suppression, especially in the attenuated leaves. Where there was sufficient room, it had formed mats, and both species were well tillered.

On September 5 the total was 389, of which 311 were *Andropogon* and 78 *Bouteloua.* The loss since August 11 was 21% and 10%, and the suppression 60% and 53% respectively. The bluestem was 12″ tall and the leaves were as broad and vigorous as in the check quadrat. It had produced many tillers and new ones were still abundant, although many shoots were dying. The grama averaged about 4″, while the control plants were 2″ taller. Though shaded considerably by the bluestem, it had formed mats in so far as the dense sod permitted.

Season of 1925—By April 7, *Bouteloua* alone had started to grow, but by June 16 there were 58 plants of *Andropogon* 6″ in height and often with 1–3 tillers each. *Bouteloua* had produced so many tillers that the individuals could not be recognized; it averaged 4″ and formed a continuous mat over portions of the quadrat. Each species was dominant in different parts of the quadrat. On July 15 both were found to be equally thrifty.

On August first the light intensity at a height of 3″ was 33%, showing that the taller bluestem cast only a light shade, due in part to the dry season. By August 18 the quadrat had the aspect of mixed prairie of the bluestem type, the grama forming a fairly continuous layer beneath its taller associate. It averaged 4″ high in comparison with 10″ for the bluestem. The two species were uniformly mixed throughout and the ground was correspondingly well covered. Many of the leaves of grama had died, but this seemed to affect its vigor but little. None of the bunches of *Andropogon* were more than an inch in diameter.

Season of 1926—By April 15 the short-grass alone had resumed growth but on June 9 *Andropogon* formed a somewhat open but uniform stand to a height of 10″. Its broad leaves threw considerable shade upon the grama; the latter had vanished under the densest growth of bluestem and was in poor condition generally. The leaves were narrow and many were dead, due largely to the reduced light, which ranged down to 20% at a height of 4″ on June 17.

By August 4, *Andropogon* was suffering severely from drouth; nearly all the leaves were dead for 1″–6″ from the tip and many to the base. The light value had fallen to 5% and *Bouteloua,* though attenuated, was thriving in consequence of its protection from drouth. As a result of the rains of late summer, *Andropogon* renewed active growth to form a stand 10″ tall, under which a layer of short-grass 4″–6″ high occurred throughout the quadrat. In some spots the grama had died and nowhere was it thriving, the leaves being very narrow and much attenuated. It was unable to blossom on account of the dense shade, and although its hold was not completely broken, it appeared that it would ultimately yield to the tall-grass.

Comparative behavior and factors—The decisive factor in this competition between tall-grass and short-grass was light, though the initial advantage also lay with *Andropogon* because of its much larger number of seedlings. The greater stature of the bluestem gave it the lead when water was abundant, while the more efficient root system of the grama was an advantage during drouth. Since both grew well during the first season, the formation of a layered mixed prairie the second year was inevitable, *Bouteloua* necessarily taking the subordinate position. The significance of the difference in stature was shown most graphically the last year, when *Andropogon* suffered much from drouth at a time when its shade protected *Bouteloua* from excessive water-loss. When late rains again permitted the bluestem to grow, its shade became too dense for the grama and the latter gave signs of increasing distress.

This alternating advantage of tall-grass and short-grass is doubtless typical of those areas of both mixed and coastal prairie where overgrazing has developed a layer of short-grass beneath the dominant *Andropogons* and their associates. It produces a characteristic annuation in which rain favors the tall-grasses and drouth the short-grasses relatively, while overgrazing often renders the former inconspicuous or even lacking.

SHORT-GRASS VERSUS SHORT-GRASS

This culture comprised *Bouteloua hirsuta* and *Bulbilis dactyloides,* and was initiated in 1925. On June 13 there was an overwhelming preponderance of the seedlings of the former, the respective numbers being 274 and 7. The average height of the stand was 2″, but some plants extended upward to 5″. Both species were thrifty, *Bouteloua* usually producing 2 tillers and *Bulbilis* 2–3 per plant. The poor stand of the latter was due to the low germination of the seed. By July 15 there were 248 plants of the one and 15 of the other. The grama averaged 3″ tall with a maximum of 7″ and tillers were abundant, some plants having as many as 8–9 each. The buffalo-grass was smaller and had tillered less, so that the plants were found only upon close examination.

On August 20, the numbers were practically the same, namely, 246 and 15; the whole area inside the frame was a solid mat of short-grass reaching a height of 3″. Flower stalks 7″–14″ tall were abundant around the margin of the frame, but were lacking in the center. Both species had tillered with equal abundance, but *Bulbilis* was too sparse to be important in the competition. The chresard on August 28 was 4% for the upper 6″, 11% in the next and 8.8% in the second foot. This was 4% and 5% less for the upper layers than in the adjacent prairie, but there was little difference in the lowermost layer. The grama was in excellent condition on September first, and by October 5 it had seeded abundantly at a height of 12″–15″.

Season of 1926—By April 10, *Bouteloua* had begun growth, but *Bulbilis* had not started. There was a total of 189 individuals or small slumps of the former and only 12 plants of the latter. *Bouteloua* formed a dense mat

2″–5″ thick over the entire quadrat, while *Bulbilis* was insignificant, though one pistillate plant was in blossom. By August 4 the grama had made good growth for a dry season and by September first it had flowered profusely. *Bulbilis* still persisted, but was unable to overcome the handicap of much larger numbers. In spite of all hazards, it had produced seed by June 30, 1927, and was evidently destined to make fair headway against the dominant *Bouteloua*.

The disparity of numbers in this quadrat was nearly 40 to 1 against *Bulbilis,* and the value lay chiefly in showing that the latter could hold its own under such a handicap. It forms a denser sod and has the great advantage of stolons, but requires more water in consequence than either *Bouteloua hirsuta* or *B. gracilis*. As a result, it ranges less widely and occurs less abundantly than the latter, but under conditions of more or less overgrazing, its vigor and stolons enable it to associate with the grama on nearly equal terms.

DOMINANT VERSUS SUBDOMINANT

In this group of cultures, mid-grass dominants were pitted against subdominant forbs in the following combinations: (1) *Andropogon scoparius* vs. *Kuhnia glutinosa;* (2) *Agropyrum glaucum* vs. *Kuhnia glutinosa* (3) *Agropyrum glaucum* vs. *Oenothera biennis*.

Andropogon scoparius vs. Kuhnia glutinosa

Season of 1924—On June 12 there was a total of 351 plants, of whic 278 were *Andropogon* and 73 *Kuhnia;* the former averaged 1″–1.5″, th latter 2″ high. By June 25 the total was 411, *Andropogon* comprising 33 of this and *Kuhnia* 80. The latter were very vigorous and formed a domi nant layer at a height of 3″, with only a few grass blades rising above i A few of the grasses were somewhat yellow and stunted from shading, bu most of them were in good condition (plate 6A).

On August 11 there were 307 plants of *Andropogon* and 66 of *Kuhni* making a total of 373; the mortality in both had been low, but 181 or 59% of the former and 18 or 27% of the latter were suppressed. The averag height of Kuhnia was 8″, the tallest reaching 11″ and the suppressed one but 4″–5″. The tallest grasses attained a stature of 14″ and enough we as tall as the forb to give a grassy appearance to the quadrat. The growt was dense, the shade deep, and competition severe. The leaves of *Kuhn* were dead to a height of 3″–4″ on the stems, and there were mar attenuated and dead grass leaves.

The number of plants on September 12 was 328, of which 274 we *Andropogon* and 54 *Kuhnia;* 205 or 75% of the former and 8 or 15% of t latter were suppressed. *Kuhnia* was very thrifty with a height of 1 and was beginning to bloom in abundance. *Andropogon* on the contra was in very poor condition, tillers were few and the plants were usual decumbent. The leaves were very narrow, the tips usually dead and ve few extended to the top of the *Kuhnia* layer.

Competition cultures of grass and forb in high prairie.

A. Culture of *Andropogon scoparius* and *Kuhnia glutinosa*.
B. Culture of *Andropogon nutans* and *Arctium lappa*.

Season of 1925—Both species were just beginning growth on April 7. On June 16 there were 47 plants of *Kuhnia,* making a loss of only 7 during the winter; 16 individuals were suppressed. The forb was 5.5″ tall with 5–7 pairs of leaves and was very thrifty except for the suppressed plants badly shaded by their companions as well as the grasses. *Andropogon* was also very thrifty, ranging from 4″–11″ tall; it was so crowded that individuals could not be recognized. Some clumps were an inch in diameter and comprised as many as 15 stalks.

The shade was dense, the dominants of *Kuhnia* taking the chief part in this, and on June 27 the value on the ground was only 1.3%–1.7%, while at 4″ it was 11%–14%. By August first the light intensity had fallen to 11% at the 3″ level. Both species were in good condition on July 15, when the chresard at the usual levels was respectively 5%, 10.6% and 12.4%.

On August 18, *Kuhnia* numbered 36; the average height was 8″, the maximum 14″. Many had a spread of 2″–3″ and were forming flower buds. Throughout most of the quadrat the forb overtopped the grass and appeared to be shading it out. *Andropogon* was 7″ tall and was quite dense. It was in much better condition on the edges where the shade was less; the stems were taller and thicker and the leaves more numerous. Under the denser stand of *Kuhnia* the grass was greatly attenuated and many plants were dead or dying.

Season of 1926—By April 15, *Andropogon* was beginning to turn green at the base, and by June 9 there was thick stand to a height of 6″–10″. The 62 plants of *Kuhnia* ranged from 2″–12″; the shorter ones were noticeably yellowish and attenuated, but the upper leaves of most were well lighted. Both competitors were growing fairly well, but the grass showed some effects of drouth.

On June 17 the light intensity at a height of 4″ varied from 3%–10%. The water-loss from the porous-cup atmometer at a height of 3″–6″ was 16 cc. as compared with 26 cc. for a bare area.

On August 8, most plants of *Kuhnia* were about 10″ tall; a few suppressed ones were only 5″, and three tall ones on the margin measured 18″ and bore flower buds. All the leaves were dried and dead to nearly the level of the grass layer at 7″–9″. The grass was fairly thrifty, since the forb cast only moderate shade, but below the grass the light intensity at 5″ was but 5%.

By September 1, *Andropogon* formed a fairly dense sod 8″–9″ high; it was flourishing and flower stalks numerous at a later date. As a result of the drouth 10 plants of *Kuhnia* had died, leaving 32, and the leaves were dead on all to a height of 8″. Some individuals were much suppressed, but nine were 10″–20″ high and bore buds or flowers. By the end of June 1927, *Andropogon* had made a rank growth, in which persisted 20 plants of *Kuhnia,* most of them in excellent condition.

Comparative behavior and factors—The more rapid growth and greater spread of *Kuhnia* gave it a distinct advantage the first season, in spite of the fact that *Andropogon* outnumbered it 4 to 1. It dominated the quadrat regardless of the fact that 32% succumbed under the intense competition and none made a normal growth. However, they attained a height of nearly a foot, and flowered and seeded abundantly. *Kuhnia* was able to obtain the larger share of soil water by virtue of its deeper root system and this permitted it to overtop the grass at all times, three-fourths of the latter being badly suppressed.

During the second season the roots of the forb extended to a depth of 4′–5′ and quite beyond those of the grass. This enabled it to absorb vigorously below the general competitive level and to place its leafy tops at a constant advantage with respect to light. As a consequence, the initial lead due to better absorption was converted into an even more decisive gain as to light, and shading proved the most critical factor for the grass.

The third season was one of severe drouth. The weaker plants of *Andropogon* had been eliminated by shading and winter-killing, and the survivors were better fitted for competition. They formed a dense sod and developed some flower stalks. Less than a third of the plants of *Kuhnia* were sufficiently thrifty to bloom, and all had assumed a position of subdominance with respect to *Andropogon.* The fourth season both species were in good condition and the quadrat exhibited the usual structure of true prairie, in which the subdominant forbs maintain themselves in greater or less abundance in accordance with the rainfall and the consequent growth of the dominant grasses.

Agropyrum glaucum vs. Kuhnia glutinosa

Season of 1924—On June 4 there were 169 plants in the quadrat; 103 of these belonged to *Agropyrum* and 66 to *Kuhnia.* The former were mostly 3″–4″ tall with 3–4 leaves, the latter were 1″–1.5″ with 4 leaves. On June 27 the total was 172, of which 106 were *Agropyrum* and 66 *Kuhnia;* the average height of the latter was 4″, with a range from 1″–6″. Beneath the forb the topmost leaves of the grass and some of the plants were dead, but in the more open places *Agropyrum* grew above the *Kuhnia* level to 6″–9″ and was beginning to tiller.

On August 12 the number of plants was 104, 40 *Agropyrum* and 64 *Kuhnia;* the mortality of the former had been over 60%, while only two of the latter had died. Suppression affected 32 of the 40 grasses and 17 of the 64 forbs. The surviving grasses were in poor condition; many had reached a height of 8″–9″ before dying, apparently from drouth. *Kuhnia* averaged 8″ with a maximum of 11″; the plants were thrifty and flower buds appearing. The shade was dense and the lower 8–12 leaves were dead in consequence.

By September 5 the total number had been reduced to 60, of which 55 were *Kuhnia* and only 5 *Agropyrum;* all of the latter and 16 of the former were suppressed. The remaining individuals of the grass were 8″ tall with 1–2 stalks, but seemed half dead. The average height of *Kuhnia* was also 8″, the tallest reaching 14″; four of the plants were well branched and about to bloom. The dense shade had caused the death of the leaves on the lower 5″ of the stems.

Season of 1925—Both species had renewed growth by April 7 and some seedlings of *Kuhnia* had also appeared. On June 13 there were 10 plants of *Agropyrum* and 73 of *Kuhnia;* the former averaged 7.5″ high, ranging from 4″–9″, with 3–5 leaves per plant. The tips of the leaves were dry for a short distance and the lowest leaf was dead on some plants. The grass had spread slightly outside of the quadrat. *Kuhnia* averaged 3.5″ tall with 1–3 stalks per plant and 6–10 leaves per stalk. The quadrat was fairly open and hence the shade was not dense.

On July 15 there were only a few plants of the grass, but the forb was abundant. Light was not critical, since the value on August first at 3″ was 25%. By August 17 there were 80 stems of *Kuhnia* and 40 individuals; the latter were much dwarfed, the average height being 10″. The leaves were dry and dead up to 6″ and had turned yellow for 2″ more. There were 8 plants of the grass, 3 of them with a single tiller about equal to the parent stalk. The height ranged from 4″–10″ and their condition was poor.

Season of 1926—Both *Agropyrum* and *Kuhnia* had started growth by April 15. On June 9 there were 20 *Agropyrum* 8″–14″ tall, usually consisting of a single stem. The 45 plants of *Kuhnia* formed an open stand 6″ high and there was but little competition for light. The surface of the soil was bare and the consequent evaporation rendered the competition for water all the more telling.

By August 4 nearly half of the grass was dead and others half dead; none were over 14″ tall, and the leaves at the base of the stems were all dead. *Kuhnia* was suffering severely from drouth and exhibited many dead basal leaves, though a few bore flower buds. On September 1 only 10 poor plants of *Agropyrum,* 6″–14″ in stature, remained. Drouth had also reduced *Kuhnia* to 22 plants, which were flowering at heights of 13″–19″. Late in June 1927 five plants of *Agropyrum* were fruiting normally and there were several other good plants. The quadrat was rather open and *Kuhnia* was also thriving.

Comparative behavior and factors—*Agropyrum glaucum* was unable to compete effectively with *Kuhnia glutinosa* during the first three seasons. It was placed at a marked disadvantage by the deeper root system of the forb, as well as by the rapid growth of its tall leafy stems. By the end of the first season its number had fallen from 106 to 5, while its competitor had decreased only from 66 to 55, and had been able to blossom. The second and third seasons were dry, and both species lost about half their number, but *Agropyrum* suffered much more severely from suppression. This situation was improved by abundant rains in 1927, and both *Kuhnia* and *Agropyrum* were thriving by midsummer, and the latter had been enabled to come into bloom.

As in the case of other associated grasses and forbs, annuation is a striking feature in their relative importance and success. The tenacious rootstock of *Agropyrum* keeps it from being completely displaced, but its less effective root system permits it to produce flower and fruit only under an adequate rainfall. It is peculiarly slow in becoming established and its behavior during a four-year period is not entirely conclusive as to its final success.

Agropyrum glaucum vs. Oenothera biennis

Season of 1924—On June 4 there were 120 plants in the quadrat, 74 of *Agropyrum* and 46 of *Oenothera.* The stand was so open that there was little competition; the average height of the plants was 1″–3″ and the number of leaves 3–4. On June 27 the total number was 121, of which 78 were *Agropyrum* and 43 *Oenothera;* the latter averaged 6″ and 7 plants were sending up flower stalks. This species overtopped the grasses, except for about a dozen long blades, and cast a fairly dense shade. *Agropyrum* averaged 4″ with a maximum of 8″; the effect of the shade was strikingly shown in the attenuated condition, the small amount of tillering and the numerous dead leaves.

By August 12 nearly half the grass had died, leaving but 34 plants, while only 3 of the forb had disappeared. The total number was 77, of which 29 of the former and 15 of the latter were suppressed. The average stature of the forbs not in bloom was 8″, but the 9 flowering stalks averaged 26″. The plants were very thrifty and leafy, and cast a dense shade. *Agropyrum,* on the contrary, was in poor condition; the leaves were short, narrow and thin, and none succeeded in rising above the level of the forb.

By September 5 there remained 64 plants, of which 23 were *Agropyrum* and 41 *Oenothera;* 11 of the former and only 2 of the latter had died since August 12, while 12 and 14 respectively were suppressed. *Oenothera* threw a dense shade, due in part to the many leaves of the large rosettes. Ten individuals were flowering and fruiting profusely at a height of 3′. The tallest grasses were on the margin as usual, where they reached 9″; there were practically no tillers and half of the plants appeared dead.

Season of 1925—Both species were beginning to grow on April 7, and on May 28 there were 14 *Agropyrum* and 13 main stems of *Oenothera.* The latter were 5″–8″ tall, while the former was represented chiefly by single stalks scattered over the quadrat, ranging from plants 4″ tall with 2 leaves to those 10″ with 4 leaves. It appeared thrifty, and was not much shaded owing to the open nature of the stand. The plants of both species, but especially of the forb, were much smaller than in the low prairie. Light measurements taken on June 27 showed the intensity on the ground to be 3%–4% and at a height of 6″ 12%–18%. On August first, due to the increased height of the forb and the death of many lower leaves, the value was greater, being about 40% at the 3″ level.

On July 15 there were 17 stalks of *Oenothera* 16″ tall and but a few plants of *Agropyrum,* which averaged 15″. On August 17, *Oenothera* comprised 13 main stems from 20″–34″ in height. These were vigorous plants for the most part with an average stem diameter of 6 mm. There were no leaves on the stems below a height of 9″–13″, but all the plants were blooming or in fruit. There were but 4 isolated culms of *Agropyrum,* 7″–14″ tall; it had not tillered and was in poor condition.

Season of 1926—*Agropyrum* had already started growth by April 15. *Oenothera* was entirely lacking at this time and on June 9; the original plants had died and there were no seedlings present, in spite of the many fruits produced the previous season. The 11 plants of *Agropyrum* consisted of a single stalk, but they were quite thrifty with an average height of 13″. By August 4 all but 6 individuals of the grass had died and the leaves of these were tightly rolled. The quadrat was bare and the loss by evaporation high. The scattered grasses were still alive at the end of the season, but in spite of the late summer rains, they made practically no growth.

Comparative behavior and factors—The failure of *Agropyrum glaucum* in competition with *Oenothera biennis* is to be ascribed primarily to the rapid growth and relatively great stature of the latter. Combined with the rosette habit and broad leaves, this insured the suppression and death of all but a few of the stronger plants. As in the preceding culture, greater success in obtaining water from the soil resulted in an even more decisive advantage in the use of light, and *Agropyrum* was almost completely eliminated by the end of the second season. The disappearance of *Oenothera* at this time gave the grass a new opportunity the third summer, but the ensuing drouth was too much for the scattered and weakened plants, about half succumbing and the others making almost no growth.

SUBDOMINANT VERSUS SUBDOMINANT

Two combinations were employed in this series: (1) *Kuhnia glutinosa* vs. *Liatris punctata;* (2) *Oenothera biennis* vs. *Liatris scariosa.* The first member of each culture is characterized by the rapid development of an efficient root system and consequent vigorous growth. Both species of *Liatris,* on the other hand, have a meager root development with a slow growth of the exceptionally tolerant shoots.

Kuhnia glutinosa vs. Liatris punctata

Season of 1924—On June 12 there were 104 plants in the quadrat, 43 of them *Kuhnia* and 61 *Liatris,* and on July 8 the respective numbers were 101, 43 and 58. The former was thriving and many individuals were branching freely near the base and higher up the stem. They were mostly 5″–6″ high, while *Liatris* was but 3″–4″ with 2–4 broad leaves. There was little or no competition as yet.

On August 12 the total was 95, 42 belonging to *Kuhnia* and 53 to *Liatris;* 9 of the former and 16 of the latter were suppressed. *Kuhnia* averaged 9″, with a range of 1″–13″; the plants were thrifty and flower buds were appearing. The lower 8–12 leaves were dead in contrast to none in *Liatris;* the latter was 3″–7″ tall with 4–5 rather broad leaves and seemed to be thrifty. About half the soil under the plants was shaded.

By September 5 there were 93 plants, of which *Kuhnia* comprised 42 and *Liatris* 51; of the one 13 and of the other 5 were suppressed. *Kuhnia* averaged 10″ in height with the tallest 13″, and was in fair to good condition. The best plants of *Liatris* were 7″ high with 4–5 leaves, but many had only one or two. The shade was not dense and the competition accordingly not severe, though the good growth of *Liatris* was due in part to its tolerance. However, 10 individuals of the latter died during the season in contrast to one of *Kuhnia,* owing probably to poor root growth.

Season of 1925—Both species were beginning to grow on April 7 and on June 13 there were 15 of *Kuhnia* and 31 of *Liatris*. The winter mortality had been 27 or 64% in the former and 20 or 40% in the latter. The one averaged 3.5″ high with 2–6 branches per plant and 10 leaves per branch, and was correspondingly thrifty. The other had an average of 5.5″, ranging from 3″–8″ and also appeared to be thrifty. The quadrat was fairly open and the effect of shade slight. On July 15 the stand was still quite open and both species were thriving, *Kuhnia* at 8″–9″ and *Liatris* at 5″

On August first the light intensity at 3″ above the soil was 46%, demonstrating that shade was not an important factor. By August 17 the 15 plants of *Kuhnia* bore 80 stems, while the number of *Liatris* had fallen to 12. The former made a uniform stand about 11″ in height; the leaves were dead on the lower 5″ of stem and yellow for an inch higher. The plants were more or less dwarfed in consequence of drouth, but were about to bloom abundantly. *Liatris* was no longer in good condition; but two plants bore flower stalks, with a height of 14″, the others having only 2–3 leaves. Many of the individuals had been uprooted by a rodent for the sake of the reserve food in the enlarged root.

Season of 1926—By June 9, *Kuhnia* was represented by 46 single stalks and *Liatris* by 15 plants; otherwise the quadrat was bare. *Kuhnia* was only 3″–8″ tall and in poor condition; the leaves of *Liatris* reached the same height, but only one plant had developed a stem. The low chresard was the limiting factor at this time. The openness of the stand is shown by the rate of evaporation; at a height of 3″–6″ it was 16 cc. as against 28 cc. in a bare area.

By August 4 six more individuals of *Liatris* had been destroyed by rodents, while *Kuhnia* had lost 4 stalks. The plants were in poor condition, none being more than a foot in height. On September first but 17 plants of this species remained, 5 of which were blossoming at a height of 14″. *Liatris* had lost another plant, leaving but 8; some were but 2″–3″ tall and with only 2 leaves, and the best at the end of three seasons were no more than 8″ in stature with 6 basal leaves. On June 30, 1927, there were still a few good plants of both species, one of *Liatris* having a stem; the quadrat was rather open and was being invaded by *Solidago missouriensis*.

Comparative behavior and factors—The culture started with fairly equal numbers of *Kuhnia* and *Liatris,* and both species made good growth the first season, owing to the adequate rainfall. The quadrat was not sufficiently dense to handicap the shorter but tolerant *Liatris,* though it suffered some loss by reason of its poorer root system. Winter-killing was fairly severe in both, and the quadrat was more open the second season. As a result, competition was almost entirely for water and the plants made growth until the drouth in July, when they became somewhat dwarfed. *Liatris* was then damaged by rodent attacks and but two plants blossomed by comparison with all those of *Kuhnia.* Little progress was made during the following very dry season, and by autumn only 17 *Kuhnia* and 8 *Liatris* were to be found in the area.

The course of competition in this culture probably represents fairly well the relative abilities of the two species. *Kuhnia* is much more vigorous and correspondingly taller, but it has neither the tolerance nor the reserve food-supply of *Liatris,* and in consequence the two usually meet on fairly equal terms in the serotinal mixed society of the true and mixed prairies.

Oenothera biennis vs. Liatris scariosa

Season of 1924—On June 12 there were 160 plants in the quadrat, 44 belonging to *Oenothera* and 116 to *Liatris;* the former had a spread of 0.5″–3″, while the latter was very small and bore a single leaf. By June 27 more than half of *Liatris* had died, leaving 47 plants; *Oenothera* had increased to 46, making a total of 93. The latter averaged 6″ tall and several were sending up flower stalks; the former was distinctly suppressed. The dense shade and moist soil had promoted rotting at the base of a number of plants.

By August 12 about half of the remaining individuals of *Liatris* had succumbed, leaving 23, but *Oenothera* maintained its number, making a total of 69 plants. All of *Liatris* and 27 of *Oenothera* were suppressed. The former were but 2″–5″ high with 1–2 narrow leaves; the latter was 26″ tall with numerous leaves and 18 flower stalks. The foliage was dense to a height of 10″ and the shade correspondingly deep.

On September 5 the total number was 62, of which 43 were *Oenothera* and 19 *Liatris;* 20 of the former and all of the latter were suppressed. The quadrat was much crowded with the rosettes of *Oenothera,* and 15 plants of the latter were flowering and fruiting, the stems reaching a maximum of 42″. *Liatris* was in very poor condition, mostly below 4″ and with a single leaf.

Season of 1925—By April 7, *Oenothera* had developed rosettes 2″–3″ in diameter, and on May 28 there were present 17 plants of this and 8 of *Liatris.* The winter mortality had been 26 in the former and 11 in the latter. Ten plants of *Oenothera* were about 10″ tall, the other 8″ or less; the leaf spread was 5″–7″, but the lower leaves were yellowish or dead, probably from drouth. None of *Liatris* was over 5″ tall and for the most part there were but 1–2 leaves per plant, though the plants seemed thrifty.

On July 15, *Oenothera* numbered 12 stems, some of them 21″ tall, while *Liatris* was only 3″–4″ in height. By August 17 all of the 14 plants of the former were in bloom. The 8 largest averaged 28″ high, the 6 smallest 18″ and the diameter ranged from 8 to 2 mm. The individuals usually had but a single stalk each. Of the 8 *Liatris* the tallest measured but 5″ with 1–2 slender leaves per plant.

Season of 1926—On June 9 no *Oenothera* was found in the quadrat; the parent plants had died at the end of the biennial period the preceding autumn, and none of the seeds had germinated. This was the more surprising in view of the fact that the quadrat was bare, except for a single plant of *Liatris* 8″ tall and with 4 leaves. By August 4 this plant too had disappeared, probably having fallen a prey to rodents.

Comparative behavior and factors—During the first season in competition with *Liatris, Oenothera* made an excellent growth with practically no loss. The former had great difficulty in becoming established and the mortality was 84%, leaving only 19 weak plants in September. The winter mortality was high in both species, and this left the quadrat too open for good growth the following seasons. By autumn of the second year the stand was reduced to 14 *Oenothera,* all in bloom, and 8 *Liatris,* none past the stage of the second leaf. The adult plants of the former died at this time and the sole survivor the succeeding spring was a single *Liatris,* to be destroyed by a rodent. The rapid growth and tall stature of *Oenothera* handicapped its competitor both as to water and light, though the latter had finally the more decisive effect.

DOMINANT VERSUS RUDERAL

In the last series of cultures on the high prairie, dominants and ruderals were combined in the following pairs (1) *Andropogon scoparius* vs. *Amarantus retroflexus;* (2) *A. scoparius* vs. *Arctium lappa minus;* (3) *A. scoparius* vs. *Verbascum thapsus;* (4) *A. nutans* vs. *Amarantus retroflexus;* (5) *A. nutans* vs. *Articum lappa minus;* (6) *A. furcatus* vs. *Arctium lappa minus.*

Root relations—The taproot of *Arctium lappa minus* develops rapidly and penetrates deeply. By August 11 of the first season plants in the control quadrat on the high prairie were 16″ tall and the taproots had reached depths of 3′ or more (fig. 7). The laterals varied from 1″–15″ in length and were most abundant in the upper 18″ of soil. The development of leaves in relation to those of the competing *Andropogon scoparius* is shown in plate 6B.

The root system of *Verbascum thapsus* differs much from that of the preceding. Although it develops a taproot, the wide-spreading horizontal branches are numerous in the upper layer of soil. Plants on the high prairie had penetrated to a depth of 3 feet by August 5. Branching was profuse and the general development of laterals excellent, making this species an effective competitor with the grasses (fig. 8).

Andropogon scoparius vs. Amarantus retroflexus

First culture, season of 1924—On June 4 there was a total of 357 plants in the quadrat, 266 of them *Andropogon* and 91 *Amarantus;* the latter was 2″

Cultures of *Andropogon scoparius* and *Amarantus retroflexus* in high prairie

A. Growth in first series, June 23, 1924.
B. Growth in second series with nitrates, August 25, 1925.

tall and dominated the grass, which ranged in height from 0.25″–1.5″. By June 25 the number had increased to 507, due to the appearance of new grasses, making 437 *Andropogon* and 70 *Amarantus*. The latter averaged 5″ high and were decidedly yellowish owing to the low nitrate-content of the soil. Some of the lower leaves were entirely yellow and a few had become dry. The grass formed an understory about 2.5″ high, only a few of the narrow leaves extending above the ruderal.

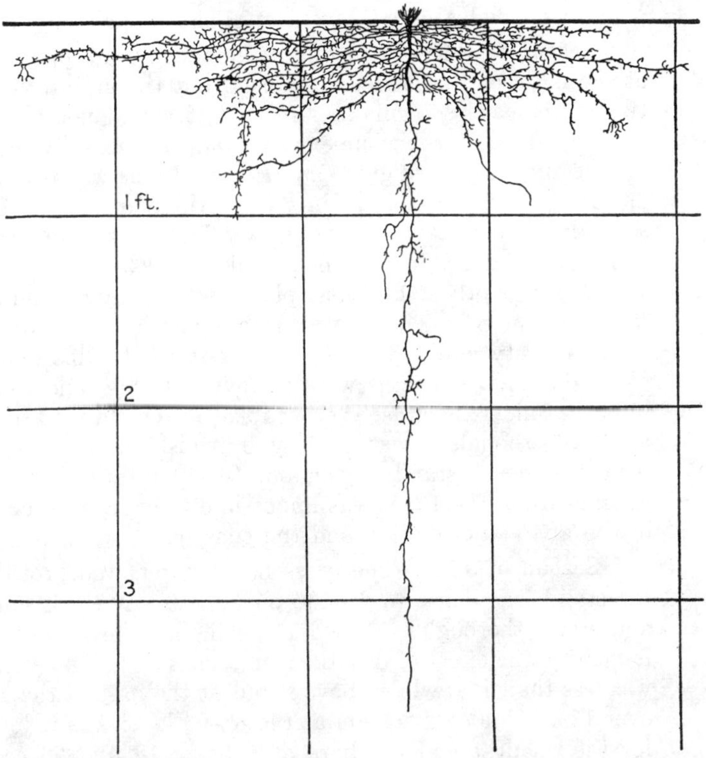

FIG. 8—Root system of *Verbascum thapsus* near end of first summer.

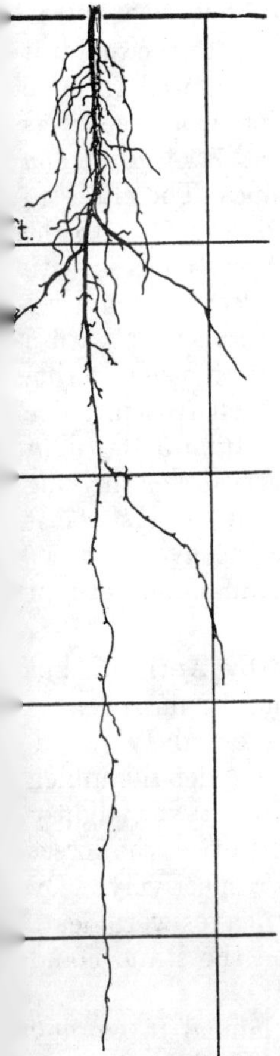

FIG. 7—Root system of *Arctium lappa* near end of first summer.

By August 11 the total had fallen to 447 plants; 57 grasses had died, leaving 380, and 3 forbs, leaving 67, while a fourth of the latter and a half of the former were suppressed. Twelve of the ruderals were 9″ tall, the remainder being about 6″; they were fruiting and all but the upper leaves had fallen. However, while the plants cast little shade, the leaves had dropped upon the grass blades, bearing them to the ground as well as shading them. The grasses made a uniform level at 7″, which was a little higher on the margins; the best plants bore 2–5 tillers, but a number of the lower leaves were dead.

On September 4 there was a total of 356 plants, of which 299 belonged to *Andropogon* and 57 to *Amarantus*. Since June 25 the mortality for the former had been 138 or about 32% and for the latter 13 or 19%; about one-half and one-sixth respectively were suppressed. However, the latter had grown scarcely any since August 11; all the leaves had fallen and some plants were dry and dead. The grass had made good growth and was now in fairly good condition, attaining a uniform level of 7″–8″. The larger plants were well tillered, but most of the suppressed plants had succumbed.

Season of 1925—*Amarantus* had entirely disappeared from the quadrat at the close of the preceding season, but the culture was followed through two more seasons to make sure that it would not develop from seeds, as well as to trace the course with *Andropogon*. By April 7, *Andropogon* was beginning to turn green and *Poa pratensis* was invading. The chresard on May 2 was the same as in the prairie, being 12%, 18% and 16% to the several depths to 2′ respectively. On May 28 the quadrat was covered with a uniformly dense stand, though about 20% was bare except for a leaf-mulch to a depth of 1″. The plants were thriving and averaged 6″ with a maximum of 8″. The leaves were somewhat narrower than usual owing to the density and to drouth, and many of the tips had been frozen. The old stalks of *Amarantus* were removed at this time. On June 5 the light intensity under the grass averaged 9.5%. On July 15 the grass was thriving at 6″–8″, though it was slightly brownish in color, and by August 17 it formed a dense stand throughout the quadrat. The height averaged 8″, ranging from 6″–11″; it was impossible to count the individuals on account of the abundant tillering and the consequent crowding.

Season of 1926—The grass had not renewed growth by April 15, but the stand was dense to 6″–10″ on June 9. A thick mulch of dead leaves and stems thoroughly shaded the soil, but the leaves were tightly rolled, indicating the effect of drouth. On June 17 the light value under the mulch was less than 1%, while above it but at the base of the grass it was slightly over 1%. The evaporation at the 8″–9″ level was reduced little, the losses for the quadrat and the bare area being 18 and 26 cc. respectively. By August 4 the height had increased little if at all and many leaves were dead. By late September a number of plants had blossomed and the stand could not be distinguished from the original prairie.

Second culture, season of 1925—This culture was planned to employ fewer individuals of each species and to favor *Amarantus* by adding Knop's nutrient solution to it, establishing another quadrat as a control. On June 9 one quadrat contained 245 plants of *Andropogon* and 25 of *Amarantus*, and the nutrient solution was added to it for the first time. In the control without fertilizer there were 225 *Andropogon* and 27 *Amarantus*. The plants in the two cultures were almost the same in development, except that those in the control were slightly taller and more leafy. The weed averaged 1″–2″ tall with 4 leaves per plant; the latter were slightly yellowish and cast little shade. The average for the grass was the same.

By June 13, *Amarantus* in the fertilized quadrat was much darker and the *Andropogon* somewhat darker than in the control, and both were in excellent condition on June 22. On June 27 the light intensity on the soil in the quadrat was 5%–7%. By July 15 there were 237 *Andropogon* and 18 *Amarantus* in the quadrat, and 207 and 21 respectively in the control. Both species were in much better condition in the treated culture. The average height of *Amarantus* was 5″ in the treated quadrat to 3″ in the control, with 8 and 2 respectively in bloom. In the one case the plants were deep green in color, in the other, yellowish. In the quadrat, *Andropogon* gave an average height of 5″ with a maximum of 10″, in the control the values were 4″ and 8″ respectively. The leaves were broader, the color deeper and the tillers more numerous in the quadrat. On August 1 the light intensity on the ground in the latter was 14% and 33% at a height of 3″.

On August 20 there were 17 plants of *Amarantus,* 1 of which was suppressed in the fertilized quadrat and 21 with 10 suppressed in the control. For *Andropogon* the respective numbers were 229 and 45, and 208 and 55. The fertilized plants of the ruderal ranged from 8″–16″, the checks from 5″–11″, while those of the grass were 7″ with 6–7 tillers per plant as against 5.5″ with fewer tillers. In both quadrats *Amarantus* was of little importance and the dead leaves were too few to handicap the grass.

Season of 1926—On May 31 no *Amarantus* was found in either quadrat; *Andropogon* averaged about 9″ tall and formed a very dense uniform stand. Many leaf-tips as well as leaves were dead, and there were likewise many suppressed or dead individuals. The dominant ones were so well tillered and made such dense mats that the separate plants could not be distinguished. Growth in the adjacent control was very similar, though the grasses were somewhat shorter and less dense; late in the season some grasses in both quadrats blossomed and set seed.

Comparative behavior and factors—As a consequence of its competition with *Andropogon scoparius, Amarantus retroflexus* was much dwarfed and consequently fruited early. This was primarily because of its inability to secure an adequate amount of water, but the lack of nitrates was a factor as well. The shade produced by the taller ruderal together with the effect of the fallen leaves was quite detrimental to the grass for a time, but late in the season it made good growth. In spite of its seed production, *Amarantus* did not reappear the second season or afterward, thus following the rule for ruderals in competition with climax dominants. When the nutrient solution was added, *Amarantus* fared better in some measure, but even then it was not a serious competitor with *Andropogon,* indicating that the holard was much more significant than the supply of nutrients.

Andropogon scoparius vs. Arctium lappa minus

Season of 1924—On June 4 the quadrat was occupied by a total of 345 plants or about 3.5 per square inch; of this number 273 were *Andropogon* and 72 *Arctium*. The latter were still very small, bearing merely cotyledons or the first pair of leaves with blades an inch long. The grass averaged 5 leaves per plant, mostly an inch long. By June 25 the total was 390,

323 *Andropogon* and 67 *Arctium;* new individuals had appeared in both species, but those of the latter had been removed to leave less than one to the square inch. At this time *Arctium* averaged 4″ in height, the largest having 4 leaves, some of them 2″–3.5″ wide. The grass showed the effect of shading by its leafy competitor in its yellowish and attenuated leaves.

By August 11 the total had fallen to 340; 12% of the grass and 18% of the forb had died, leaving respectively 285 with 139 suppressed and 55 with 25 suppressed. *Arctium* averaged 7″ tall and shaded the entire quadrat, the largest leaves measuring 5″× 3.5″. The grasses extended a slight distance above this level, but most of them were attenuated and sickly, since the competition for light was severe.

On September 4, *Andropogon* numbered 235 plants and *Arctium* 47, giving a total of 282. The number had been reduced from 3.9 to 2.8 per square inch, the burdock suffering a loss of 35% to 27% for the grass. Of the latter, 36% had been suppressed and of the former 47%, nine of these consisting of but a single living leaf. *Arctium* had deteriorated greatly during the month; there was a large number of dead fallen leaves and many blades had been badly eaten by grasshoppers. The best grasses were 10″ high, but the stand was far from uniform, and tillering was much less than in the controls.

Season of 1925—On April 7, *Arctium* had started to grow, and *Andropogon* was beginning to turn green. On May 28 there were 155 *Andropogon,* of which 87 were suppressed and 39 *Arctium* with 16 suppressed; during the winter, 80 or 33% of the former and 8 or 17% of the latter had died. The forb overtopped the grass in about half of the quadrat; it averaged 5″ in height with leaf blades 3″× 2″. The grass also was 5″ tall with a maximum of 8″; it was in fair condition except under the taller plants of *Arctium.* The light intensity at this time under *Arctium* ranged from 4.5% to 11%; on June 27 it was but 3% below the layer of dead leaves and 9% at 4″ above the soil beneath the large leaves of this species.

On July 15, *Arctium* averaged 6″–7″ in height and many of the leaf blades were 5″ long, some being rolled from drouth. *Andropogon* was 6″–8″ tall and slightly brownish in color. By August 3 many leaves of the ruderal had been mostly eaten or completely destroyed by insects. On August 17 there were 18 plants of *Arctium* but four alone were of any importance; these were 6″–7″ high with 2–3 large leaves. The damage caused by grasshoppers had increased, and even the petioles in many cases had been eaten down to mere stubs. The growth of the grass was far from uniform; the best plants were well-developed clumps as tall as 11″, on which flower stalks were appearing. Many were suppressed and attenuated with a stature of 1″–4″, and in the areas densely shaded all the individuals were dead. Shade now seemed less important than earlier, owing to the fact that the living leaves were at a height of 6″–9″. As usual, the dead leaves of *Arctium* weighted down and shaded the blades of grass. By October 5, *Andropogon* bore flower stalks 18″ tall on the edges of the quadrat.

Season of 1926—*Andropogon* was beginning to turn green at the base, but *Arctium* had not yet appeared. By June 9 a dense sod occupied one-half of the quadrat, while the center was somewhat open; the height of the grass was 6″–10″. Of the 7 plants of the ruderal, all but one were to be found in the open center; with two exceptions, none had more than a single leaf, the second leaf having died from drouth. The living leaves were about 6″ high and with blades 2″×3″ that were badly rolled as a result of drouth.

On August 4, *Arctium* was represented by 4 remnants, none over 5″ in height; grasshoppers had riddled the leaves. The grass had not increased its stature. The chresard at the several depths to 2′ was 6.4%, 7.5%, and 5.2% respectively. By September 1, numerous flower stalks were appearing on the grass and later these blossomed. Five plants of *Arctium* had sprung from the old bases during the late summer rains; none were over 6″ high with blades 1″–4″ long, and they were without effect upon the grass. On June 30, 1927, 3 small leaves of the ruderal were found in the grass sod.

Comparative behavior and factors—*Arctium lappa minus* proved to be a much more vigorous competitor of *Andropogon scoparius* than *Amarantus*, owing naturally to its greater duration and broader leaves, as well as to its greater capacity for growth and persistence. During the first season, the grass was seriously hindered in its growth by the shade cast by *Arctium* and by the dead leaves of the latter that fell upon it. On the other hand, the forbs withstood the effect of the grass with great difficulty because of the lesser efficiency of a taproot system by comparison with one consisting of fibrous roots. *Arctium* remained a formidable competitor during the first half of the second year, but it suffered much more than the grass during the drouth of July and was almost destroyed by insect attacks. Only a few plants appeared the third season and these were so badly damaged by drouth and insects as to have no effect upon the grass. The latter lost most of its suppressed individuals through winter-killing, but formed a good sod elsewhere, flowering late in the season.

Andropogon scoparius vs. Verbascum thapsus

Season of 1924—On June 12 there was a good stand of *Andropogon*, but *Verbascum* was far from uniform; it was thinned to leave one plant to each 2 square inches, making a total of 40. By July 8 there were 291 plants, 247 of them *Andropogon* and 44 *Verbascum*, forming a fairly open stand. The latter ranged from seedlings to plants with 4 leaves; these were about 1.5″×2.5″ and completely covered the grass in places. The grass had a mean height of 3″, frequently with 1–2 tillers. The culture seemed to be on the eve of intense competition (plate 8).

By August 11 the total number had been reduced to 266, with 235 for *Andropogon* and 31 for *Verbascum;* 100 of the former and 14 of the latter were suppressed. The largest plants of the forb had a total spread of 8″ and occurred at the margins. They produced a dense shade, causing the death of the grasses beneath. The best grasses were thriving at an average height of 10″.

On September 12 the total was 231, with 206 belonging to the grass and 25 to the forb; 114 of the one and 7 of the other were suppressed. The leaves of the *Verbascum* had an average size of 4″×1.5″ and the plants were thrifty, except where crowded to the extent of 3 or more to the square inch. The average height of the dominant mulleins was 8″ and of the suppressed plants 6″; where they were close together, the grass exhibited the effect by low stature and narrow leaves. Where the grass was dense, on the other hand, the mullein was similarly reduced in height and size of leaf. The root and shoot relations at the end of the first season are shown in the bisect, plate 9.

Between July 8 and September 12, *Andropogon* had lost 41 or 17% and *Verbascum* 19 or 43%. The grass showed the effect of the competition during this time chiefly in the attenuation and stunting due to shading, while the much higher mortality of the forb bore witness to the greater ability of the grass to obtain the major share of the holard under drouth conditions.

Season of 1925—By April 7, *Verbascum* had already produced 4–5 leaves per plant, while *Andropogon* was just beginning to grow in the more exposed parts of the quadrat. The chresard on May 26 was but 4.7% in the upper 6″ or 3% less than in the prairie, while it was rather uniformly 13% at greater depths to 3′. As a consequence, there was intense competition for water in the upper layer of soil.

On May 27 there was one mullein 6″ tall in one corner of the quadrat, with a spread of about 8″; the dead leaves around this had completely destroyed the grass beneath. There were also 7 smaller plants with leaves nearly erect owing to the crowding by the grasses. The latter were 5″–8″ tall, but it was impossible to recognize the individuals because of tillering and density; most of the plants appeared as small clumps (plate 9).

On June 5 the light intensity on the ground under the dead leaves of *Verbascum* was 1.5%, under the green leaves 2%. By July 15, *Andropogon* was thriving at 9″–12″ tall and *Verbascum* had developed 3 flower stalks, one of which produced seed. On July 29, determinations of the chresard showed that there was but 4.7% in the first and second 6″ layers, while the amount was 8% and 10.6% in the second and third foot respectively. On the whole, these readings were slightly higher than in the adjoining undisturbed prairie.

On August 17 there were but 7 plants of *Verbascum*, ranging from those with 4 partly developed leaves and a width of a third of an inch to one with 12 such leaves and a spread of 7″. The leaves were somewhat erect and there were three flower stalks, two 10″ and one 22″ tall. The plants covered only about a fifth of the total area, but the dead leaves were as effective as the living in reducing the light and hence no grasses grew under them. The grasses averaged 6″ with a maximum of 12″; they were well developed, except for the suppression near the mulleins.

The striking features of the course of competition during the second year were the heavy mortality in *Verbascum* due to winter-killing, but 8 plants surviving out of 25, and on the other hand the early start of this species in the spring, and the shading and weighting down of the grasses by its broad leaves, both living and dead. By October 5 the grasses on the edge of the quadrat had produced flower stalks 18″ in height.

Season of 1926—*Andropogon* made excellent growth during this year, partly owing to the fact that *Verbascum*, like practically all the annuals and biennials used in the cultures, failed to develop from self-sown seed. However, the dead leaves exerted a marked effect on the grasses close to them, through the exclusion of light. The stand was open, but the grasses were 10″–12″ high and not greatly affected by the existing drouth. The evaporation relations at the surface were determined by means of cans filled with moist soil; in the bare portions of the quadrat the loss was 14.2 cc., in a denuded quadrat 16.1 cc., and under the mulch of leaves but 9 cc. On August 4 many of the leaves were dead as a result of the drouth, but the grass produced numerous flower stalks and blossomed late in September. At this time it could not be distinguished from the native sod.

Comparative behavior and factors—During the first season *Verbascum thapsus* did quite well in competition with *Andropogon scoparius*. Where abundant, its broad leaves enabled it to suppress the grass more or less readily, but in a dense grass sod it was naturally suppressed in its turn. The light relations were clearly to its advantage, but it depended largely on the upper and drier soil layer for its water supply, while the grass obtained a more adequate amount from greater depths. The loss from winter-killing reduced the mullein to the point where its competition was effective in but one or two small areas. It dominated these and the grass flourished elsewhere, both species being able to produce flowers. *Verbascum* vanished at the end of the second season of its biennial period and left the quadrat to its competitor. The grass made fair growth the third summer in spite of drouth and finally became merged into the native prairie sod.

ANDROPOGON NUTANS VS. AMARANTUS RETROFLEXUS

Season of 1924—On June 12 there were 567 individuals in the quadrat; 492 of these belonged to *Andropogon* and 75 to *Amarantus*, the number of the latter being planned so as not to exceed one to the square inch. Both species were flourishing and many new plants were appearing; the grass averaged about an inch tall and the forb 2.5″. By June 25 there was a total of 717 plants, 648 of the former and 69 of the latter. *Amarantus* ranged from 1″–7″ tall with an average of 5″; although yellowish in color, it dominated the stand, shading the grasses. A few of the grass blades projected above the forb, but in general they were narrower and paler than those in the control quadrat. On the margins the grass was not shaded, and the plants were correspondingly more vigorous and the blades broader and larger.

By August 11 the total number had been reduced to 635, of which 569 were *Andropogon* and 66 *Amarantus;* since June 25 the mortality had been 12% and 4%. Of the grasses 306 were suppressed, while all the ruderals showed the effect of insufficient nitrates, 13 of them being especially reduced. By September 12 the total had been reduced to 451, *Andropogon* comprising 392 and *Amarantus* 59; the losses during the preceding month were 31% and 11% respectively. More than two-thirds of the grasses were suppressed and all the ruderals were in poor condition, nearly all of the leaves having fallen from the stems. The average height of the latter was 9.5″ with a range of 1″–16″; nearly all had blossomed, the spikes averaging an inch long. The grass had grown very poorly and the plants were mostly dead in the center; they were about 7″ tall around the edges

The competition had been severe, owing to the large number of plants and the excessive shading. The falling of the leaves of the ruderal upon the grass blades was especially disastrous to the latter. Of the maximum number of 648, the grasses had suffered a mortality of 40% by September 12. The ruderals fared better, only 21 having died, but all were greatly stunted, owing to the inadequate supply of nitrates and probably of water as well. The grass roots had proved far more efficient in absorption than those of the weed.

Season of 1925—On April 7, *Andropogon* was starting to grow on the margin of the quadrat, but no *Amarantus* was present, not even a seedling appearing during the two following seasons. By June 16, *Andropogon* numbered 138 with an average height of 5″; the individuals could still be distinguished, since the competition of the preceding summer had reduce tillering. The stand had attained a height of 9″ by July 15, but was no very dense at this time, and the following month saw little change. The plants were thriving and well developed, but they were scattered in ope tufts and none produced flowers.

Season of 1926—By June 9, *Andropogon* formed a good but ope growth over the quadrat, 7″–13″ in height; the leaves were brown and drie at the tips owing to drouth. On August 4 the grass was 10″ high, bu most of the leaves were rolled and many were dead. As a member of th low-prairie community, *Andropogon nutans* suffered much more from drout than *A. scoparius* with its shorter stature and bunch habit. However, i recovered rapidly in consequence of abundant rains, and by late Septembe constituted a uniformly dense mat 10″ high and not to be distinguishe from the native sod.

Comparative behavior and factors—As in the culture with *Andropogo scoparius,* a mid-grass, *Amarantus retroflexus* grew well in competition wit the much taller *A. nutans* earlier in the season and handicapped the gras in its growth. The dense grass stand in its turn was even more detriment to the forb, so that it flowered early and only a few plants attained a fa height. Owing to its annual habit, this species vanished at the end of th summer and no seedlings were found the following season. As a cons quence, the grass made a good growth in spite of the effect of the previous competition, which was seen also in the high loss from winter-killing. The third year was one of drouth and *Andropogon* was unable to grow well until after the rains of late summer; it then made a dense sod but too late to permit flowering.

Andropogon nutans vs. Arctium lappa minus

Season of 1924—On June 4 there were 373 plants of *Andropogon* and 69 of *Arctium;* the latter bore 3–4 leaves and overshaded the grass, which was but 1.5″–2″ tall. By June 25 the grass had increased to 590; new plants of the burdock had also appeared, but many of these were removed, reducing the total to 67. The largest were 4″ tall and bore 4 leaves, the latter about 2″ wide and long. Only a few of the older grass seedlings stretched above the forb and all showed distinctly the effect of the shade; the more recent seedlings were about an inch high, and very delicate and slender. Few of the grasses were in as good condition as those in the control.

By August 8, *Andropogon* had decreased to 427 and *Arctium* to 31; of these 329 and 14 respectively were suppressed. The general level of the forb was 8″, while the dominant plants of the grass were 2″ higher. Many of the former had died, but there were still present 22 leaves 4.5″×3″; the dead leaves were also an important factor in suppressing the grasses. Many of the latter were likewise dead, and tillering was poor in the living ones (plate 6B).

By September 12 nearly all the plants of *Arctium* were dead; but 3–4 remained with leaves 8″ long, the other 9 being very much suppressed. Almost half of the grasses had died during the month, leaving 241 of which 145 were suppressed. However, many new tillers were appearing and the plants at the edges had attained a height of 8″. After the grasses had died down in the autumn, the lack of competition enabled the forb to renew growth to some effect.

Season of 1925—By April 7, *Arctium* had produced 2 leaves per plant and some seedlings had appeared; at this time *Andropogon* was barely starting and then only on the margins. On June 13 there were 104 plants of the latter and 13 of the former; the grasses averaged 5″ with 2–5 tillers, but they were very uneven in height and density and many were suppressed. The forb was about the same height; the leaves were about 4″ long and nearly as wide, so that they formed a nearly continuous cover over a large part of the quadrat and shaded the grass considerably. The light intensity under this layer was but 1%–4%. The seedlings of the burdock were removed at this time.

On July 15, *Arctium* was about 6″–7″ tall, with many leaf blades ″ long; the grass was scattered but the growth good. On August first the light value below the leaves of the burdock was 12%. On August 18 there were 12 plants of *Arctium,* 7 of them fairly large with 3–5 leaves; the latter had blades 3″–5″×2″–3″ on petioles 5″ long. This species was important in only one part of the quadrat and even here the leaves were badly eaten by insects. In spite of this, the fallen leaves were sufficiently abundant t bring about much suppression of the grass. *Andropogon* was fairly we developed along the margins and as a single bunch in the center, and ave aged 9″ in stature.

Season of 1926—By April 15, plants of *Arctium* were an inch hig though the grass had not begun to grow. By June 9 the grass formed a open irregular stand, consisting of small clumps and individuals; there w little mulch and two-thirds of the quadrat was bare. Ten plants of *Arctiu* were present, ranging from 4″–7″; none bore more than 2 leaves and, on hal one leaf was dead or badly wilted. The grasses were also suffering fro drouth, and the quadrat was so open that practically all the competitic was for water.

By August 4, *Arctium* had vanished and the grass was suffering bad from drouth. It took on renewed growth late in the summer, reaching height of 10″, but did not flower. A single small plant of *Arctium* reviv in September, and on June 30, 1927, a single live leaf was present in tl dense stand of the grass.

Comparative behavior and factors—*Arctium* proved a successful con petitor of *Andropogon nutans* during the first season, suppressing the gra by virtue of the shade cast by its large leaves. However, the grass ma good growth in autumn after the ruderal had dried up. During the fir half of the second season, competition continued to be severe, but tl *Arctium* was so much damaged by drouth and insects in late summer th it exerted little effect, except by its dead leaves. *Andropogon* had large disappeared from the center of the quadrat, but flourished along the edg The drouth of the third year brought about the almost complete disappea ance of *Arctium*, and the grass formed an uneven stand that was not ab to produce flower stalks.

ANDROPOGON FURCATUS VS. ARCTIUM LAPPA MINUS

First culture, season of 1924—On June 4 the total number of seedlin in the quadrat was 368, of which 303 were *Andropogon* and 65 *Arctium;* t former bore about 4 leaves per plant, while the latter exhibited cotyledo only or 2 small leaves in addition. By July 25 there were 485 plan 424 *Andropogon* and 61 *Arctium;* the average height of the latter was 4″ a the largest individuals bore 4–5 leaves 3″×4″ in dimension. It formed good stand, casting considerable shade and protecting the soil, and some the leaves weighted down the grass blades. As a result the grass was mo or less attenuated; its average height was slightly less than that of the for but some blades exceeded the latter by 6″–8″.

On August 12 there were 401 plants, 359 belonging to the grass a 42 to the ruderal. Since July 25 the mortality had been 15% for the o and 31% for the other; about half of each were suppressed, namely, 169 a 23. On the whole, *Arctium* was in very poor condition; the average height w 8″ and the average size of the leaves 3″×2.5″, the largest being 4.5″×3. *Andropogon,* likewise, was not flourishing; the average height was 12″, b many of the leaves were discolored and attenuated. By September 5 the total number had been reduced to 249, of which but 4 belonged to *Arctium*. The loss since August 12 had been 32% for the grass and 90% for the forb, while 98 and 3 respectively were suppressed. *Andropogon* was 11″–12″ tall; it had been and was still suffering severely, as indicated by the poor color, narrow blades and dead leaves.

The auspicious start of *Arctium,* followed by a rapid decline to insignificance, was the striking feature of the season's development. From a total of 65 there had been a reduction to 4, making a mortality of 94%, by comparison with 42% for *Andropogon.* Shading by the ruderal was the primary cause of suppression and death in the grass, while the latter overcame its competitor by virtue of its better system for absorption.

Season of 1925—By April 7 the plants of *Arctium* usually bore 2 leaves each, except for some new seedlings, while those of *Andropogon* were just starting. On May 26 there were 151 clumps of the latter, of which 63 were suppressed and 19 of the former, of which 8 were suppressed. Several of the ruderals that appeared dead the preceding September evidently retained enough vigor for a new start. The four largest ruderals were 5″ tall with 2–3 leaves per plant, the blades measuring 2″×3″ and hence casting only a light shade. The stand of grass was thin but uniform throughout, averaging 4″ high. There were no large clumps, but individuals could not be distinguished with certainty.

On July 15 the growth of *Andropogon* was uniform and good, but *Arctium* was sparse and suffering from drouth. By August 17 but 2 plants of this species remained; they possessed but one leaf each, badly eaten by insects, and the taller measured only 6″. The grass formed a dense stand throughout the quadrat with a level of 8″–10″; the lower leaves were dead to a height of 2″–3″. The shade under the grass was not heavy and drouth was obviously the major factor in the disappearance of *Arctium.*

Season of 1926—By April 16 one plant of *Arctium* showed renewed growth. On June 9, *Andropogon* had an average height of 9″, but covered only about a half of the quadrat. There were but 3 individuals of the ruderal, one of them bearing only two leaves. The leaf blades were about 3″×2″; they were slightly rolled and had no effect upon the grass. During the ensuing drouth all three seemed to be dead and even the grass was half-dead and brownish. Two ruderals showed new growth following the rains in September and the grass stand became as dense as that of the native prairie. A single dwarf of *Arctium* with 2 leaves was present on June 30, 1927, in a good growth of *Andropogon.*

Second culture, season of 1925—This culture was seeded less abundantly, with the plan of securing a better opportunity for the ruderal. On June 9 there were 137 plants of *Andropogon* and 15 of *Arctium;* the latter usually had 2 leaves 1″×0.25″ and the spread of each plant was about 3.5″. The ruderal was beginning to overtop the grass, but most of the blades of the latter projected above the weed to a distance of 1.5″–2″. Both species were thrifty.

On July 21 there were 124 *Andropogon* and 16 *Arctium;* the first averaged 4″ with a maximum of 7″ and usually had 1–3 tillers per plant. The second averaged 3″ tall, with 2–3 live and 1 dead leaf per plant as a rule. The leaf blades were 2″–3″×1.5″–2″ and slightly yellowish in color. The shade was not dense and many of the grass leaves still extended above the ruderal. The drouth was sufficiently severe to cause large cracks in the soil of the quadrat.

On August first the light intensity on the ground under a live leaf of *Arctium* was 17%. By August 20 there were 127 plants of *Andropogon* and 18 of *Arctium.* The former was very thrifty, averaging 6″ and with a maximum of 11″, and had tillered readily to form small clumps. The forb was about 4″ high with 3 leaves per plant; the blades were 2.5″×2″ and covered the grass to some extent. The plants showed good color and apparently were thrifty. Neither species appeared to be suffering from the effects of competition, owing chiefly to the thinner stand, and the mortality was much less than in the first culture during 1924.

Season of 1926—By April 10, seedlings of *Arctium* were abundant and the plants of the previous year were starting to grow. By May 31, *Arctium* numbered 27 and *Andropogon* 136; the latter formed an open but good stand 6″–8″ tall. Many of the ruderals were small and suppressed, only 13 being of fair size; the largest possessed 2 leaves about 2″×2″. By August 4, *Arctium* had disappeared except for the remnants of a single plant. *Andropogon* occurred chiefly in bunches with most of the lower leaves dead, and hence the quadrat was quite open. By September first 6 plants of *Arctium* were found, having come up during the late rains, but they were of no significance. The grass was in fair condition, but was unable to produce seed. Three small plants of *Arctium* were found in the dense sod of *Andropogon* on June 30, 1927.

Comparative behavior and factors—When pitted against *Andropogon furcatus, Arctium* flourished and became dominant until the drouth of late summer, when all but 4 plants disappeared. The effect of over-shading had been to suppress the grass very greatly. Although the few survivors of the ruderal were recruited in number the second season by seedlings, they had little or no effect upon the grass, which formed a dense sod that limited absorption by *Arctium.* A few of the latter persisted to the end of the culture, but they were too much dwarfed to produce any visible effect.

In the second culture both species grew fairly well the first season, as a consequence of the much smaller number of plants. However, the drouth of the following summer greatly favored the grass, and the forb was reduced to a few insignificant survivors.

GENERAL SUMMARY

The general series were essentially the same on the high as on the low prairie, though the cultures were somewhat more numerous. The same life-forms and species were employed, and the chief differences in the course and outcome of competition are to be ascribed to the lower chresard and generally higher transpiration, and to the lower stature of the mid-grasses and the consequent effect upon light. Winter-killing was often less severe on the high prairie, since the greater dryness enabled plants to mature before the oncoming of winter. On the other hand, conditions were more trying during drouth intervals, and these naturally affected the tall-grasses most.

Competition between dominants—When the germination of *Stipa spartea* and *Andropogon nutans* was more or less equal, the rapid growth of the former, its greater hardiness and early start in the second spring proved decisive, the suppression of the bluestem being due mostly to light. *Sporobolus asper* enjoyed practically the same advantages over *Andropogon,* except that it germinated more poorly and began growth rather later in the spring. It tillered the best of the three and was better able to withstand drouth. In the contest between two mid-grasses, *Andropogon scoparius* and *Agropyrum glaucum,* the amount of germination was decisive as to the result, each species winning when its number of seedlings was several times that of the other. Moreover, a wetter season favored the bluestem when it was most numerous, and a dry one the wheat-grass when it outnumbered its competitor, while the one had the lead in tillering and the other in hardiness. Water was in consequence more critical than light, though the latter had much effect as usual.

Owing to a poor stand of both, the first culture of *Agropyrum* and *Bouteloua racemosa* was inconclusive. In the second the germination of the former was eight-fold greater, and this permitted it to maintain the lead through the first two seasons. As a boreal species, *Agropyrum* appeared earlier in the spring and hence overtopped its rival. During the fourth year, both species grew vigorously and appeared to occupy the ground on equal terms. In the case of *Elymus canadensis* and *Andropogon nutans,* germination favored the latter, but the rye-grass grew more readily, withstood winter better and resumed growth more promptly in the spring. Water was the decisive factor, drouth affecting both tall-grasses seriously, but preventing blooming in the bluestem and giving the final dominance in 1927 to *Elymus.*

Competition between tall-grass and short-grass—When the germination of *Agropyrum* exceeded that of *Bulbilis* by 25 times, it became the dominant, the short-grass persisting as a lower layer. In the first culture where the number of seedlings the first year was much the same, *Bulbilis* became dominant the first summer by virtue of its better absorption and lesser transpiration. *Agropyrum* began to grow earlier in the spring and this with its stature enabled it to hold a place even in the dominant sod of short-grass. Although *Stipa viridula* outnumbered *Bouteloua gracilis* 6 to 1 at the outset, its greater need for water practically excluded tillering and led to an excessive mortality the first season, while the drouth of the third summer eliminated it entirely. The initial advantage lay with *Andropogon nutans* in competition with *B. gracilis,* and this was enhanced by its stature. As a result the decisive factor was light, though drouth compensated the short-grass for this handicap from time to time, and the outcome was a culture of two layers in accordance with the respective life-forms.

Competition between dominant and subdominant—In competing with *Andropogon scoparius* and with *Agropyrum, Kuhnia* possessed the three advantages of rapid growth, greater spread of stems and leaves and a better root system, though it germinated less well. It held the lead over *Andropogon* for two seasons and over *Agropyrum* throughout, but became subordinate to the former in the third year and was reduced to equality with the wheat-grass in **1927**. These cultures afford experimental demonstration of the relation between dominance and subdominance in grassland. The normal control of the dominants may be greatly modified at any time by the course of annuation or aspection, and it may be destroyed by overgrazing. As in previous cultures of *Oenothera biennis* and grasses, the forb profited by its rapid growth, rosette of broad leaves and tall stems to dominate *Agropyrum* completely for two seasons, after which it disappeared entirely, leaving the space to its competitor.

Competition between subdominants—In both these cultures, a vigorous and rapidly growing perennial, *Kuhnia,* or biennial, *Oenothera,* was pitted against the slow-growing *Liatris,* respectively *punctata* and *scariosa*. The taller more leafy species dominated for the first two years in both cases, but was unable to supplant the more tolerant *Liatris* completely. The latter would have made a much better showing had it not been for the fondness of rodents for its fleshy rootstocks.

Competition between dominants and ruderals—The competition of *Andropogon scoparius* followed the same general course and gave much the same results with three different ruderals. These were an annual, *Amarantus retroflexus,* and two biennials, *Verbascum thapsus* and *Arctium lappa minus*. In all cases, the forb suppressed the grass as a result of the shade cast by its broad leaves during the first or second season. In its turn, it was handicapped later in the summer by the better absorption of the grass and it disappeared at the end of its life period without leaving a trace, in spite of its abundant seed production. These cultures and those discussed in the preceding chapter all prove the fact drawn from studies of climax and succession in the field, namely, that annuals and biennials are unable to compete with the climax grasses during the proper season for the latter. In consequence, they never occur in the climax except in disturbed places, or in relatively open ones, such as are to be found in the desert-plains and bunch-grass associations, during the season of winter rainfall in particular. The cultures of the taller *Andropogons, nutans* and *furcatus,* gave practically the same results, though the greater stature was even more decisive than in the case of *Andropogon scoparius*.

4. SUPPLEMENTARY STUDIES OF COMPETITION IN THE PRAIRIE

SOD TRANSPLANTS

General plan—In contrast to the method of seedling cultures, mature individuals of selected species were transplanted into the low prairie to permit studying the course of competition with the grasses already in possession. For this purpose the following species were transferred to an unmown area of low prairie on March 31, 1924: *Agropyrum glaucum, Bouteloua gracilis* and *Bulbilis dactyloides.* Pure stands with an area of one, one-half, and one-eighth square meter were cut to a depth of 6″ and reset in the midst of tall-grasses after an opening had been made of the exact size and depth to fit them. This was done at a time when the soil in general and the holard in particular were especially favorable to the transplanting.

Season of 1924—By May 13, *Agropyrum* constituted a thick even stand about 6″ high, which was well in advance of the low-prairie sod and consequently was not shaded by the latter. Both *Bouteloua* and *Bulbilis* were also thriving, at a level of about 3″. By early June the surrounding grasses overtopped *Agropyrum,* which was now 10″ tall, except in the case of the eighth-meter quadrat, which was 15″ in height. This was due to the shading effect of the overlapping tall-grasses, which stretched above it to a height of 17″–26″. By the middle of July the native stand was 3′ high; although the wheat-grass was deeply shaded, especially in the smaller quadrats, it still showed good growth. In September the tall-grasses were 4′ high and overlapped the transplants to a distance of about 6″ on all sides. The areas of *Agropyrum* were badly invaded by tall-grasses; the leaves were narrow and none of the plants had produced flower stalks.

By the middle of June the light intensity in the smallest quadrat of *Bouteloua* had been reduced to 8.8% and a month later the grasses were greatly attenuated in this, as well as on the overshaded edges of the other two transplants. By September 10 the grama in the smaller areas was almost dead as a result of shading, but it had flowered sparsely in the largest. *Bulbilis* grew distinctly better, though it too was greatly shaded by midsummer. Blossoming occurred in all three areas, but the leaves were markedly attenuated in the two smaller ones. The prairie was not mown and during the winter much debris from the fallen grasses accumulated over the ground to a depth of 4″–6″.

Season of 1925—By May 26, *Agropyrum* had made a sparse growth in the largest quadrat to a level of 10″. *Poa pratensis* and *Spartina cynosuroides* had invaded to a considerable degree and much dead material had collected about the margins. As a result the growth in the half-meter quadrat was still poorer, while in the smallest one only a few attenuated stalks 10″ high could be found. By July first the surrounding grasses were 2′ high and the invasion was still more pronounced. The meter area exhibited only scattered plants and the smallest contained only a few blades. By September first the invaders had attained a height of 15″ and only a close examination revealed remnants of wheat-grass in the two larger quadrats, none of them being able to form flower stalks. In the smallest area the shade was very dense and no wheat-grass could be found.

Bouteloua had made only a scattered growth to a height of 7″ by May 26, owing to the shade of the adjacent plants and the debris present. *Spartina* had invaded to a small amount. The growth in the middle quadrat was much the same, but the smallest was covered to a depth of 6″ by fallen material and no grama grass was to be discovered. By midsummer the plants in the meter quadrat had reached a height of 13″ and a few in the best lighted portion were blossoming. The half-meter was densely shaded throughout; the plants were much attenuated and occupied less than one-third of their original territory. On June 27 the light intensity at a height of 4″ where the grama had died was only 10%.

By late summer the grasses about the meter quadrat, chiefly *Spartina*, were over 3′ tall and cast a dense shade. Invading grasses occupied much of the quadrat, the grama being limited to a good stand 20″x23″ in the middle. Here the foliage was 13″ tall and some of the flower stalks reached 26″, flowering and fruiting rather abundantly. The half-meter area was much more badly shaded and invaded, only three bunches of grama persisting in the middle; these attenuated plants were 15″ high, but had not blossomed.

Bulbilis was represented at the outset by a rather open and poor growth about 5″ tall. The soil was dry and cracking, the fallen debris on the edges was weighting down the grass, and both *Spartina* and *Poa* were invading to a considerable extent. In the half-meter area conditions were similar, though the growth seemed slightly better. The smallest quadrat was covered with 4″ of debris and but two remnants of buffalo-grass remained. By the first of July these two were dead as a consequence of the very dense shade. At this time the stand in the largest quadrat had reached a level of 6″–7″, but the plants were very slender owing to the invasion of *Spartina*. Late in the summer *Bulbilis* had largely disappeared from the margins, the tall-grasses overlapping well toward the middle of the quadrat. The growth in the latter was fairly good, but there were no flowers or fruits and the attenuated plants were clearly yielding to the invading *Spartina* and *Poa*.

Season of 1926—*Agropyrum* had been strongly invaded by *Poa* and *Solidago* by June 5 and much debris was scattered throughout both the quadrats. The wheat-grass was obviously losing in the struggle, occurring but thinly in one-half of the area. By August first the soil was so dry that the leaves of *Spartina* were rolled and the transplant grasses looked dead and dry, although the shade was less than usual and the invasion correspondingly reduced. On September first the wheat-grass remained as a few remnants; no leaves extended above a foot and the area was badly invaded by prairie grasses 40″ tall. By the first week in July, 1927, a scattered growth remained in the center of the large quadrat; it was in poor condition and none had blossomed. The shade was dense, the quadrat thoroughly invaded and the surrounding grasses 3.5′ high. Several remnants of *Agropyrum* still persisted in the half-meter quadrat.

Bouteloua covered about half the area of the larger quadrat with a poor growth on June 5. *Solidago* and *Spartina* with a stature of 2′ cast a dense shade around the margins and were invading the area. In the half-meter sod only a few stalks of grama remained and none of these had blossomed. The tall-grasses were rapidly closing over it. By September there was a small area of grama about a foot square in the middle of the large quadrat. The much attenuated leaves reached a height of 16″, but in spite of this flower stalks 2′ tall had developed. Only an occasional isolated plant was found by July of 1927, as a result of the very dense shade cast by *Solidago* and *Spartina,* which were 2.5′–3.5′ tall. No *Bouteloua* remained in the half-meter area, where *Spartina* was 5′ high.

Bulbilis in the half-meter quadrat was invaded almost throughout by *Spartina* on June 5 of the third season. The latter was already 2′ high and a layer of debris 2″–4″ thick covered all but the few relics of buffalo-grass, which were found only in the best-lighted portion. In the larger quadrat a few staminate spikes had appeared, but the growth was very scattered. By September *Bulbilis* had vanished from the half-meter area, which was covered with *Spartina* over 3′ tall, a condition that still obtained in July, 1927. The buffalo-grass persisted in the large quadrat as a stand 12″x16″ in the center, but on July 3, 1927, only a few isolated plants remained.

Summary and conclusions—The three species, *Agropyrum glaucum, Bouteloua gracilis,* and *Bulbilis dactyloides,* passed through practically the same course in their competition with the tall-grasses and forbs of the low prairie. They became well established at the outset, but even in the first season were suppressed along the margins by the overhanging prairie grasses and as a consequence they were readily invaded by the latter. This process went forward with increasing momentum during the next two seasons, and all three suffered a similar fate, being reduced to a few weakened shoots destined to ultimate extinction. *Agropyrum* fared somewhat better than the short-grasses, owing to its greater stature and vigorous rhizomes. The size of the transplant quadrat naturally played a large part in the rate of disappearance, since both shading and invasion could take place only at the edges. The eighth-meter area had practically been merged in the low prairie by the end of the first season, and the same fate overtook the half-meter in the third. This was reflected in the behavior of the large quadrat, where the transplant made its final stand in the center during the last season.

The above results are in harmony with those secured in the low prairie in studies of ecesis from 1918–1921 (cf. "Experimental Vegetation," pp. 44, 53, 84). The sods employed in the latter were more nearly the size of the eighth-meter quadrat and their persistence was in fairly close agreement, varying somewhat with the season. However, when species typical of the low prairie were employed, they were able to establish themselves with little loss, chiefly because they matched their competitors in rate of growth and height, but partly because of their being accustomed to the habitat. Both series of experiments, as well as those involving sowing and seedling transplants, reveal the extraordinary difficulty species have in entering and establishing themselves in closed communities of the climax or subclimax type.

DENUDED QUADRATS

General plan—The method of the denuded quadrat was employed to obtain a measure of the effect of competition upon various dominants and subdominants of the prairie. Meter areas were cleared in several ways to leave usually one individual or one species, so that it was possible to compare its growth and behavior with the same species in the undisturbed prairie sod. This procedure was employed with dominants, subdominants and combinations of the two, as well as reciprocally by removing one or more species from one quadrat and leaving these alone in another, thus providing a double control.

On April 8, 1924, a series of quadrats was practically denuded on the high prairie in accordance with the above plan. The plants selected for study were left undisturbed, but the soil around them was cut away to the depth of a half-inch. The sod was then inverted and cut into small pieces. All invaders were removed as they appeared, and a loose mulch maintained throughout the two or three years of the experiment. This naturally afforded an extra supply of the three major factors—water, light, and nitrates—to the plants remaining in the area.

Dominants

Two quadrats were installed on a gravelly knoll near the high-prairie station. From one of these all the vegetation was removed with the exception of the bunches of *Stipa spartea,* while in the other denuding affected everything but *Koeleria cristata.* During the first season no differences were to be noted until late in the summer, when the plants in the quadrats exhibited a better growth and remained green longer than those in the prairie.

Stipa spartea—By June 3 of the second year the plants in the quadrat had made much better growth than those under the full competition in the prairie. This was the evident result of the greater water-content available to the scattered bunches in the quadrat. Thus the chresard within the denuded area was 1.5%–3.5% higher to a depth of 3′ than that on the outside. In the prairie most of the grasses had 4″–8″ of the leaf-tip dead, while the leaves were normal on the inside. In addition, the bunches within the denuded area were much greener and bore fewer dry basal leaves. Few of the plants on the outside of the quadrat had bloomed, and these possessed fewer and smaller flowers than those within (plate 10A).

Koeleria cristata—By May 11, 1925, the plants in the quadrat were pushing out their spikes, well ahead of those in the native prairie. This

Comparative growth of grasses in prairie and in denuded quadrat.

A. *Stipa spartea.* B. *Koeleria cristata.*

more rapid growth was due to temperature as well as to the more abundant holard, since the bare soil was several degrees warmer than that of the prairie, owing to the absence of the screen afforded by the grass cover. On June 3 the plants of *Koeleria* in the prairie had rolled their leaves as a result of drouth and gave other signs of wilting. Rolled leaves also occurred within the quadrat, but they were much greener and there were fewer dead ones. The number of spikes to each bunch was much larger on the inside and the plants were much better developed in every way. The spikes were larger and broader, often being twice as long as those in the prairie (plate 10B).

By this time the quadrats and adjacent prairie had been much trampled in measuring the plants and disturbed by the excavation of specimens, and therefore the experiment was not followed further. However, similar quadrats on the high prairie during succeeding years yielded results in every way comparable. It was repeatedly found that the plants in the denuded quadrats bloomed about as freely during dry seasons as wet ones, while those in the prairie sod were much dwarfed and bore few or no flower stalks.

Andropogon nutans and *scoparius*—In another quadrat all the plants were removed except a single bunch of each of these two species. During the first season, drouth in the spring brought about decided differences between the plants of *Andropogon nutans* in the quadrat and the prairie. The bunch without competition bore foliage to a height of 29″ and flower stalks to an average of 43″ with a maximum of 58″; there were 39 such stalks with panicles that averaged 10.5″ long. The plants in the prairie had foliage 26″ high, flower stalks 33″ and 39″ respectively and panicles 9″ long. On the other hand, *Andropogon scoparius* showed little or no difference in the two situations, owing to the fact that it was at the edge of the denuded area and was consequently subject to considerable competition from the prairie grasses.

During the following season, the behavior of *Andropogon nutans* was essentially the same. By late autumn *A. scoparius* was flowering abundantly in the quadrat at a height of 2′, while flower stalks were sparse and scarcely over 18″ tall on the plants outside. Similar differences were observed for both species during the third year of growth, and in October the grasses on the inside were 4″ taller on an average than those on the outside. On July 6 the chresard of the quadrat was considerably higher than for the prairie, as follows: 0″–6″, 8.7:6.1%; 6″–12″, 14.1:11.4%; 1′–2′, 13:10.1%.

From another quadrat all the plants were removed with the exception of one bunch of *Andropogon nutans* and the subdominant forbs. Little difference was observed until after drouth began about August 15, when the bluestem on the inside forged ahead, as did also several plants of *Helianthus rigidus* and *Solidago missouriensis*. Late in May of the following year the forbs had gained still further over those plants in the undisturbed prairie.

Subdominants

Kuhnia glutinosa—A single mature plant of this species was left in the center of a meter area; by the middle of the summer it had made vigorous growth and was considerably larger than those in the prairie. On September 10 the differences between the two were marked; the plant inside had developed 18 stems with an average height of 2′. In the absence of competition for light, the stems spread almost horizontally, some nearly sprawling on the ground, the mass having a diameter of 2′. After many measurements, a typical branch was selected from this and a plant outside for comparison. The flower stalks were respectively 12″ and 6″ long, the stem diameter 5 mm. and 3 mm.; the individuals on the outside were often more than 2′ tall and erect, but possessed few branches. Determinations of the chresard on September 13 gave higher values for the denuded area as usual: 0″–6″, 8.7:7.5%; 6″–12″, 11.8:7.3%; 1′–2′, 11.1:6.9%; 2′–3′, 11:9.3%.

By May of 1925 the plants had made a vigorous start, but were frozen nearly to the ground by late frost, in spite of which they were as vigorous and as different by late autumn as during the previous season. On May 2 the chresard in the quadrat at the several levels to 2′ was 13.5%, 17.3% and 16.4% as compared with 11.8%, 17.4% and 14.3% in the prairie.

On May 29, 1926, *Kuhnia* was in better condition in the quadrat than on the prairie, and by September this plant had 25 stems and considerably surpassed any of those in the prairie. During 1926 the chresard in the *Kuhnia* quadrat as compared with that in the prairie outside was regularly much higher, as the following table shows:

Table 5—*Chresard of quadrat and prairie*

Date	Depth in feet	Kuhnia quadrat	Prairie	Difference
1926		%	%	%
May 4....	0.0–0.5	16.1	5.4	10.7
	0.5–1	17.2	10.0	7.2
May 29....	0.0–0.5	5.1	6.1	—1.0
	0.5–1	17.3	11.3	6.0
	1–2	20.0	9.5	11.0
	2–3	16.7	5.6	11.1
June 29....	0.0–0.5	6.7	3.5	3.2
	0.5–1	13.9	6.9	7.0
	1–2	14.0	7.9	6.1
	2–3	12.5	8.4	4.1
	3–4	11.2	7.1	4.1
	4–5	8.9	8.3	.6
July 17....	0.0–0.5	13.0	4.6	8.4
	0.5–1	10.7	6.1	4.6
	1–2	12.0	5.7	6.3
	2–3	10.2	5.9	4.3
	3–4	8.7	7.2	1.5

The nitrates were likewise slightly more abundant in the upper layer of soil in the quadrat than in the prairie; in the latter it was usually 1 ppm. or less, by contrast with 3.4 ppm. for the 6″ layer on June 19 and 2.3 ppm. for the 8″ on July 20. On clear warm days the soil temperatures were higher in the denuded quadrats than in the grassland. For example, at 4 p.m. June 6, the temperature at 1″ below the surface was 38.8° C. in the quadrat as against 25.4° in the prairie, and at a depth of 4″ it was 29.5° and 23.5° respectively.

Helianthus rigidus—A number of plants of this species were left in the quadrat after denuding and by August 10 they were distinctly ahead of those in the surrounding vegetation, the contrast being marked by September. The average height was respectively 37″ and 29″, and the stem diameter 6 mm. and 3 mm.; the leaves were thrice as broad in the quadrat and the number of nodes in the first foot of stem 9 as against 5. There was an average of 4 heads per stalk inside as against 2 on the outside and the corresponding diameters of heads were 21.6 mm. by contrast with 16 mm. Flowering took place within and without the quadrat at practically the same time. The semi-wilted condition of the plants in the prairie at periods of stress showed clearly that these differences were due to the chresard. On May 26, 1925, the available water-content in quadrat and in prairie was as follows at the usual depths: 9.5:8.2%; 8.7:9.8%; 12.5:12.2%; 7.1:14.1%. On July 6 the values were the following: 6:6.1%; 13:11.4%; 10.5:10.1% and 10.3:9.4%, and on August 28: 6.7:7.9%; 11.3:13.7%; 6.6:9.3%; 4.6:7.2%. The differences in chresard favored the prairie plants about as often as those in the quadrat, but the contrast in growth and size is readily explained by the fact that the water available had to be shared by many more individuals. During the following seasons similar results were observed, but they were never as pronounced as in the first year.

Solidago missouriensis—Little difference was found until August 10, 1924, when the dry conditions of late summer brought about marked effects. The 10 plants in the quadrat were still blooming on September 10, while those in the prairie were through; the former bore much larger thyrses and the leaves were alive nearly to the base of the stems instead of being dead to a height of 4″–6″. The plants inside averaged 18.5″ as against 16″ in height. Rosettes of the current year had many broad leaves and averaged 7″ high, while those in the grassland bore fewer, narrower and more erect leaves.

The measurements taken during 1925 showed that the plants in the quadrat were again a third larger and better grown than those in the prairie, in spite of the fact that drouth might have been expected to offset the advantage of the denuded area by drying out the bare soil. The chresard for quadrat and prairie on July 6 was as follows: 0″–6″, 13.3:6.1%; 6″–12″, 12.5:11.4%; 1′–2′, 13.3:10.1%. However, the plants in the quadrat did not prosper the following year, growing scarcely better than those on the outside, a result to be ascribed to the effect of severe drouth on a bare area as contrasted with one bearing cover.

Amorpha canescens—Although the plants of this species made a good growth in 1924 and put forth several new shoots from the underground parts, they were no better developed than those in the prairie. They were followed closely during the season of 1925, but no essential differences in stature, flowering, fruiting, etc., were discerned. This unexpected result is apparently to be explained by their deep root system, which enables them to escape severe competition with the grasses for water and nutrients. The plants continued to thrive during the next two summers, but without exhibiting any advantage over those in the prairie.

Brauneria pallida—This quadrat was given attention every two weeks to prevent invasion and to keep the surface well mulched. During the latter part of the season, the plants in the denuded area grew more vigorously and began to flower before any of those outside, but by October there was little observable difference between the two sets. Late in May of 1925, flower heads began to form in advance of those on the outside and the plants themselves were more thrifty. This difference continued throughout the season and by autumn representative plants exhibited striking differences in vigor and leafiness, and in the number and size of the heads (plate 11A).

Petalostemon candidus—During the first year the plants in the quadrat grew well in the early season, but late summer drouth led them to fruit at a height of 14″–24″, exceeding only slightly in vigor those in the prairie. In the quadrat, leaves occurred nearly to the base of the stem, while in the grass the lower leaves were shaded and hence usually turned yellow and dropped off. During the year of 1925 no significant differences could be discovered, owing to the unusual weather conditions, but in May, 1926, the plants in the denuded area were much better than those in the prairie. They were 10″–19″ tall, bore many more stalks and branches, and possessed leaves nearly to the base. By September they were once and a half to twice as large as those in the surrounding grassland.

Aster multiflorus and *Solidago rigida*—On June 8, 1925, two additional quadrats were denuded, only the plants of *Aster* being left in one and of *Solidago* in the other. At the end of the first season, *Aster* showed no advantage in the absence of competition, but the golden-rod was exceptionally large and robust, being almost as thrifty as individuals on the low prairie and much better than any of those on the high prairie. The stems were 18″ tall with many rosettes and the thyrses 4″–6″ wide. On May 29, 1926, both species were much taller and further advanced in the quadrat than on the outside. Chresard determinations on May 4 and August 13 gave the following values:

Comparative growth of forbs in prairie and in denuded quadrat.

A. *Brauneria pallida.* B. *Solidago rigida.*

TABLE 6—*Chresard of quadrat and prairie*

Date	Depth in feet	% in the quadrat	% in the prairie	Difference in favor of quadrat
May 4....	0.0–0.5	10.8	5.4	5.4
	0.5–1	12.1	10.0	2.1
	1–2	13.3	12.7	0.6
	2–3	11.1	9.3	1.8
Aug. 13....	0.0–0.5	13.0	18.5	−5.5
	0.5–1	12.9	10.0	2.9
	1–2	8.3	4.0	4.3
	2–3	6.5	5.9	0.6
	3–4	5.0	4.3	0.7

By September first marked differences occurred in the case of *Solidago* (plate 11B). The plant in the quadrat comprised 14 branches, though the maximum for those outside was five; it was 21″ tall and just beginning to blossom. It was much better than those in the prairie with respect to stature, size and number of leaves, size of thyrses and number of flower heads. At this time the plants of *Aster* were nearly 2′ tall, averaging a third larger in every way than those in the prairie.

SHORT-GRASS AND MID-GRASS

On April 8, 1924, four representative areas a meter square were carefully selected in the high prairie; in these *Bouteloua gracilis* was present to a considerable degree as a result of mowing, often twice a year. In two of these quadrats, *Bouteloua* was removed, leaving the mid-grasses, while in the other two all plants but those of grama were taken out. However, it was only by repeated visits and persistent effort that these two quadrats were kept free of other grasses during the three growing seasons. By August 10 the mid-grasses in the quadrats were somewhat better developed than in the prairie, but the stand was sufficiently dense that the removal of the short-grass reduced competition much less than expected. The following year, not even this difference was noted, the grasses having taken possession of the small areas formerly occupied by *Bouteloua*. In the area adjacent to the quadrats, grama was very scarce by the end of the second season after mowing ceased. During the third summer it entirely disappeared, the mid-grasses shading it out completely just as the tall-grasses had done in low prairie.

In the quadrats with the mid-grasses removed, *Bouteloua* made an excellent growth, soon outstripping its companions on the outside and reaching a height of 9″ by midsummer. The flower stalks were 16″ tall, but none occurred on the outside, where the leaves also were greatly attenuated. This was largely due to overshading by the mid-grasses, which formed a level 17″ high. Late in May of 1925, *Bouteloua* was extending its territory about the quadrats and was far in advance of the isolated clumps under the unmown mid-grasses. The latter covered the short-grass to a depth of 4″–7″ with fallen leaves and stems, which played a large part in the disappearance of grama from the general area.

By July first only those plants of *Bouteloua* in the quadrats were blossoming, at a height of 10″–13″, and by autumn the areas resembled greatly the pure short-grass cover so often found in the mixed prairie. The grama spread still further the following year, making an excellent growth with tall flower stalks, in striking contrast to its complete disappearance outside. The chresard values in the grama quadrats and in the prairie were as follows in 1925:

TABLE 7—*Chresard in Bouteloua quadrat and prairie*

Date	Depth feet	% in the quadrat	% in the prairie	Difference in favor of quadrat
July 6....	0.0–0.5	1.8	9.0	— 7.2
	0.5–1	8.9	5.2	3.7
	1–2	11.9	11.0	0.9
	2–3	12.9	9.4	3.5
Aug. 18....	0.0–0.5	18.5	10.6	7.9
	0.5–1	9.4	19.6	—10.2
	1–2	10.1	15.7	— 5.6
	2–3	14.9	8.5	6.4

DENUDED QUADRATS WITH CENTRAL SOD

In this experiment, five quadrats on the high prairie a meter in exten were treated in the following fashion on June 8, 1925. An area of a squar foot was reserved in the center and the remainder of the sod was loosene overturned and pulverized to a depth of 3″. By September the grasse were notably better developed in the central square, around which th soil had been cultivated and kept free from weeds. *Andropogon furcatu* bore flower stalks with a height of 30″, *Panicum virgatum* with 29″, *Bou teloua racemosa* 26″ and *Andropogon scoparius* 24″. The average heigh of the foliage in the various quadrats was 11″, while in the prairie ou side it was but 8.5″ and there were practically no flower stalks on an of the grasses.

At this time the cover of an area 10″ square was removed from ea of the quadrats and from an equal extent of the prairie, placed in sack thoroughly air-dried and weighed. The yield from the quadrats wa 206.1 gm., from the prairie 107.9 gm.; this increase of 91% came part from the better growth of foliage but chiefly from that of the flower stalk

Early in the spring of 1926, five additional quadrats were treated the same manner. On June first the grasses in these were hardly bett developed than those in the suroundin prairie, owing to April drout but those of the quadrats of the previous year were practically twice well developed. The latter had profited greatly by the lack of compet tion for a year, while drouth had prevented the new quadrats from obtai

Denuded quadrats in high prairie with central relicts.

A. Growth on June 16, 1926, in quadrat denuded in the spring.
B. Growth on same date in quadrat denuded in spring of 1925.

ing a similar advantage. During both seasons, chresard determinations in the bare soil of the quadrats showed a considerable excess over that for the natural prairie (plate 12).

Samples for the determination of nitrate-content were made near the edge of the central square on July 20, 1925. This revealed the fact that the roots of this central group occupied the bare area fairly thoroughly, reducing the nitrates in the surface 6″ to 1.5 ppm. On June 1, 1926, the nitrate-content in the bare areas was slightly higher, namely, 1.6–2.4 ppm. in the upper 6″ level and 1.2–1.6 ppm. in the second. However, this was not much higher than in the prairie sod, where the value was 1.2 ppm.

Owing to the drouth, the prairie grasses were drying and were dead halfway to the ground by July 20, 1926, but those in the central squares were in excellent condition. Even the grama was turning brown in its partially cleared quadrats and many prairie species were wilting. The sod of *Bouteloua racemosa* was flowering at the usual height, though none was found outside.

On August 15 the central squares of both the 1925 and 1926 quadrats were harvested. There was a rank growth of *Andropogon, Stipa* and *Bouteloua racemosa* in the 1925 quadrats to a general level of 12.5″, but these were beginning to dry badly. In the 1926 quadrats the level was 10″; both *Panicum virgatum* and *Bouteloua racemosa* bore a few flower stalks at heights varying from 18″–26″. The grasses outside the bare areas had been clipped early in the spring before growth was resumed, in order to remove the debris of the preceding year. The general level of the foliage was only 6.5″; it looked dead and dry and bore the general appearance of autumn vegetation. The average yield of the 1925 squares was 56.2 gm., and that of the 1926 37.8 gm., while the undisturbed prairie produced only 21 gm. The yield of the former was 63% and of the latter 44% more than that for the prairie; the difference between the 1-year and 2-year quadrats was 49% in favor of the latter.

Significance of results—One of the major objects of the preceding experiments with denuded quadrats was to permit the separation of competition and climatic effects in the prairie. As a consequence of previous studies (Clements and Weaver, 1924), as well as a study of the course of annuation in the prairie, it had been recognized that the optimum development of dominants and subdominants far exceeded that normally observed. In years of excessive rainfall, such as 1915, the stature of the tall-grasses, e.g., *Andropogon furcatus, A. nutans* and *Spartina cynosuroides,* reached 10′–12′ or more than twice the usual height in the low prairie. This optimum was reflected in the subclimax prairie with a rainfall of 35″–40″, and was suggested also in roadsides and other disturbed areas in which competition was reduced. The results obtained with both grasses and forbs in the present instance demonstrate that their stature in the prairie is largely an outcome of competition, the latter being the immediate limiting factor within the much wider limits set by the climate of the true prairie. The elimination of competition led to an increase in stature and volume ranging from 30% to 100%, and produced more or less striking changes in form. The increase was naturally less the first season, but gradually reached the maximum as a consequence of the effect of larger amounts of water, light and nutrients in augmenting the storage in underground parts. Furthermore, it appears probable that the development secured in the absence of competition does not represent a final optimum, but that still greater amounts of the three primary factors will produce even larger plants.

The results obtained by the reciprocal removal of mid-grasses and short-grasses throw further light upon conditions in both the mixed and coastal prairie, where the association is characterized by two layers of grasses. The excellent growth of *Bouteloua gracilis* in the quadrats thinned to leave only this species attests the ability of grama to thrive in a fairly humid climate, quite apart from the relict areas of the xeroclines, which represent a very different local climate. In areas repeatedly mown it was able to maintain itself but not in a thrifty condition, while in those mowed twice a year the handicap to the mid-grasses enabled it to persist in fair amount. When mowing ceased, the grama became very scarce in two seasons and disappeared completely in the third, as a result of shading by the mid-grasses. The effect of grazing upon the relative success of short-grass and mid-grass is essentially the same, as the next section shows, but overgrazing always gives the former a much more decisive advantage.

Thus, it is evident that the original migration of short-grasses into the mixed prairie and to a certain degree their persistence in it could have taken place only in a climate with recurring dry periods that reduced the stand of mid-grasses in both height and density to the point where they no longer shaded out the short-grasses. During the historical period, this reduction and the consequent spread of *Bouteloua* and *Bulbilis* has been greatly promoted by grazing and especially by overgrazing. The latter supplies the explanation of the layer of short-grass in the coastal prairie, which is found in a rainfall of 30″–40″. Species of *Bouteloua, Bulbilis* and *Hilaria* are able to enter and to remain only where the taller grasses have been kept down by grazing, and they disappear in a few seasons whenever the latter are sufficiently protected to grow to their normal stature with the consequent effect upon light intensity and water-content.

The denuded quadrats with central sod aid also in the explanation of the open spacing of plants in semi-arid and arid regions. In many such communities the open area around each individual shrub or perennial herb may be several times greater than the extent of the top, but the uncovering of the root system shows that it takes tribute from most if not all of the area concerned. This condition is reflected in a humid climate by the much better growth of the plants in the central square within the denuded meter, an increase made possible by the extent to which the roots entered the denuded area.

COMPETITION IN EXCLOSURE

General plan—During the past decade the method of the fenced area in the form of both exclosure and enclosure has been developed and applied on a large scale to the grassland formation of North America (Clements,

1920, 1928). It has proved fundamental to an understanding of climatic control and competition effects, since these have everywhere been modified and often profoundly altered by the effects of grazing under fencing or herding. In addition to the changes wrought by the larger mammals are those caused by rodents; the two effects regularly exist together and can be separated only by means of rodent exclosures, which are now a constant feature of such installations. In short, the exclosure, supplemented under favorable conditions by the enclosure, and with its full complement of quadrats, transects, etc., is indispensable to the analysis of vegetation that seeks to distinguish the effect of climate from that of competition or of animals, and to separate the effects of different life-forms or species of animals.

Long-continued overgrazing of the original tall-grass meadow or low prairie on the flood-plain of Salt Creek had resulted first in a fairly pure sod of *Bulbilis dactyloides* with mixed areas or relicts of *Agropyrum glaucum* or *Poa pratensis* with various ruderals, and had finally caused the destruction of much of the short-grass sod itself. A badly overgrazed area of this character was fenced on April 2, 1924, the area exclosed being 42′x20′. Somewhat less than half of it was occupied by a thicket of *Symphoricarpus occidentalis, Amorpha fruticosa* and *Prunus americana,* ranging from 4′–8′ high. The exclosure was divided into permanent squares 5′ on a side and the outlines of the thicket, blue-grass, wheat-grass and buffalo-grass communities accurately traced. In addition, four permanent quadrats 1 meter in size were located at strategic points, and these were charted and photographed from time to time. Quadrats 1 and 2 were placed in areas where *Agropyrum* and *Bulbilis* were competing with each other, No. 3 in a community so badly overgrazed that it was now more or less denuded and largely occupied by ruderals, and No. 4 in an area where *Symphoricarpus* was actively moving into the grassland.

Although the spring was late, quite marked differences in the growth inside and outside of the exclosure had occurred by April 18, and after a year of protection a fine stand of grass occupied the area in general. For the sake of clearness, the course of development in each quadrat is discussed in order, followed by that of the area as a whole.

Quadrat 1

The buffalo-grass, *Bulbilis dactyloides,* formed a dense sod throughout this entire quadrat on June 11, 1924. *Agropyrum glaucum* had been greatly reduced by grazing and was now sprinkled lightly over the sod, being more abundant at the lower end, where there were 67 stalks by contrast with 62 in the major portion of the quadrat. Under protection this species made a rank growth by September 10 and produced seed in abundance at a height of 18″ (fig. 9).

By May 29 of the following spring considerable change had taken place in the quadrat. The mat of buffalo-grass was 3″ high over the area and had already blossomed. The wheat-grass averaged a foot tall and had

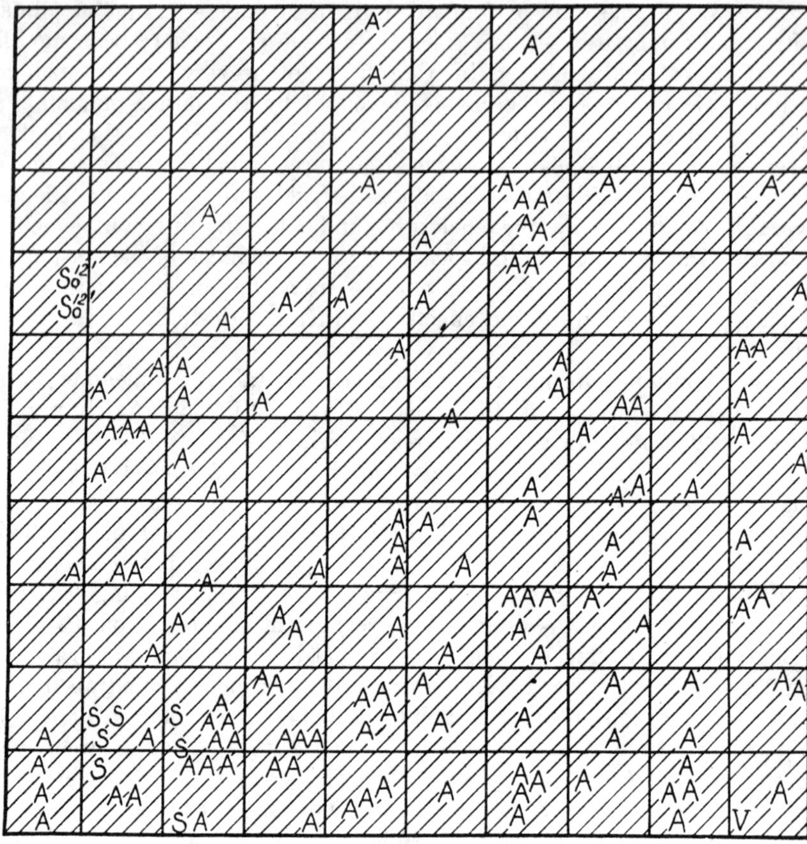

FIG. 9A—Quadrat 1 in exclosure: chart for 1924

//// or B, *Bulbilis dactyloides;* exponents indicate number of stems or unit clumps in this and next.
A or unhatched area, *Agropyrum glaucum.* *(Cont. on page opposite)*

TABLE 8—*Light intensity in the exclosure*

Date	Above Bulbilis under thin Agropyrum	Above Bulbilis under thick Agropyrum	Under dense Agropyrum	Under Symphor. Grass absent	Under Symphor. Poa present	Under Symphor. soil bare
1925	%	%	%	%	%	%
Aug. 30..	49	...	20	10	...	5.7
June 13..	49	...	20	10–14.3	...	5.5
July 12..	..	...	...	......	...	...
July 19..	..	38.1	...	......	15.5	8.3
July 23..	..	20	6.5	......	8.4	4.3
1926						
Aug. 7..	..	22	6	......	9.5	7.5

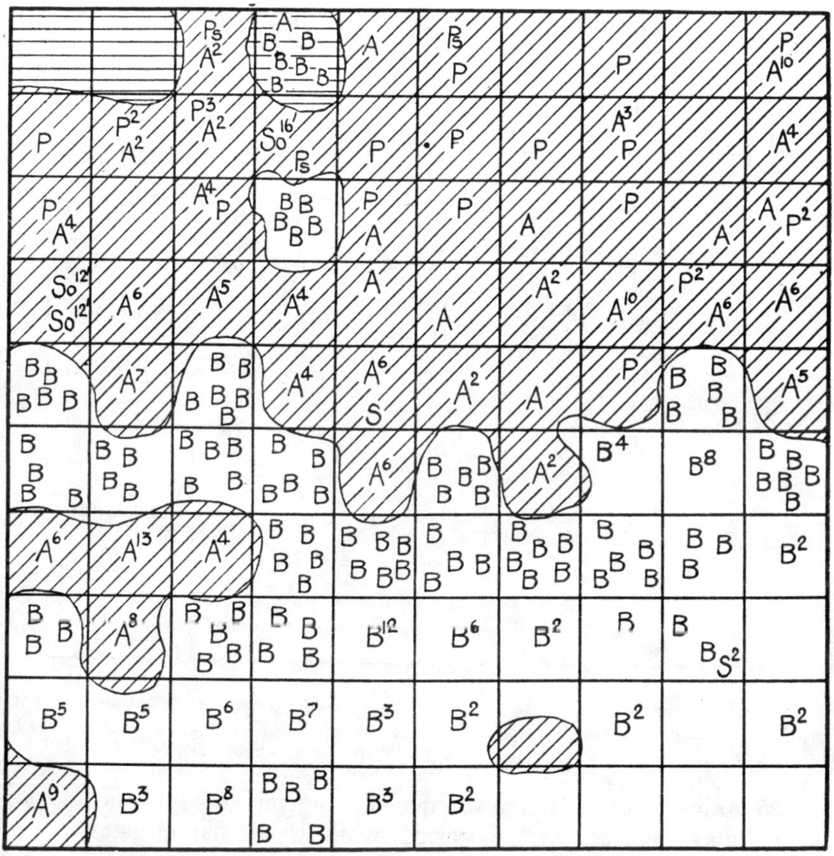

FIG. 9B—Quadrat 1 in exclosure: Chart for 1926.

≡ or P, *Poa pratensis;* So., *Symphoricarpus occidentalis;* exponent is height in inches; V, *Veronica peregrina;* S, *Sporobolus asper;* Ps, *Panicum scribnerianum.*

reatly increased its territory; in the lower strip the stalks numbered 302, n increase of 351%, while in the rest of the quadrat the gain was 290%. n the former this grass was now sufficiently dense to shade seriously its horter competitor and the latter was beginning to lose its hold (cf. table 8). 'he competition for water was evidently less severe than that for light, nce the root systems were fairly well matched as to depth and intimate ontact with the soil. However, the soil frequently became very dry dur- g the season and this undoubtedly exerted some effect upon the struggle able 9). In spite of this, *Agropyrum* reached a stature of 18″–28″ and roduced spikes, even when growing in the dense sod of *Bulbilis* (plate 13).

By June 4, 1926, *Agropyrum* had made still further gains; it now dom- ated 28 of the 30 sq. dm. at the lower end, where *Bulbilis* was now much tenuated and rapidly disappearing. In fact, it had vanished from 6 of e squares and was poorly represented in the others. The wheat-grass so dominated 18 other squares and in the two upper decimeters there were

TABLE 9—*Holard and chresard in the exclosure*

Date	Depth in feet	Bulbilis		Agropyrum		Symphori-carpus	
1925		%		%		%	
June 12...	0.0–0.5	9.2	—0.1	11.0	1.7	16.0	6.7
	0.5–1	8.3	0.1	9.0	0.8	11.9	3.7
	1–2	6.7	—0.3	7.1	0.1	9.9	2.9
	2–3	8.4	—1.6	9.1	—0.9	10.5	0.5
Aug. 19...	0.0–0.5	7.1	—2.2	8.3	—1.0	...	..
	0.5–1	8.0	—0.2	8.1	—0.1	...	..
	1–2	5.3	—1.7	6.5	—0.5	...	..
	2–3	7.4	—2.6	9.2	—0.8	...	..
1926							
July 21...	0.0–0.5	6.8	—2.5	8.5	—0.8	10.3	1.0
	0.5–1	6.6	—1.6	7.9	—0.3	9.0	0.8
	1–2	5.4	—1.6	6.5	—0.5	7.2	0.2
	2–3	7.2	—2.8	8.9	—1.1	8.5	—1.5
	3–4	13.6	3.6	...	..	15.0	5.0
Aug. 7...	0.0–0.5	...	..	7.8	—1.5	...	..
	0.5–1	...	..	8.1	—0.1	...	..
	1–2	...	..	6.2	—0.8	...	..
	2–3	...	..	9.7	—0.3	...	..

The values were not determined with the usual regularity, since they were essentia like those obtained for the low-prairie station but a mile distant on the same flood-pla The hygroscopic coefficients for the several depths were 9.8%, 9.0%, 7.0% and 9.3%.

now 25 stalks in place of the original 3. It averaged 14″ tall and a fe flower stalks were already appearing, while the buffalo-grass was ever where attenuated and but 7″ high. Had it not been for the very d summer, the diminution of the latter would have been much more pr nounced. By July 6, 1927, *Bulbilis* had vanished from half of the qua rat and had become interrupted in the other, where it had bloomed. T wheat-grass had flowered throughout at a level of 32″ and was destin to dominate.

QUADRAT 2

As in the preceding, *Bulbilis* also formed a dense sod over this quad on June 11, 1924. *Agropyrum* was scattered rather uniformly througho to the number of 90 single stalks; in addition, there were a few plants *Poa pratensis* and a relict bunch of *Andropogon furcatus* in the upper p tion. By the summer of 1925 the mid-grasses had increased greatly number, although *Bulbilis* was still rather uniformly present to a height about 4″, with numerous flower stalks 5″–6″ tall. *Poa* had also flowe at 15″–22″, while the wheat-grass was about a foot tall. The latter h accomplished an effective regeneration, although it was not yet sufficien dense to completely replace the short-grass. The 90 stems of the preced year had increased to 337, and there were 204 stalks of blue-grass pres

Effect of protection on *Bulbilis* pasture.

A. *Bulbilis* community before exclosure, 1924.
B. The same community June 10, 1926, *Agropyrum* becoming dominant.

also. *Agropyrum* made further gains by June 4, 1926, with an increase from 337 to 527 stalks, or 57%; it was scattered through the quadrat with a continuous layer of buffalo-grass beneath. The latter was 3″–4″ tall, the former 12″–15″; *Poa* had decreased to 22 stalks and no other species were of any importance. The isolated shoots of *Symphoricarpus* in this and the preceding quadrat had made but slight growth. On July 6, 1927, *Bulbilis* was still present, but was being more and more handicapped by the wheat-grass.

Quadrat 3

When this quadrat was first charted on May 20, 1924, more than half of the surface was bare. *Grindelia squarrosa* was dominant throughout with a height of about 7″; the other ruderals that occurred more or less frequently with it were *Verbena stricta, Vernonia baldwini, Oxalis stricta, Capnoides montanum, Veronica peregrina, Alopecurus geniculatus, Hordeum pusillum, Festuca octoflora,* etc. No *Bulbilis* was present, but *Agropyrum* occurred throughout, most of the decimeter squares containing 2–5 stems. By September it was seeding abundantly at a height of 18″–26″. The ruderals had made a vigorous growth and *Grindelia* had flowered profusely at 3′–4′.

By the following spring *Agropyrum* formed a very dense growth from which the weeds had disappeared as if by magic; of these only a few suppressed plants of *Verbena stricta* remained. The wheat-grass was in practically complete possession; as a rule 5–12 stalks were found in each decimeter square, but 15–20 were common and only a single square had not been entered. On June 4 of 1926, every square was dominated by a dense growth of *Agropyrum,* which was already 12″–18″ tall. The early and rapid growth of this species was obviously an important feature of its success in competition. An occasional plant of *Poa pratensis* or *Panicum scribnerianum* occurred here and there. On July 6, 1927, *Agropyrum* remained in complete possession, forming a rank stand with leaves 2′ and flower stalks 3′ tall.

Quadrat 4

This quadrat was located in the ecotone between an invading tongue of *Symphoricarpus* from the thicket and an area of *Bulbilis.* In 1924 the shrub was represented by 41 new shoots and 18 a year or more old. The entire area was covered with an undergrowth of *Bulbilis,* except for a few squares where *Agropyrum* dominated or where the shade of the shrubs permitted *Poa* to grow. The ruderals indicating overgrazing were similar to those of the preceding quadrat, but were less abundant. *Agropyrum* occurred rather uniformly but sparingly throughout, except where the shrubs were densest. By September many shoots of *Symphoricarpus* of the current year had made a growth of 21″–39″ and had produced an abundance of fruits.

By the middle of June 1925, the taller shrubs cast more shade so that *Bulbilis* was disappearing rapidly from one-half of the quadrat. In many squares it was already dead and its place was being taken by the much more tolerant blue-grass. However, the death of *Bulbilis* was due in large part to the dense growth of wheat-grass, which attained a stature of 12″–15″ by June first. The spring was very dry and the soil hard, making conditions very unfavorable for the spread of the shrubs, only 2 new shoots being found.

The following spring was also unfavorable for the growth of shrubs; not only were no new shoots found in the quadrat, but there were practically none along the whole front of the thicket. Owing to the drouth the foliage was more open than normal and the shade less dense, with the consequence that the grasses easily held their own. *Agropyrum* was somewhat attenuated in the shade, its stature being 21″ by contrast to 15″ in the open. Beyond the shrubs *Bulbilis* flowered profusely at 4″–7″ and *Poa* at 20″–28″; *Agropyrum* was but 12″–15″ high because of the drouth and produced very few spikes. By August 7 the buffalo-grass had dried on the ground, while the wheat-grass was still green. Plants of *Andropogon* were partly wilted and the leaves rolled, but the deep-rooted forbs of the prairie were still thriving.

When the quadrat was observed on July 6, 1927, *Symphoricarpus* had not increased its territory, but a rank growth was still disputing the area with a dense stand of *Agropyrum,* which grew up to the edge of the shrubs but yielded to *Poa* beneath them. A small amount of buffalo-grass still persisted under the dense cover of wheat-grass, but it was very much attenuated.

Exclosure Maps

In addition to charting the quadrats, the vegetation of the entire exclosure was mapped each year and notes made of the growth and behavior of all the important species (fig. 10). The major features of the secondary succession from the ruderal stage due to overgrazing were as follows:

(1) The practical disappearance of the weeds by the end of the second year of protection.

(2) The replacement of *Poa pratensis* over large areas by *Agropyrum glaucum;* it was shaded out in other areas by the advance of the shrubs.

(3) The replacement of *Bulbilis dactyloides* wholly or in part by *Agropyrum;* at no point was the former able to extend its area (plate 14).

(4) Isolated individuals or groups of shrubs grew somewhat in height and often produced a few shoots, but little more than held their own against the wheat-grass and its associates. The main shrub area extended its front only slightly and this occurred chiefly during the first year of protection, the other seasons being unfavorable for the growth of shrubs.

(5) The appearance of *Sporobolus asper* and especially of the low-prairie tall-grasses, *Andropogon furcatus* and *nutans,* together with their thrifty growth, suggested that *Agropyrum glaucum* would gradually yield to these, and low prairie would develop. This indication was supported by the

reappearance of two important subdominants after two years of protection, namely, *Kuhnia glutinosa* and *Salvia pitcheri.*

In 1927 the area gave further evidence of passing from the wheatgrass stage into low prairie. *Andropogon furcatus* was becoming fairly abundant and another species of the prairie, *Panicum scribnerianum,* had reappeared. In addition to *Kuhnia* and *Salvia,* other perennial forbs such as *Erigeron ramosus* and *Asclepias verticillata* were present. *Symphoricarpus*

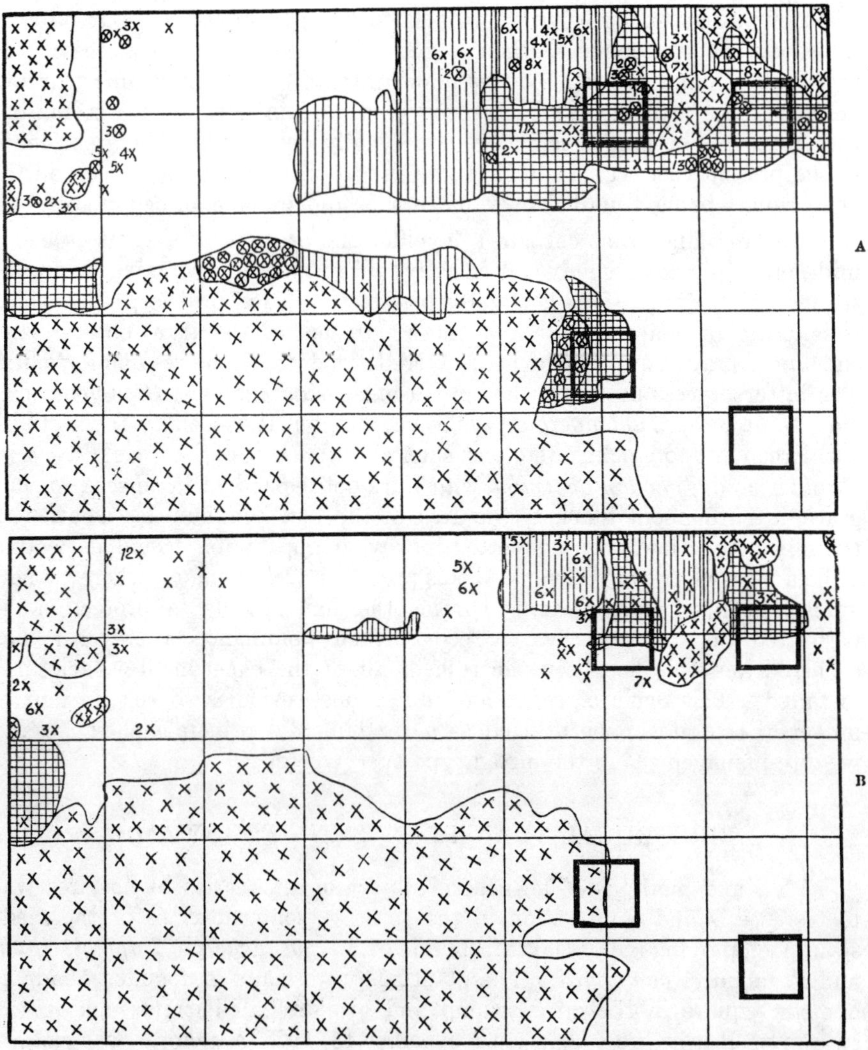

FIG. 10—Maps of exclosure in overgrazed pasture: A, mapped in spring of 1924 at time of fencing; B, at end of 3rd year of protection, August 1926. Unshaded, *Agropyrum glaucum;* squares, *Bulbilis dactyloides;* lines, *Poa pratensis;* x, shrubs, mostly *Symphoricarpus.* The heavy squares are the four quadrats.

was making little progress, having spread into the wheat-grass in two places with a total of only 15 shoots, which were too scattered to make headway against the grasses, especially of the tall-grass type.

Significance of results—This exclosure experiment is an epitome of the course of development in overgrazed mixed prairie wherever it has been given adequate protection against stock and rodents. ("Plant Indicators," 1920:305–309; "Plant Succession and Indicators," 1928; 381–385). If the taller grasses had completely vanished over a considerable area, the regeneration of prairie might be so difficult as to become impossible, but no such areas have been found. Although a careful examination is necessary to reveal the suppressed mid-grasses under intense overgrazing during drouth periods, they have always been discovered, and in normal or wet seasons they regularly become conspicuous and usually dominant. The area fenced in the present case was in much worse condition than has been seen on the open range, being equalled only in corrals and other confined places.

All grasslands that contain a layer of one or more short-grasses below mid- or tall-grasses behave essentially alike under overgrazing and subsequent protection. This is equally true of the mixed prairie proper which is climatic in character, and of those portions of the true, coastal, and subclimax prairies, in which the 2-layered structure is the result of grazing. The latter process promotes the spread of all short-grasses, *Bouteloua gracilis, B. hirsuta, Muhlenbergia gracillima,* etc., but it especially favors those with creeping rootstocks or stolons, such as *Bulbilis dactyloides, Hilara cenchroides,* and *Cynodon dactylon,* which are not dependent upon seed for migration. Throughout much of the area of the four communities mentioned, the mixed type had been converted into what appears on casual inspection to be a pure short-grass cover and this process is still going on wherever new areas are subject to overgrazing. On the other hand, incidental protection has permitted the taller species to recover their dominance wherever it has occurred and this has been the rule in all of the experimental exclosures installed. This cycle of reduction and reconstitution is so certain and so universal that it can be utilized as a fundamental principle of range and pasture management in the mixed type everywhere.

RECIPROCAL TRANSPLANTS IN REED-SWAMP

Plan and methods—The general object of this series of cultures was to test ecesis and competition in the three typical consocies of the reed-swamp, employing reciprocal transplants of *Scirpus lacustris, Typha latifolia* and *Phragmites communis.* The plantings were made in denuded quadrats a meter square in the open swamp and also in half-barrels sunk in the swampy soil near the edge. On November 13, 1923, a quadrat was cleared and most of the rhizomes removed in a *Phragmites* swamp about a pond, and similar ones prepared in *Typha* and *Scirpus* areas along a sluggish stream. Each quadrat was marked by means of a stake a meter long at each corner and the outline of the area indicated by a wire fastened to the stakes. The latter projected about 4″ above the soil level and the wire was attached one inch from the top of the stakes. Pieces of sod a foot square and 1′–2′ deep were cut from the respective communities and transplanted into the other two, in accordance with the following system:

Scirpus Swamp	Typha Swamp	Phragmites Swamp
Phragmites Typha	Phragmites Scirpus	Typha Scirpus
Typha Phragmites	Scirpus Phragmites	Scirpus Typha

The seral substages in the reed-swamp are as follows: (1) *Scirpus*, typically in shallow water; (2) *Typha*, in soft mud; (3) *Phragmites*, in rather firm but saturated soil, the upper two inches sometimes merely moist. The *Scirpus* area was covered with water from a depth of 2″–3″ to over a foot in the small pools between the plants; the soil in the *Typha* community consisted of soft wet slime with scattered shallow pools, while *Phragmites* grew in water-logged soil with few or no small pools. The clearing away of vegetation, the removal of rhizomes and the placing of the quadrats and transplants entailed many difficulties in the case of both *Scirpus* and *Typha*, but was not much more difficult with *Phragmites* than in grassland.

On April 2, 1924, before the spring rains had begun and while the soil still had a frozen crust, two holes were dug in the *Phragmites* community, and half-barrels well bored with holes on sides and bottom were sunk in them. These were filled with muck and several rhizomes of each of the three dominants were planted in them. The same procedure was carried out in the *Scirpus* and *Typha* consocies. To avoid confusion through many details, the course of development is traced separately for each community.

Scirpus Swamp

Season of 1924—On May 4, no *Typha* was found in the quadrat and only one clump of *Phragmites* was growing, comprising 6 shoots 2″–12″ tall. Seven stems of *Scirpus* had invaded the area, which was open and covered with water 2″ deep; the surrounding bulrushes were 6″–24″ tall. By June 5 the latter were about 3′ tall; many appeared as invaders in the margin of the quadrat and one clump with 7 stalks occupied the center. The individuals on the inside gave no signs of blooming, while those on the outside were in full blossom. *Phragmites* was represented by 3 stalks 16″–30″ high. By September 4 the bulrushes on the outside were 6′ tall, and there were 11 stalks of *Phragmites* on the inside, varying from 2′–6′ high, but none had produced seed. A dense layer of *Leersia oryzoides* was found at 2.5′ (plate 15).

Season of 1925—On March 27 a vigorous clump of *Phragmites* and one of *Typha* was transplanted into a thick growth of *Scirpus*, while the soil was still frozen. On June 5, *Scirpus* was 4′–5′ high and beginning to bloom. The light intensity below it was 5% on the ground beneath a dense growth and ranged from 7%–15% where it was more open, while at 2′ it was respectively 25% and 38%. The nitrate-content in the various swamps varied from 1.2–2.5 ppm., showing that there was nowhere an excess. The 1925 transplant of *Typha* did not grow, but that of *Phragmites* had produced 15 stalks by August 15. The tallest was 5′, but none had blossomed. In the quadrat at this time there were 11 stalks of this species 7′ tall; it had spread for a foot and was thriving, the flower clusters being on the point of appearing.

On May 14, 1926, *Phragmites* was thriving in the quadrat at its normal stature, in spite of the fact that water stood on the surface of the soil. *Scirpus* was also thriving, while the clump of *Phragmites* transplanted in 1925 was 1′–4′ tall and vigorous.

Barrel culture—On May 4, 1924, this contained one plant each of *Typha* and *Phragmites* 8″ high, and 16 of *Scirpus* ranging from 4″–24″. The water level was 3″ below the soil surface. The bulrushes surrounding the quadrat were fairly thick and 2′–2.5′ tall. By June 5 these formed a dense stand 5′ high and in blossom, which cut off the light from the south side of the culture. *Typha* was 24″ and *Phragmites* 27″ tall, but slender, while *Scirpus* was represented in the barrel by 40 stems 42″ or less in height. These occupied a third of the area solidly and were increasing their extent; seven were in bloom. Water stood within an inch of the top of the barrel. Two *Phragmites* planted just outside the barrel in the dense stand of bulrush were 42″ tall.

By August 15 a single stalk of *Typha* 4′ tall remained in the barrel; *Phragmites* had died, in spite of the fact that those outside were thriving. On May 14, 1926, the single dwarf cat-tail still persisted, but the drying of the swamp during the summer permitted it to be overrun by cattle.

Of the four original sods of *Typha* only this one poor plant survived; nor were there any volunteers or invaders from the outside. The three clumps of *Phragmites* transferred to the quadrat became established, increased their area and blossomed; the plant in the barrel was the only one to succumb, apparently because of deficient aeration due to the layer of water.

Typha Swamp

Season of 1924—By May 4 only one of the two clumps of *Phragmites* had developed, producing 9 stems 6″–18″ high, but later the other also began growth. Both sods of *Scirpus* were growing, comprising 22 stems that were 5″–30″ tall. Water stood an inch deep on the surface. *Typha* formed a dense growth around the quadrat, but it was only 2′ high and consequently diminished the light very little. It had begun to invade, seven stems occurring within the quadrat.

By June 5 the stand of *Typha* about the area was 44″–48″ tall and the six invading stems were of the same height. Water stood on the surface of the quadrat, and the light intensity was good. There was a total of 38 stems of *Scirpus* 10–13 mm. in diameter and with an average height of a meter. The clumps were thrifty and spreading, the largest measuring 7″x12″. *Phragmites* was represented by 16 stalks ranging from 6″–36″ in height; these were dwarf and pale, the general condition being poor.

By September 4 a dense growth of *Typha* 7.5′ tall surrounded the quadrat, and there were 31 vigorous invaders on the inside. The surface soil was saturated but firm enough to stand upon, and the shade was deep. *Scirpus* numbered 59 stems, which were as well developed as in its typical area, being 10–16 mm. thick and 4.5′–7′ tall. In one corner they had spread 14″ beyond the original sod and many had blossomed. *Phragmites* comprised 20 stalks with a maximum of 6′; they had not blossomed and were for the most part in poor condition.

Season of 1925—On June 5 a fine stand of *Typha* 4′–6′ tall was growing from the water-covered soil; many leaves had been frozen back for 3″–12″. The quadrat was in general well-lighted, and both *Scirpus* and *Phragmites* were thriving. Under a dense growth of both old and new *Typha* the light intensity near the soil was 3%–4% and at a height of 2′, 20%–30%; under the new plants only, the respective values were 10%–13% and 26%. On April 30 and June 5, the pH values were determined for all three swamps, which yielded almost identical results, namely, *Scirpus* 7, *Typha* 7, and *Phragmites* 7.05.

On August 15 nothing remained in the quadrat but a dense growth of *Typha*. These were 6′–7′ tall and growing in running water 2″ deep, the stream having changed its course to cover the quadrat. The light values below the cat-tails was 10% at 1′, 25% at 3′ and 62% at 5′. Again on May 14 of the next year no traces were found of *Scirpus* or *Phragmites*, *Typha* forming an excellent pure stand.

Barrel culture—On May 4, 1924, 3 shoots of *Scirpus* had developed a height of 3″; the area was well lighted and stood 2″ above the water level. By June 5 there were 5 stalks of bulrush 10″–17″ tall, but in rather poor condition; the 7 stems of *Phragmites* were 10″–22″ and the 4 of *Typha* about 10″ tall. By September 4, all the individuals of *Typha* had died and there were only remnants of *Phragmites*, the low light intensity and poor aeration having affected these more than *Scirpus*. There were 31 stalks of the latter, but they had been eaten back to 3′–5′. On August 15, 1925, five stems of bulrush but 28″ high persisted in the barrel, and the reed-grass had vanished. Several plants of *Typha* had reappeared and grown to a height of 6′ and considerable *Leersia* was present. By May 14, 1926, everything had disappeared from the barrel but *Leersia*.

Second quadrat, 1925—On March 27, new sods of *Scirpus* and *Phragmites* were transplanted to the *Typha* swamp. By August 13 one clump of reed-grass had died and the other was in poor condition, with 5 stalks 3.5′ tall. One sod of bulrush also had died, but the other had made an excellent growth and enlarged its area. The light intensity under the bulrush was 14%, that under the reed-grass and adjacent cat-tail 16%. On May 14, 1926, the clump of *Scirpus* was thriving, but no *Phragmites* could be found.

Of the five clumps of *Phragmites* planted in the *Typha* swamp during two different years, none prospered and all finally died. In general, *Scirpus* grew much better, though a single clump alone persisted, apparently owing to its location in deeper water and a more open situation. Light appeared to be the most important factor in the actual competition, though this was determined largely by the greater vigor and stature of the *Typha* which was in possession.

Phragmites Swamp

Season of 1924—By May 4 the two clumps of *Scirpus* were in excellent condition with a stature of 2′ and *Typha* was nearly as flourishing. There were a few plants of the latter native to the area on the margin of the quadrat. The circumjacent *Phragmites* was 2′–3′ high up to the edge of the area. Water stood an inch deep and the shade cast by the old stems of the reed-grass was also a considerable factor (plate 15).

On June 5 the layer of water on the surface had decreased a half-inch. The quadrat was practically surrounded by a dense new growth of *Phragmites* and the light intensity was greatly reduced. Under the stand of old and new reed-grass it was 2.7% at the level of a foot and under the new alone 8% at the same level; in the more open center of the quadrat it was 24%. The clumps of bulrush were spreading and had put forth many new shoots; there was a total of 97 stalks of which 27 were in blossom at about 3′. The bunches of cat-tail were thrifty at a height of 2.5′–3.5′. No *Phragmites* had invaded, but there was a low scattered growth of subdominants.

On September 4 the layer of water had disappeared for the most part, but the soil was still saturated. *Scirpus* now comprised 94 stalks, the largest being 6.5′ tall; it had extended its area, one rhizome having migrated for 12″. About half of the plants had borne fruit, but many were dead, partly as a result of insect damage. *Typha* had attained a stature of 5′–5.5′, but also had been badly eaten by insects. Only one plant had fruited, but 7 well-developed stems had arisen from rhizomes. A single stalk of *Phragmites* was found at this time.

Season of 1925—Growth had not started by March 25, but on June 5 the new stems of *Phragmites* about the quadrat were 3′–5′ tall. *Scirpus* was thriving, some individuals being already in bloom. *Typha* was 2′–3′ high, but 4″–12″ of the tips had evidently been frozen. Light values near the surface ranged from 6%–15%, while at 3′ they were about 30%. They decreased greatly during the next two months and on August 15 the intensity in the forenoon was 0.4% near the soil under *Phragmites,* 2.5% at the 2′ level, 12.9% at 3′ and 35% at 6′. There was no standing water, but the soil was very wet. *Leersia, Heleocharis* and other subdominants formed a layer 22″ high. The original clumps of *Typha* were both dead, but 5 of the later shoots 4′–5′ tall still persisted. Both bunches of *Scirpus* had widened their area 2′ or more; some were 5′ high and had seeded, though they were not very thrifty on account of the dense shade.

A

B

Competition in the reed-swamp.

A. *Scirpus* and *Typha* transplanted into *Phragmites* consocies.
B. *Phragmites* transplanted into *Scirpus* consocies.

On May 14, 1926, *Scirpus* was the only one of the three dominants to ›e found in the quadrat. The dry summer hindered the growth of the bul-ush especially and the shoots from both clumps were only poorly developed n September first. The plants of the cat-tail were very vigorous and the nvaders of the reed-grass fairly well developed, the drier soil handicapping hese species much less. *Leersia* had been particularly favored and now nade a dense layer 3′ high through the swamp.

Barrel cultures—On May 5, 1924, there were 8 stalks of *Phragmites* ″–24″ high, 16 of *Scirpus* 16″–20″, and 1 of *Typha* 6″ tall. Water stood n the surface, and the light values were good. *Phragmites* had made a ense new growth about the barrel where the old stalks had been tramped own. By June 5 there were 30 stalks of *Scirpus* 28″–44″ tall, with six f them in bloom, 5 of *Phragmites* 2′–2.5′ and 2 of *Typha* 15″–24″ high. he barrel was surrounded by a dense growth of reed-grass, which reduced he light value to 5%. On September 4, but 9 stalks of bulrush remained live, most of them eaten to a few inches of the soil; *Typha* was dead and *hragmites* also, except for one poor stalk.

On August 15, 1925, a half-inch of water covered the soil, which was ractically bare of vegetation. Neither reed-grass nor cat-tail was present, nd only one clump of much dwarfed half-dead bulrush was found. The hade cast by *Phragmites* about the area was so dense that only a sparse ndergrowth could develop. On May 14, 1926, no plants were growing in he barrel and on September first nothing but *Leersia oryzoides*.

In barrel No. 2 there was an excellent growth of *Scirpus* and *Phragmites* n May 5 and one good *Typha*. The light was good, and the water level as 3″ below the surface. By June 5 the new growth of *Phragmites* about he barrel was 6′ tall. The 37 stalks of *Scirpus* occupied one-half the area, ut only 4 were of normal diameter and 10 had seeded at 2.5′–4′. There ere 6 stalks of reed-grass 1′–2′ high and 3 of cat-tail at 18″–33″. The ght value at the level of a foot was 12%. On September 4 there were 3 stalks of bulrush, nearly all of which were eaten back but still alive. *hragmites* consisted of 2 stalks 4′–5′ tall, while *Typha* was dead, except for ne new 8″ shoot. None of the three dominants was found in the culture he following year.

Sod transplants—On March 27, 1925, four clumps each of *Scirpus* and *ypha* were transplanted directly into the natural stand of *Phragmites,* and plants of *Typha* already present were marked by stakes. On August 15, *hragmites* was 9′ high and the shade was consequently very dense; water ozed readily from the surface. The bulrushes were thriving at 4′–8′ high nd many had blossomed. Two transplanted individuals of the cat-tail nd two of the original ones had died; the others ranged from 2′–6′ in height.

On May 15, 1926, *Phragmites* was already 3′ tall and *Scirpus* had made n excellent growth, greatly extending its area. The plants were 2′–3′ high nd some had begun to bloom. The 5 clumps of *Typha* were in fair con-ition, some reaching 2′ and more. By September first *Phragmites* had ormed a dense stand 7.5′ high, in spite of the relatively dry soil. All clumps of *Scirpus* had made good growth and some had quadrupled their area, though many of the stems were slender, owing to the dense shade. They had fruited, but were now badly eaten back. Two more plants of *Typha* had died and but one of the remaining three had blossomed; all the live plants were badly damaged by insects.

Summary—Although both *Scirpus* and *Typha* are occasionally found scattered through the *Phragmites* consocies in nature, they are always to be regarded as relicts. The holard of such areas is too low for them, especially in late summer or dry years, and the dense and continuous layer of rhizomes of *Phragmites* practically prevents their invasion. Owing to the water layer, their typical habitats are less subject to frost after growth begins in spring, and this constitutes a further disadvantage in the *Phragmites* habitat. However, the paramount factor in their elimination is shade, the greater stature and broad spreading leaves of the reed-grass reducing the light intensity to a decisive degree.

The barrel cultures were intended to prevent root competition and thus make it possible to determine the respective importance of light and water, but in spite of the measures taken, the access of oxygen was rendered more difficult and the accumulation of harmful gases increased to the point where none of the three dominants was able to survive.

GENERAL SUMMARY

Occupation—The results of transplanting sods of grasses and of reeds into low prairie and swamp respectively emphasize the great advantage enjoyed by plants and species already in possession. This is chiefly and often wholly a matter of competitive ability arising from having shoots and roots at the optimum level for the species concerned. Even when care is taken to select the most vigorous transplants and to give them every advantage so that initial ecesis is successful, the occupants practically always emerge victorious in the course of three or four years (cf. "Experimental Vegetation," 147–151). Moreover, when the habitat factors seemed definitely to favor the invader, the plants in possession were rarely much disturbed and the final outcome was in their favor. This is in full accord with the behavior of climax dominants in nature, as is attested by innumerable natural experiments. It is practically impossible for alien species to invade and ecize in the face of the control exerted by the occupants, except where disturbance has destroyed these or weakened their weapons of competition in some respect. A corollary of this of far-reaching importance is the fact that random migration into climax communities is rarely if ever effective, and that mass migration alone, under the disturbing action of a climatic change, can bring about a modification or replacement of an entire formation or association. All of the usual evidence as to the effectiveness of migration is drawn from the observation of disturbed areas, especially roads and railways, or from migration along water-ways, where occupation is slight or lacking and ecesis direct.

The study of denuded quadrats and of the exclosure reveals two important facts; the first is, that competition between dominants, dominants and subdominants, and between subdominants is always more or less reciprocal, each leaving an impress upon the other in growth-form and often in habitat-form (Clements and Clements, 1915). The second is that disturbance, notably overgrazing, regularly gives an appearance of occupation that is fictitious. The plants in possession are there chiefly or only because of the coaction exerted by animals, and the real competitive relations can be revealed only by eliminating these effects. Such influences exist wherever communities are laid under tribute by animals, and hence are wellnigh universal, since even forest reproduction is affected by them (Clements, 1910; Pearson, 1920).

5. COMPETITION IN THE ECOTONE BETWEEN WOODLAND AND PRAIRIE

Scope of experiments—The chief object of this group of studies was to throw light upon the relations between trees and shrubs on the one hand and grasses on the other, particularly with respect to the contact between the deciduous forest and the prairie, and the climatic processes involved. The ecesis of several species of trees and shrubs had been followed in connection with studies in experimental vegetation (Clements and Weaver, 1924), but only incidental attention was paid to competition. The success of tree plantations in the prairie has led to the assumption that even the true prairie owes its persistence to fire and that the climatic relations are not controlling The fact is overlooked that such groves have been artificially aided in a number of ways, such as the destruction of the grass cover, the use of mulches or actual watering and the employment of young trees tall enough to escape overshading. In short, the most critical time in the whole process, that of germination and establishment, is avoided by the use of transplants, while the physical factors and competitive relations are profoundly modified to the advantage of the tree. The distinction between true prairi and subclimax prairie has been based upon the climatic relations of mid grasses and tall-grasses, for which there is an extensive body of evidenc drawn from natural experiments. The purpose of the present investiga tions was to supplement this by an actual experimental attack upon th competition between woody plants and grasses in the two prairi communities.

COMPETITION OF TREES AND GRASSES IN LOW PRAIRIE

Plan and methods—This series of experiments was initiated in the sprin of 1924. The species of trees selected were those commonly grown fc shade and wind-breaks in Nebraska and adjacent prairie states, and com prised *Acer saccharinum, Gleditsia triacanthus, Ulmus americana, Ace negundo* and *Fraxinus lanceolata.* Seeds were employed in the case of th first three and seedling transplants with the last two. The plan was t secure four degrees of the competition with the grasses in increasing severit Four long parallel trenches 4″ wide and deep were made by removing th native sod; they were filled with loose soil free from roots and the seed sow water being added from time to time to insure germination and establishmen

In the case of the first row, the sod was overturned to a depth of 4″ a distance of 6″ on either side, and was then thoroughly pulverized to con stitute a good mulch. By frequent shallow hoeing this denuded area wa kept free from invaders during the following seasons, while the overhan ing grasses along the edges were repeatedly clipped to insure good lighti and thus confine the competition to the soil. Even here for a time at lea

Growth of tree seedlings in low prairie.

A. Mulched and clipped rows. B. Excavation of root systems of tree seedlings.

there was little or no competition for water or nutrients, since the roots of the dominants of low-prairie spread but little and penetrate nearly vertically (Weaver, 1920).

The grass along the second trench was kept clipped to the ground for a distance of 6″ on either side. In consequence there was practically no competition for light, and the demands of the clipped cover for water and nutrients was moderate and the corresponding competition not severe. Along the third trench the grasses were watered rather freely from time to time during the first year and especially in periods of drouth. Care was exercised to walk along the clipped row when watering and thus avoid trampling the grass about the trees. Since sufficient water was present at all times, the competition in this row was chiefly for light and to some extent for nutrients. The trees in the fourth trench were flanked by the grasses of the prairie and were entirely unaided in competition with them (plate 16).

Season of 1924—During the spring and summer, conditions for growth were very favorable, although the precipitation during August and September was slight. The trees in the mulched row made much better growth than in the clipped one, while those in the others did much more poorly than in the latter, there being little difference between the watered and the unaided row. The chresard in the mulched row on August 4 was 12.2% in the upper 3″, which was 4.5% higher than in the clipped one, although at a depth of 3″–6″ it was practically the same for both. On September 5 the chresard was as follows:

TABLE 10—*Chresard in tree rows*

Depth in feet	Mulched row	Clipped row	Prairie	Excess in mulched row over prairie	Excess in clipped row over prairie
	%	%	%	%	%
0–0.5	13.7	10.2	10.6	3.1	–0.4
0.5–1	16.4	12.4	9.4	7.0	3.0
1–2	19.1	15.7	8.0	11.1	7.7
2–3	18.6	18.6	11.0	7.6	7.6

The water-content of the clipped row was greater at all depths than in the unbroken sod, except for the upper level, but was much less than in the mulched row.

As a whole, the mortality among the trees increased with the degree of competition. The general growth and height were greatest in the mulched row, next in the clipped and least in the two remaining rows, the watered in some cases being slightly higher than the unaided. Watering caused the grasses to grow more vigorously, with the consequence that the tree received even less light than in the unaided row. In both they were now surrounded by a grass stand 3.5′–4′ tall (plate 17).

TABLE 11—*Condition of trees on September 10*

Species	Mulched row	Clipped row	Watered row	Unaided row
Gleditsia triacanthus	Ave. 8″ Max. 12″ Lvs. 15	6″ 9″ 9	3″–4″ 4–5	....
Acer saccharinum..	Ave. 14″ Lvs. 16	8″ 13 Several dead	8″ 8″–10″ Several dead	....
Acer negundo	Ht. 8″–13″ Lvs. 12	5″–11″ 10	6″–8″ 8	4″–7″ 6
Fraxinus lanceolata	Ht. 2″–12″ Lvs. 15	All dead	4″ Eaten	3″–5″ Eaten
Ulmus americana ..	Ht. 1″–12″ Lvs. 16	1″–8″ 10 Few dead	2″ 4 Many dead	2″ 3–4 Very many dead

Season of 1925—On April 7 the plants of *Acer negundo* and a single individual of *Ulmus americana* bore new leaves in the mulched row, and those of the first alone in the clipped row, while even this species was putting forth only a few leaves in the watered and unaided rows. The vicissitudes of the season were great, but they served to exemplify the further dangers that tree seedlings meet in grassland. Many of the latter had been cut off at heights of 2″–9″ by rabbits, though to a less extent where they were concealed by the tall-grass. The spring itself was late and cold and also quite dry; severe freezing occurred twice in the latter part, doing much damage to leaves, except again to those well protected in the grasses. The water-content in the mulched row on May 4 was considerably higher than for the clipped; this amounted to an excess of 8% and 6% for the first and second 6″ levels respectively, with little difference at the greater depths. Even in the clipped row the chresard was 12%, but by May 26 this had fallen to 9%, or 5% less than for the mulched series. The water-content of the unmown prairie was much lower than this.

By June 27 the grasses had reduced the light about the unaided trees to 16% at a height of 3″. On August first the intensity at the level of the topmost leaves in the watered row was 50%, in the unaided 67%, while at the height of 6″ the values were reduced to 26% and 37% respectively. On June 30 the soil temperatures in the mulched row were slightly higher than those in the clipped one; an inch below the surface they were 100° and 98° and at a depth of 8″ where the roots were abundant, 80° and 78° respectively. Just under the loose surface of the mulched row the value was 125°.

On June 3 the chresard in the surface soil was 17% in the mulched, 12% in the prairie and but 7% in the clipped row; even in the second 6″ level the differences were still marked, being respectively 28%, 17% and 11%. The value of the loose soil as a mulch was further demonstrated by the exceptional development of roots in the first foot of soil in this row.

Growth of tree seedlings in mulched, clipped, watered and unaided rows.

A. *Gleditsia triacanthus.* B. *Fraxinus lanceolata.* C. *Ulmus americana.*

Similar results were obtained on June 22 and July first, and on July 15 the mulched row contained twice as much available water in the first foot as did the clipped one. These differences were accentuated by the fact that the chresard of the clipped row and unmown prairie was drawn upon by the grasses as well as the trees, giving the seedlings in the mulched row a still

TABLE 12—*Chresard on the low prairie during 1925*

Date	Depth	Mulched maples	Clipped maples	Prairie	Mulched boxelders	Clipped boxelders	Unaided boxelders	Mulched elms	Clipped elms	Prairie
		%	%	%	%	%	%	%	%	%
May 4	0.0–0.5′	20.8	12.4							
	0.5–1	22.0	16.6							
	1–2	14.4	20.4							
May 26	0.0–0.5′	14.4	9.1	4.4	20.3	...	...	...	...	4.4
	0.5–1	12.7	12.6	8.9	18.6	...	...	...	...	8.9
	1–2	20.2	18.0	12.4	18.4	...	...	...	...	12.4
	2–3	18.8	20.4	13.7	21.7	...	...	...	...	13.7
June 3	...	...	...	...	...	...	...	...	...	...
	...	...	...	...	...	...	...			
	...	...	...	...	...	...	...			
June 22	0.0–0.5′	27.8	22.1	20.7	...	...	...	16.9	7.4	12.1
	0.5–1	28.8	19.5	19.5	...	...	...	27.9	11.5	17.1
	1–2	21.4	19.7	19.1	...	...	...	22.3	16.6	25.7
	2–3	...	...	15.8	...	...	...	19.7	20.8	22.3
July 1	...	...	...	...	22.7	12.5	19.4			
	...	...	...	...	15.7	21.3	21.6			
	...	...	...	...	21.0	20.7	19.4			
July 8	0.0–0.5′	...	11.8	5.5						
	0.5–1	...	4.2	4.7						
	1–2	...	9.9	9.7						
July 15	0.0–0.5′	14.0	7.1	5.5						
	0.5–1	23.1	11.0	8.5						
	1–2	20.0	18.2	12.6						

TABLE 13—*Development of trees on August 26*

Species	No. leaves	Ave. length leaf, cm.	Ave. width leaf, cm.
Gleditsia			
Mulched	207	10.5	2.0
Clipped	37	9.0	2.1
Watered	9	5.0	1.5
Unaided	7	6.0	1.0
Acer s.			
Mulched	201	8.0	6.6
Clipped	35	4.3	4.3
Watered	4	1.5	2.0
Unaided	2	3.0	2.0
Acer n.			
Mulched	80	12.0	11.0
Clipped	38	6.5	7.5
Watered	15	4.3	4.2
Unaided	10	5.0	7.0
Ulmus			
Mulched	129	5.0	2.5
Clipped	35	3.7	1.7
Watered	4	2.6	1.4
Unaided	5	5.0	1.5

TABLE 14—*Behavior of trees during 1925*

Species	Mulched	Clipped	Watered	Unaided
Gleditsia				
Ave. ht. in.	13.7	6.6	4.5	5.0
Greatest growth, in.	17.0	10		
No. winter-killed	0	3		
No. died during summer	4	9	2	16
Acer s.				
Ave. ht. in.	21.7	9.1	6.5	7
Ave. diam. mm.	9	4	2.5	2
Greatest growth, in.	12	7		
No. winter-killed	4	1		
No. died during summer	1	12	13	12
Acer n.				
Ave. ht. in.	21.5	12	7	11
Ave. diam. mm.	11.5	5.7	3	3.2
Greatest growth, in.	20	17		
No. winter-killed	0	0	0	1
No. died during summer	0	0	2	5
Fraxinus				
Ave. ht. in.	5	0	3	
Greatest growth, in.	6	0	0	
No. winter-killed	3	0	0	4
No. died during summer	1	0	6	
Ulmus				
Ave. ht. in.	14.8	5.1	4	4
Ave. diam. mm.	...	...	1.4	1.4
Greatest growth, in.	18	7		
No. winter-killed	4	11		
No. died during summer	1	10	8	5

greater advantage. This is revealed by a comparison of the growth, made by means of average specimens determined by careful measurements. During this season no water was given the trees in the row watered the previous year and in consequence they made a poorer growth in most cases than those in the unaided row, where the grasses had grown less rankly the first summer (pl. 18).

Season of 1926—The dry weather resulted in much winter-killing, while the spring was both late and cold, and with the summer was dry until about the middle of August. The mortality was least in the mulched, next in the clipped and unaided, and greatest in the watered row. Many of the twigs were frozen, the terminal bud or a few inches of the stem being dead.

The chresard values for the several rows at the usual depths are given in table 15.

TABLE 15—*Chresard for tree rows, 1926*

Date	Acer saccharinum		Acer negundo		Unaided trees
	Mulched	Clipped	Mulched	Clipped	
	%	%	%	%	
May 4....	12.8	8.6	24.0	11.1	
	23.6	16.8	20.2	19.9	
	22.2	22.4	19.6	19.8	
	24.3	24.9	22.5	20.5	
May 28....	10.2	9.9			
	16.8	11.5			
	20.5	15.5			
June 18....	9.8	8.2	...	...	10.0
	10.6	8.6	...	...	9.6
	15.5	11.3	...	...	16.7
	23.9	16.4	...	...	17.5
June 29....	7.3	6.9	...	...	9.2
	10.5	8.6	...	...	9.5
	13.5	11.7	...	...	13.6
	7.2	14.6	...	...	18.3
July 17....	9.0				
	8.5				
	11.4				
	15.3				
July 20....	...	1.2			
	...	4.9			
	...	8.8			
	...	13.7			

In the surface foot especially there was always more water present in the mulched than in the clipped row. *Poa pratensis* formed a dense mat about the trees in the latter, and much water was lost through transpiration, in spite of the fact that the grasses were kept short. It is somewhat surprising to find that the trees in the unaided and the originally watered row were as well supplied with available water as those in the clipped one. In spite of drouth there was at least a small amount of water to be had at all depths, and much in the deeper soil where the trees were well rooted, at least the mulched ones. The average day and 24-hour humidity as well as the daily evaporation from June 10 to August 27 are shown in table 16.

TABLE 16—*Temperature, humidity and evaporation in low prairie, 1926*

Date	Ave. day temp.	Ave. daily temp.	Ave. day humidity	Ave. daily humidity	Ave. daily evap. corrected full week
	° *F.*	° *F.*	%	%	*cc.*
June 10....	76.5	69.7	44.8	55.9	27.6
June 17....	77.8	75.1	72.0	74.7	19.9
June 24....	80.6	75.6	64.1	71.5	18.3
July 1....	82.4	76.6	55.7	66.0	29.3
July 8....	84.1	79.9	64.3	72.5	27.1
July 15....	79.1	73.8	52.2	63.5	18.1
July 22....	93.4	88.4	50.0	59.5	44.4
July 29....	90.4	87.2	51.3	61.4	26.0
Aug 6....	86.1	81.7	43.6	52.6	32.6
Aug. 13....	85.9	81.0	42.8	51.5	23.8
Aug. 20....	77.7	74.0	70.5	78.0	19.6
Aug. 27....	86.7	82.3	71.2	80.1	18.0

Determinations of nitric nitrogen were made for the surface foot of soil on June first. The results were as follows: 1.4, 1.1 and 0.9 ppm. for the mulched, clipped, and unaided rows of *Ulmus americana* respectively and for *Acer negundo* in the same order 1.2, 1.0 and 1.0 ppm. In the upper 6 inches on June 19 the mulched row of *Acer saccharinum* had at its disposal 4.7, the clipped 1.1 and the unaided 0.9 ppm. Hence the plants were using all of the nitrates as rapidly as they were made available, with the exception of those in the mulched row, which did not have to share their supply with the grasses.

The rate of starch and photosynthate accumulation was determined on June 30. *Acer saccharinum* bore very thin leaves in the shade of the unaided row and the photosynthate present was 0.0991 gm. per 100 sq. cm. *Ulmus americana* under the same conditions exhibited starch only in the uppermost of the four leaves, where the amount was moderate; the photosynthate was 0.0598 gm. per unit. *Gleditsia triacanthus* in the unaided row, although the leaves were folded on account of drouth, gave a good starch test, with a photosynthate value of 0.0763 gm. per unit. On August 27 with adequate chresard and good illumination, the leaves of *Ulmus* and *Gleditsia* yielded but 0.0335 and 0.01125 gm. per unit area respectively, probably on account of the lateness of the season.

Even during the driest part of the summer, which was about August first, the trees in the mulched row were in good condition generally, though the leaves of *Gleditsia* folded in the forenoon and those of *Ulmus* rolled and withered somewhat. In the clipped row the latter was dying, but the former was prospering. On August 9 the condition of the trees had become worse, some of those in the unaided and watered rows being badly wilted.

Growth of *Acer negundo* in mulched, clipped and unaided rows.

A. 1924 seedlings on August 26, 1925.
B. 1925 seedlings on August 15, 1926.

Acer saccharinum had died in these same rows and *Fraxinus* survived in the mulched row only.

TABLE 17—*Development of trees on August 9, 1926*

Species	% loss in 3 years	Ave. ht. in.	Ave. diam base, mm.	Ave. max. spread, in.	Dry wt. ave. plant, gm.
Mulched					
Gleditsia	13	18.0	8.3	11.1	22.62
Acer s.......	40	24.6	13.2	11.9	50.32
Acer n.......	20	33.0	17.5	20.2	87.21
Fraxinus	67	12.6	8.9	7.0	...
Ulmus	14	19.4	8.9	8.8	28.27
Clipped					
Gleditsia	44	9.0	5.2	6.0	2.25
Acer s.......	92	15.0	7.3	4.5	6.00
Acer n.......	10	14.8	8.7	7.7	5.84
Fraxinus	100	...	...	...	...
Ulmus	68	6.4	3.8	2.9	0.84
Watered					
Gleditsia	50	4.8	1.8	1.5	0.65
Acer s.......	100	...	...	...	...
Acer n.......	55	9.8	3.7	3.0	0.91
Fraxinus	100	...	...	...	...
Ulmus	90	3.6	1.2	1.0	0.21
Unaided					
Gleditsia	93	9.8	2.2	5.2	0.65
Acer s.......	100	...	...	...	...
Acer n.......	90	5.5	5.3	3.0	0.72
Fraxinus	100	...	...	...	...
Ulmus	75	3.5	0.8	1.2	0.17

The average loss of all trees in the mulched row was 31% and in the clipped 62%, while it was 79% and 92% respectively for the watered and unaided rows. With respect to the most favorable method of planting, i.e., the mulched row, the losses by species were as follows: *Gleditsia* 13%, *Ulmus* 14%, *Acer negundo* 20%, *Acer saccharinum* 40% and *Fraxinus* 67%. Taking all four methods into account, *Acer negundo* had the least with 44%, *Fraxinus* the most with 92%, while the others were *Gleditsia* 50%, *Ulmus* 62% and *Acer saccharinum* 83%. Of all the species, *Acer negundo* made the best growth as measured both by height and dry weight (fig. 11).

New series—In 1925 the preceding experiment was duplicated to the the extent of employing the same species and methods. Owing to late frost and drouth, replanting proved necessary and a large amount of water was required to permit the seedlings to become established. As in the previous series, the watered row was aided during the first season only. The growth of the trees was naturally poorer than for the preceding year, but the results were of value in exhibiting the effect of the dry years and drouth periods that constantly recur in the prairie climate.

Season of 1925—By June 27 the light intensity at the height of 4″ in the unaided row was only 45%. On August 18 the chresard in the mulched row of *Acer saccharinum* was 21% or more at all levels to 3′. The effect of replacing the grass by a mulch was shown by the fact that the second foot of soil contained 7% more available water than the prairie and the third foot 12% more. Owing to the dry year the grass in the watered row made a much better growth than that in the unaided one. The chresard for the mulched, clipped and unaided rows on August 28 was as follows:

TABLE 18—*Chresard on August 28*

Depth in feet	Mulched	Clipped	Unaided
	%	%	%
0.0–0.5....	12.9	13.8	9.2
0.5–1	20.3	18.4	10.4
1 –2	21.1	16.8	7.7
2 –3	22.0	15.0	12.9

The development of the trees by August 26 is well illustrated by *Acer saccharinum,* which was representative (table 19). The results for maple and ash at this time were as follows:

TABLE 19—*Growth of trees, August 26*

Species	Row	Ave. no. leaves	Ave. length leaf-blade	Ave. width leaf-blade
			cm.	*cm.*
Acer saccharinum.	Mulched	30	9.3	9.1
	Clipped	27	5.5	5.4
	Watered	13	3.2	2.9
	Unaided	12	4.5	4.2
Fraxinus lanceolata ..	Mulched	22	7.0	3.5
	Clipped	13	5.0	2.7
	Watered	7	3.0	1.0
	Unaided	8	2.5	1.0

Season of 1926—During the winter of 1925-26 many of the individuals died. The losses were greatest in the watered row where the wood had ripened least before the oncoming of winter, next in the unaided and clipped respectively, and least in the mulched row, where the conditions for ripening were better and the trees were larger. On the other hand, the tips of the latter were frozen back further in the spring, owing to the greater exposure resulting from their height, and the damage by rodents was also greater.

On June 29, 1926, the chresard in the clipped row was reduced to 2.3%, 8.1% and 10.5% at the several levels to 2′ respectively, but was 2%–3% higher in the mulched row. The season continued dry and on July 20 the values were further reduced to 0.8%, 5.6%, 6.9% and 4.1% for the clipped row and to 4.9%, 8.0%, 9.8% and 17.3% for the mulched one, the latter being about 3% higher for the three upper levels and 13.1% for the lowest. On account of the drouth, the trees grew poorly and by August first some even of *Gleditsia* and *Acer negundo* were dying. A week later it was

TABLE 20—*Development of trees at end of summer, 1925*

Species	Ave. height in.	Ave. diam. mm.	Per cent mortality
Gleditsia			
Mulched	10.6	..	0
Clipped	5.0	..	25
Watered	4.0	..	10
Unaided	3.0	..	10
Acer s.			
Mulched	15.0	5.6	0
Clipped	9.4		43
Watered	6.3	2.4	6
Unaided	7.6	2.5	36
Acer n.			
Mulched	14.6	6.5	0
Clipped	7.0	3.6	67
Watered	6.3	3.1	0
Unaided	4.3	2.0	40
Fraxinus			
Mulched	7.3	..	20
Clipped	6.3	..	79
Watered	2.6	..	13
Unaided	2.2	..	38
Ulmus			
Mulched	12.2	..	41
Clipped	3.8	..	89
Watered	3.0	..	44
Unaided	2.5	..	73

TABLE 21—*Development of trees planted in 1925, on August 9, 1926*

Species	% loss in 2 years	Ave. ht. in.	Ave. diam. base, mm.	Ave. max. spread at crown, in.	Dry wt. of ave. plant, gm.
Mulched					
Gleditsia	0	17.6	6.6	10.8	7.11
Acer s.	14	17.8	10.2	6.9	11.75
Acer n.	0	25.9	9.6	13.7	28.26
Fraxinus	27	12.9	6.3	6.5	9.29
Ulmus	41	13.7	7.5	8.6	6.98
Clipped					
Gleditsia	31	6.7	3.2	6.1	0.86
Acer s.	71	10.2	4.3	2.4	2.96
Acer n.	89	15.0	6.3	7.0	6.43
Fraxinus	93	8.0	4.3	1.5	1.04
Ulmus	95	4.0	2.9	2.0	0.35
Watered					
Gleditsia	20	5.3	2.0	2.9	0.61
Acer s.	61	6.3	2.2	0.9	1.85
Acer n.	80	7.0	3.7	2.0	0.57
Fraxinus	80	3.3	1.2	0.5	0.06
Ulmus	69	2.8	0.9	0.1	0.06
Unaided					
Gleditsia	0	4.1	1.5	2.0	0.07
Acer s.	93	8.0	2.3	0.5	0.32
Acer n.	60	6.0	2.0	1.5	0.18
Fraxinus	81	3.5	1.1	1.5	0.06
Ulmus	93	2.0	1.0	1.0	0.04

evident that the trees of this series were suffering much more than those planted a year earlier. The grass level stood at 8″–10″, and practically all the trees in it were wilting or dried. The relative development of the species under the several conditions is shown in table 21, p. 163, and representative individuals are portrayed in plates 16–18.

The average losses were least, 16%, in the mulched row and greatest in the clipped row, 76%, due both to the lack of water-retaining mulch and the presence of transpiring grasses. The respective losses in the watered and unaided rows were 62% and 65%, watering having been discontinued during the current season. It is significant of their greater resistance to drouth that neither *Gleditsia* nor *Acer negundo* suffered loss under the mulch.

Summary—Mortality increased as the conditions of competition became more severe. Watering the trees in the grasses during the first season did not compensate for the water removed by the latter, since the additional moisture caused them to grow taller and hence to shade the trees more. The respective percentages from the mulched to the unaided row were 24, 69, 72 and 75.

The trees made progressively better growth and development as less and less competition was encountered, except that those in the watered row were often slightly poorer than those in the unaided one, owing to the condition stated above. Considering only the most favorable method of planting, the sequence of losses was as follows: *Gleditsia* 6.5%, *Acer negundo* 10%, *Acer saccharinum* 27%, *Ulmus* 28% and *Fraxinus* 47%. When all methods of planting are taken into account, the losses were as follows: *Gleditsia* 31%, *Acer negundo* 51%, *Ulmus* 68%, *Acer saccharinum* 73% and *Fraxinus* 81%.

The best explanation of the success of some species and the failure of others is to be found in the development of the root system. In fact, the rate and degree of development and the fineness of branching were regularly correlated with survival. In consequence, the root systems were examined in detail and the following account for each species throws much light upon its behavior in the four degrees of competition.

Root Development Under Competition

Method—The root system of each species under the various degrees of competition was examined in careful detail during August of the third season of growth, 1926. A long trench 3′ wide and 6′ or more deep was dug parallel to the mulched row of trees, but at some distance from it. The root systems were then excavated by digging into the wall of the trench with minute care. After several individuals of each species had been examined and a typical root system drawn, the walls of the trench were cut back until the second or clipped row was encountered. Smaller trenches were employed in the excavation of plants from the watered and unaided rows, the trees in both these being so nearly of the same size that one lot from either sufficed for both. In the drawings the roots are shown

in a single plane, while a single representative branch is drawn to the natural size and all other drawings to scale.

The soil profile shows an upper 18″ consisting of a rich nearly black silt-loam, very granular in structure and slightly lighter in color below the first foot. Below this upper layer the soil became still lighter and it was a dark gray at the 2.5′ level. At greater depths it was darker in color with more clay, a condition that extended to many feet in depth. This soil is distinctly columnar and contained fissures 2′–3′ deep or even more, the larger ones being nearly one-half inch in width.

Gleditsia Triacanthus: Honey Locust

Mulched row—This species possesses a strong taproot with many large laterals and penetrates the soil deeply. A representative taproot was 20 mm. thick at the surface, 13 mm. at a foot deep and 3 mm. at 2 feet; it reached a depth of over 6′. As a whole the system consisted of two somewhat distinct portions, namely, a taproot with numerous short fine branches and a few major laterals in the deeper soil, and a layer of many wide-spreading and much branched horizontal laterals in the upper 16″. The latter ranged from 2–8 mm. in diameter and arose almost wholly at a depth of 3″–8″, a single plant bearing about a dozen. They extended laterally from the taproot for a distance of 14″ to nearly 5′ and throughout were densely clothed with branches often 8″ long. Large and small branches arose from these at the rate of 5–8 per inch, the larger being again branched to the same degree. This mass of laterals completely occupied the upper 12″–14″ of soil, forming an intricate network of branches (fig. 11).

The root system of the honey locust was the most profusely branched of all and readily explains the ability of this tree to successfully endure drouth. As with all the trees examined in the mulched row, the roots exhibited a marked preference for the mulched area, the maximum lateral development always being found here and not in the sod on either side. Roots frequently extended into the sod and then curved back again into the loose moist soil beneath the mulch, though many also came into competition with the grass roots. It was infrequent that a root originating in the first foot turned downward, but several of those from the second foot took this course and long branches often ran parallel with the taproot. These in turn were so profusely branched as to considerably increase the absorptive area in the deeper soil.

Clipped row—The representative plants in this row had taproots 6–8 mm. wide near the surface but tapering to 2 mm. at a depth of 18″; the taproots reached depths of about 5′. The root system resembled closely that of the mulched trees, but was very much less extensive. There were usually about 7 major horizontal branches and numerous smaller ones in addition; the maximum spread was only 3′ and branching was largely confined to the first 2′. These laterals were but 1–2 mm. wide and bore about 8 branches per inch, which were 2″–8″ long for the first 20″, decreasing rapidly in length and then ceasing altogether for the last 6″. The branches extended in all directions just below the dense sod, being sufficiently vigorous to penetrate the sod and the compact soil below. The difficulty the roots experienced in penetrating this hard soil was indicated by the flattened tips. The lateral spread in the surface soil was hindered by the density of the sod and in consequence depth of penetration was relatively greater than lateral extent (fig. 12).

Watered row—Six small trees with a stem diameter of only 2–3 mm. at base and a height of 4″–6″ were excavated from the dense sod of this row. The roots were relatively no better developed than the tops and reached depths of but 18″–20″. The most striking feature was the almost complete absence of strong horizontal roots. For the first foot the taproot gave rise to finely divided branches at the rate of about 8 per inch, and these rarely exceeded 3″ in length. A single horizontal branch extended about a foot laterally, but clusters of smaller branches filled the soil to a distance of 3″. The branching of the larger laterals and of the deeper portion of the root system was profuse. The absence of widespreading laterals was evidently due to the competition afforded by the dense masses of grass roots on either side of the narrow trench.

Acer Saccharinum: Silver Maple

Mulched row—Typical plants 20″–26″ tall exhibited strong taproots and many large branches, nearly all of which arose in the upper 8″. One taproot measured 24 mm. at the surface, 6 mm. at a foot and 4 mm. at 4′, attaining a depth of about 7′. There were about 10 major laterals, ranging in diameter from 3–10 mm. Approximately a third of these pursued a horizontal course in the mulched soil to distances of 4′–6′, entirely within the upper 6″ and sometimes within 1″–2″ of the surface. The others grew obliquely for 3″–18″ and then turned downward to penetrate deeply, frequently paralleling the course of the taproot at a distance of 4″–24″. The superficial roots were profusely branched throughout, the larger laterals being about 1 mm. thick and 3′ long, and covered with minute branches at the rate of 15–22 per inch.

The taproot and the larger roots that entered the deeper soil were often rather poorly branched beyond 6″ to 18″–20″, but below this the branching was profuse, forming dense networks of rootlets for 4″–6″ on all sides. In comparison with the honey locust, the superficial portion of the maple root system extends more widely, but does not occupy the soil to so great a depth or so thoroughly. The deeper portion is better developed in the maple, but this corresponds with the greater development of the shoot (fig. 13).

Clipped row—The maples in this row were very much dwarfed and the single survivor at the time the roots were studied was but 12″ high with 3 small branches and 18 badly eaten leaves. The root system was correspondingly poor, the taproot being only 7 mm. thick at the surface and reaching a depth of 3′. The main laterals were sparsely developed no branches occurring near the soil surface or in sufficient number to fill the soil. Below the 6″ level, branches 3 mm. to 3″ long were found at the rate of 2–5 per inch, and only the longer ones were again branched. In comparison with the honey locust in the clipped row, the root system was hardly half as well developed, and in fact hardly equalled that of the watered row, except in penetration. However, no direct comparison with the latter could be made, since no maples survived in the watered or unaided rows.

ACER NEGUNDO: BOXELDER

Mulched row—The boxelder, like the honey locust, was represented at the end of the three years under all four degrees of competition, and ranked next to it in the percentage of survival.

The taproot was 30 mm. thick at the top, but gave off so many large laterals within 1″–2″ of the surface that it was only 15 mm. wide at 6″ deep and no larger than some of the major branches in the second foot. The root system penetrated as deeply as 7′–9′, the branches sometimes exceeding the taproot in length. As a whole it was well branched throughout and occupied more soil to a depth of 7′ than either of the preceding species. Large branches, sometimes 15 mm. thick at the point of origin, spread widely and often horizontally in the surface soil, some to nearly 7′. They were abundantly supplied with both long and short rootlets at the rate of 8–16 per inch. More of the larger branches spread widely and then ramify more deeply than in either the honey locust or maple. This is correlated with the fact that the roots were required to furnish water for a shoot development far more extensive than that of the honey locust in particular. Like the latter, the resistance of boxelder to drouth and its consequent range far to the westward are connected with the efficiency of its double root system.

Clipped row—An examination of 5 of the 10 boxelders that survived in the clipped row showed taproots penetrating to 7′–8′, but with laterals spreading only 30″. One taproot produced 12 laterals 1–3 mm. thick in the first 7″ of its course; seven of these extended horizontally and five rather vertically downward. Most of the former were 2′ or less in length; in fact strong horizontal roots were not a feature of growth in the sod. They grew in all directions from the base of the plant and showed no tendency to follow the row, as did the roots under mulched conditions. All of the roots were well branched and gave rise to laterals at the rate of about 12 per inch. The taproot pursued a rather vertical course, giving off short branches a few inches long at the rate of 6–12 per inch. At irregular intervals much longer ones occurred; these spread about 6″ from the taproot and then followed it downward. In spite of its restricted spread, only 30″ in the surface soil and much less in the deeper levels, this root system was well developed and profusely branched, with a high efficiency in competition with the grass roots.

Ulmus Americana: Elm

Mulched row—The root system of the elm in this row was extensive. The taproot was 18 mm. thick at the top, branching profusely above and tapering to 3 mm. at the 5′ level; it reached a depth of 8.3′. The large laterals spread more or less widely and then extended downward for 5′–7′. Some of these spread in the surface foot for 4′–5′. All the roots were profusely branched, the superficial ones bearing about 8 to the inch, 4″–8″ but often 14″ long. The taproot was most abundantly supplied with absorbing rootlets below 5′, where they formed dense networks comparable to that of the honey locust (fig. 14).

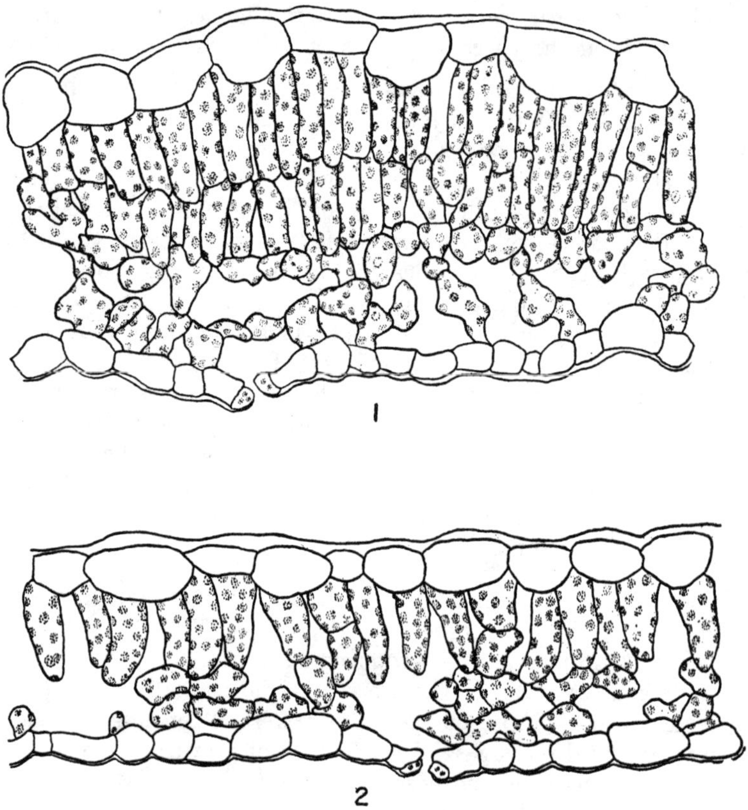

Fig. 14—Structure of leaf of *Ulmus americana* from clipped and unaided rows.

In correspondence to the poor growth of elms in the watered and unaided rows, the taproots were scarcely 2 mm. thick at the top; at the 8″ level they were but 0.5 mm. and they stopped at 17″. Their growth was extremely slow, as shown by the absence of the white smooth root-tips so characteristic of the mulched row, the branches extending to the very tip. Laterals occurred at the rate of 12 per inch.

FRAXINUS LANCEOLATA: ASH

Mulched row—The taproot had a diameter of 13 mm. at the top; it gave off as many as 5–7 branches in the upper 6″ and tapered rapidly, measuring 3 mm. at the 3′ level and 1.5 mm. at the maximum of 61″. Roots rarely grew in a horizontal direction and the greatest lateral spread was reached at 18″ deep, where it was but 2′. The main laterals grew outward and downward for only 5″–10″ and then penetrated deeply;

TABLE 22—*Development in 1924*

Date	Prairie	Transition	Dense scrub
Tilia americana			
May 5..........	4; good	2; good	2; good
May 18..........	3; good	2; good	2; good
June 11..........	2; good	1; small	1; good
July 12..........	2; leaves yellow	1	0
Aug. 10..........	2; poor	0	
Sept. 10..........	2; poor		
Acer negundo			
May 5..........	4; good	6; good	6; good
May 18..........	4; good	6; good	6; good
June 11..........	4; 2–4 in., good	5; 4–8 in.	4; leaves thin, broad
July 12..........	4; 4–5 in., yellow	5; 8–11 in., thriving	1
Aug. 10..........	4; 5–8 in., good	5; 6–13 in., good	0
Sept. 10..........	4; 8 in., good	5; 7–12 in. excellent	
Glēditsia triacanthus			
June 11..........	0	0	0
July 12..........	5; thriving	2; small	
Aug. 10..........	3; good	3; 1 good	
Sept. 10..........	3; 5 in., good	2; poor	
Ulmus americana			
June 11..........	Many; small	Many; smaller	2
July 12..........	Many; 2 in.	About 36	2; good
Aug. 10..........	About 36; poor, 2 in.	......	4; 2–3 in., fair
Sept. 10	14; fair–poor	4; 5 in., good	4; fair
Acer saccharinum			
June 11..........	27; 2–6 leaves	27; fair	2; poor
July 12..........	27; 5–8 in., good	......	14; leaves eaten
Aug. 10..........	15; fair	......	12; 1–8 in., eaten
Sept. 10..........	13; 5–8 in. fair, some dying from drouth	7; 5 in., fair	8; fair

lepths of 47″–56″ were attained by several of these. Nearly all were lothed with small mostly horizontal rootlets at the rate of 8 per inch, nd these were usually not over 2″–6″ long. The root system as a whole vas characterized by the more or less vertical growth of the main roots nd the almost complete absence of a superficial portion so pronounced n the honey locust, maple and boxelder. This seems to furnish the best xplanation of the fact that the ash made the poorest record in competition vith grasses of all the species tested.

TREE TRANSPLANTS AND SEEDLINGS IN LOW PRAIRIE

In the further analysis of competition in the prairie with respect to the establishment of trees, transplants and sowings were made on April 24 1924. Six natural seedlings of *Acer negundo* and 4 of *Tilia americana* were transferred to the low prairie area utilized for the exclosure study (p. 138), the same number to the transition between the latter and low scrub of *Symphoricarpus,* and 9 each to the dense shade of the scrub community At the same time seeds of *Gleditsia triacanthus* were sown in the same sites and seeds of *Ulmus americana* and *Acer saccharinum* on May first. The detailed results for the three successive years are given in tables 22 and 23

TABLE 23—*Development in 1925 and 1926*

Date	Prairie	Transition	Dense Scrub
1925			
Tilia americana			
May 29..........	1; 2 in.		
Sept. 1..........	1; 2 in.		
Acer negundo			
May 29..........	4; 2–10 in., fair	4; 10-15 in., thriving	
Sept. 1..........	4; 4–11 in., fair	4; 14–21 in., thriving	
Gleditsia triacanthus			
May 29..........	3; 5 in., poor	2; fair	
Sept. 1..........	3; poor	2; fair	
Ulmus americana			
May 29..........	12; 2–4 in.	0	0
Sept. 1..........	2; poor		
Acer saccharinum			
May 29..........	5; 6–10 in., poor	0	0
Sept. 1..........	5; 6–11 in., fair to dying		
1926			
Tilia americana			
June 4..........	1; poor		
July 23..........	1; 1 in., fair		
Sept. 1..........	0		
Acer negundo			
June 4..........	4; 3–8 in., good	4; 11–20 in., fine	
July 23..........	4; 4–12 in., good	3; 14–20 in., fine	
Sept. 1..........	2; 11, 16 in., fair	3; 14-20 in., fine	
Gleditsia triacanthus			
June 4..........	3; fair	2; fair	
July 23..........	3; 6 in., lvs. folded	2; good	
Sept. 1..........	3; 6–11 in.	2; 7, 16 in., good	
Ulmus americana			
June 4..........	2; 2–3 in.		
July 23..........	2; wilting		
Sept. 1..........	0		
Acer saccharinum			
June 4..........	0	0	
July 23..........	5; 7–10 in., poor-dying		
Sept. 1..........	0		

Summary of Results

Gleditsia triacanthus:
Prairie: 5 established, 2 died the first season; 3 survived in 1926.
Transition: 3, one died first season, 2 survived.
Dense scrub: None became established, owing to deep shade.

Acer negundo:
Prairie: 4, 2 died of drouth in 1926, 2 good plants survived.
Transition: 6, one died first year, 1 winter-killed, 1 died 1926; 3 fine plants survived.
Dense scrub: 6, all died by July 12, the shade too dense

Acer saccharinum:
Prairie: 27, 14 died first summer, 8 winter-killed; 5 died summer of 1926.
Transition: 27, 20 died first summer, remainder the following winter.
Dense scrub: 14, 6 died first summer, remainder following winter.

Tilia americana:
Prairie: 4, 2 died first summer, 1 in winter; 1 of drouth 1926.
Transition: 2, both died by August, probably from drouth.
Dense scrub: 2, both died by July 12, owing to dense shade.

Ulmus americana:
Prairie: 36, 22 died first season, 2 in winter; 10 in 1925, 2 in 1926.
Transition: 36, 32 died first season, remainder in winter.
Dense scrub: 4, all died in first winter, weakened by dense shade.

Survival at the end of three years was as follows:

	Gleditsia	*Acer n.*	*Acer s.*	*Tilia*	*Ulmus*
Prairie	60%	50%	0%	0%	0%
Transition	66	50	0	0	0
Dense scrub	0	0	0	0	0

It is significant that *Gleditsia* and *Acer negundo* also had the highest percentage of survival in the competition rows on the low prairie. Actual growth was generally better in the transition, where the shade handicapped the grass rather more than the trees and also conserved water in some measure.

SHRUB TRANSPLANTS IN LOW AND HIGH PRAIRIE

Plan—A study was made of the competition relations of several species of shrubs that occur as outposts in the prairie or along the margin of woodland. This was for the purpose of determining the course and outcome of the competition itself, and as well the actual status of such scrub communities with respect to extension or shrinkage. The species employed were *Symphoricarpus occidentalis, Rhus glabra,* and *Corylus americana;* the first two occur in lowland and ravine throughout subclimax and true prairie, while the last is restricted to the edges of the deciduous forest. These were transplanted in the form of rhizomes, as it is chiefly by means of propagation that these species spread outward from their respective areas. The rhizomes were placed in parallel trenches 4″×4″ in section, cut in the prairie sod, about 6 being planted in each row. The operation was the same for both low and high prairie, except that *Corylus* was not used in the latter, owing to its greater demands, and the treatment of the rows was the same as for the trees of the preceding sections. This experiment was repeated in both low and high prairie in 1925 (plate 19A).

Season of 1924—The physical conditions were very favorable during this first season and the plants flourished in most cases. However, the shade in the watered and unaided rows was so dense that several died, being only 13% on July 25. The excellent condition of the plants in the mulched and clipped rows of the low prairie is shown in figure A, as well as the rank growth of grasses on August 10. A representative specimen from each of the rows of *Rhus*, with the exception of the watered where a single individual remained, is portrayed in figure B. As the detailed tables indicate, growth on the high prairie was less, but this was compensated in considerable degree by the fact that the shade was less dense. As a consequence, the growth was better in some respects and none of the plants died.

Season of 1925—On May 7, of the three species on the low prairie, *Symphoricarpus* alone had begun growth. Excessive temperatures as high as 95° intervened and were followed by severe frost late in May; *Rhus* was frozen at both stations, but *Symphoricarpus* was not injured. By May 26 it became evident that several of the shrubs had been winter-killed. *Corylus* had suffered most, *Symphoricarpus* least, while *Rhus* was dead in the watered row of the low prairie. The second species had increased greatly in number, in spite of the fact that rabbits had cut many of the stems or eaten the buds, the damage being greatest where the plants were not protected by the grass.

The chresard for the native grasses of the low prairie has already been given (table 12), and the following tables furnish the values at the usual depths for the transplant rows in low prairie and the grassland of the high prairie:

TABLE 24—*Chresard in low-prairie rows, 1925*

Date	Rhus glabra			Corylus clipped	Prairie sod
	mulched	clipped	unaided		
	%	%	%	%	%
July 15..	6.8	5.8	1.6	...	5.5
	9.7	10.0	7.7	...	8.5
	13.3	15.3	13.1	...	12.6
	19.7	13.7	11.0		
July 29..	...	...	...	13.7	4.7
	...	...	...	15.4	7.4
	...	...	...	13.5	8.8
	...	...	...	...	14.9
Aug. 28..	18.1	12.4			
	15.3	16.2			
	14.8	13.0			
	16.5	11.9			

TABLE 25—*Chresard in high prairie, 1925*

Date	0′–0.5′	0.5′–1′	1′–2′	2′–3′
	%	%	%	%
May 2..........	11.0	17.4	14.3	
May 26..........	8.2	9.8	12.2	14.1
June 3.........	5.4	8.3	11.1	12.1
Rhus mulched.	7.6	15.2	15.4	16.2
clipped..	3.4	7.6	17.4	14.3
June 22.........	25.5	19.2	13.8	
Rhus mulched.	17.0	19.1	16.7	
July 6..........	9.0	5.2	11.0	9.0
August 18.......	10.6	19.1	15.7	8.5
August 28.......	7.9	13.7	9.3	7.2

Light values under the unaided shrubs at a height of 6″ were only 14%–18% on June 27. On August first, *Rhus* and *Symphoricarpus* in the watered row on the high prairie yielded respective intensities of 20% and 31% at a height of 3″, while in the unaided row these were 14.6% and 45%. On August 25 the values at 6″ were respectively 6.8% and 38% for *Symphoricarpus* and *Corylus* in the watered row in low prairie, and in the unaided one 17.6% and 23.7%. The effect of the shade cast by the grass was decisive in both these rows, as shown by the development of the shrubs, while in the mulched and clipped row the differences in height and vigor were caused largely by water relations.

On August 9, tests for starch in the shrubs on the low prairie gave the following results:

TABLE 26—*Starch values for low-prairie shrubs, 1925*

Species	Row	Leaf ht.	Amount
Symphoricarpus 1924	Clipped	18″	Moderate
	Watered	18″	None
Rhus 1924...........	Clipped	12″	Much
	Unaided	12″	¼ as much
Corylus 1924........	Clipped	3″, sun	Much
	Watered	3″, shade	¼ as much
Symphoricarpus 1925	Watered	Top	Very much
		Base	¼ as much
Rhus 1925..........	Watered	Top	Much
		Base	½ as much

The growth of the several species under the various conditions was recorded on July 26, and the measurements are given in table 27:

At the end of the season of 1925, *Rhus* had formed new clumps in the low prairie and had sustained no losses, except in the unaided row. So great was the tolerance of *Corylus* for shade that it lost no plants, while *Symphoricarpus*, the least tolerant of the three species, suffered some loss in both the watered and the unaided rows. On the high prairie some individuals of *Rhus* and *Symphoricarpus* died in both the mulched and unaided rows, and some of the latter in the watered row also, largely

TABLE 27—*Growth of shrubs on July 26, 1925*

	No. leaves	Ave. leaf length	Ave. leaf width	Size of leaflet
Low Prairie				
Rhus glabra		cm.	cm.	cm.
Mulched	55	26.0	13.0	7.7×2.6
Clipped	27	25.5	13.0	7.3×2.7
Unaided	8	11.0	7.0	3.7×1.3
Symphoricarpus				
Mulched	240	5.0	3.6	
Clipped	40	4.0	2.5	
Watered ...	16	2.6	1.5	
Unaided	6	3.5	1.6	
High Prairie				
Rhus glabra				
Mulched	24	32.0	24.0	7.5×2.7
Clipped	17	30.0	17.0	7.5×2.8
Watered ...	7	10.0	7.0	3.5×1.3
Unaided	18	9.5	8.0	4.5×2.0
Symphoricarpus				
Mulched	131	4.6	1.8	
Clipped	103	4.0	2.7	
Watered ...	13	2.5	1.6	
Unaided	11	2.5	2.0	

a result of its lesser stature. The maximum development of *Rhus* on the high prairie was 16″, or about a half of that on the low prairie. The differences in the case of *Symphoricarpus* were not so marked, but practically all made better growth in the low prairie as a result of the higher chresard, except where the shade was too dense. Propagation by rhizomes occurred in nearly all cases at both stations (plate 19B, C).

TABLE 28—*Chresard for Rhus glabra, 1926*

Date	Low Prairie				High Prairie		
	Rhus glabra			Prairie	Rhus glabra		Prairie
	Mulched	clipped	unaided		Mulched	clipped	
	%	%	%	%	%	%	%
May 28...	6.7	1.9	...	6.1 (Aug. 4)	9.4	12.4	13.0
	9.0	5.8	...	11.3	11.3	14.2	14.4
	13.2	10.1	...	9.0	14.0	14.3	14.1
June 18...	8.6	6.9	6.8	4.3	12.8	11.0	14.1
	10.4	10.5	8.6	7.1			
	12.5	14.8	11.5	8.4			
	15.3	13.3	8.9	8.8			

Season of 1926—There was little winter-killing in 1925-26, apart from the disappearance of *Rhus* in the unaided row and most of the group

Growth of shrubs in low prairie.

A. *Rhus glabra* in mulched and clipped rows.
B. 1924 plants of *Rhus* on August 27, 1925.
C. 1924 plants of *Symphoricarpus* on the same date.

actually increased in number of stems. The light values on June 27 at a height of 3″ under the unaided shrubs of the low prairie were only 3%–4% and at a foot 10%–12%. On August 27 the topmost leaves of *Rhus*, *Corylus* and *Symphoricarpus* gave photosynthate values of 0.02562, 0.2670 and 0.0288 gm. per 100 sq. cm. On June 19 the nitrate-content of the soil for the first 6″ was 4.7 ppm. in the mulched *Rhus*, 1.0 ppm. in the clipped and 1.5 ppm. in the unaided. The chresard under the various conditions for *Rhus* at the usual depths in low and high prairie is given in table 28, p. 174.

By August 6, 1926, the drouth had become so severe that all of the shrubs exhibited wilting and rolling of leaves. At this time the plants were cut off at the surface of the ground, taken to the laboratory and the following data secured:

TABLE 29—*Growth in high prairie, 1926*

No. plants	No. plants alive	Ave. ht.	Ave. stem diam. at base	Ave. dry wt.	Ave. no. leaves	Ave. length comp. leaves	Ave. width comp. leaves
Symphoricarpus		*cm.*	*mm.*	*gm.*		*cm.*	*cm.*
Mulched	29	28.9	3.4	2.61			
Clipped	21	20.9	3.1	2.69			
Unaided / Watered }	16	19.3	1.8	0.73			
Rhus glabra							
Mulched	10	46.1	11.1	46.87	41.5	22.5	12.3
Clipped	6	38.1	10.4	40.03	27.0	22.4	12.4
Unaided / Watered }	8	24.0	4.3	5.19	11.7	11.0	11.0

TABLE 30—*Growth in low prairie, 1926*

Species	% gain (+) or loss (−) in 3 years	Ave. ht.	Ave. diam. at base	Ave. spread of tops	Dry wt. ave. plant
Rhus glabra		*in.*	*mm.*	*in.*	*gm.*
Mulched	+200	20.4	9.1	16.9	45.95
Clipped	+ 67	16.0	7.9	17.1	31.22
Watered	−100				
Unaided	−100				
Symphoricarpus					
Mulched	+ 44	20.4	11.2	11.4	17.68
Clipped	— 45	14.3	5.7	6.1	6.45
Watered	— 86	11.0	2.0	3.0	1.22
Unaided	— 69	17.0	3.9	4.0	3.09
Corylus					
Mulched	— 63	21.0	11.3	11.0	8.65
Clipped	— 60	14.0	11.6	12.0	4.25
Watered	— 67	14.5	5.3	6.5	6.28
Unaided	— 67	12.0	5.4	4.0	1.95

The average for the mulched row was a gain of 60%; the average loss for the clipped row was 13%, from the watered 84% and from the unaided 79% (fig. 15).

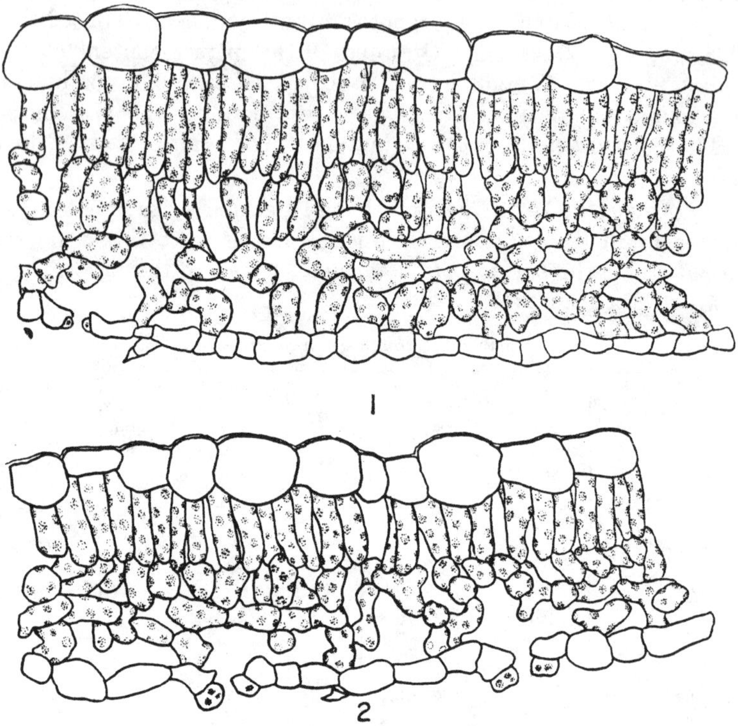

FIG. 15—Structure of leaf of *Corylus americana* from mulched and unaided rows.

Second series—The second planting of rhizomes of the three species was made on April 5, 1925. The spring was cold and dry, but a period of exceptionally warm weather brought out the leaves and they were then frozen late in May. As in the previous series, *Symphoricarpus* was not injured. On June 27 the light intensity under the unaided shrubs in the high prairie was 36%–40% at a height of 6″,. and on August first it was respectively 19% and 26.6% at a height of 3″ under the watered *Rhus* and *Symphoricarpus*. Beneath the same species in the adjacent unaided row, the light values ranged from 25% to 35%, while in the low prairie it was 21% and 19.5% in the watered row and 25% and 34% in the unaided row at 6″.

On August 26 the plants were measured with the results given in table 31, p. 177.

Season of 1926—On August first the plants of *Rhus* in the clipped row were somewhat wilted in both stations, as were also those in the watered and unaided rows, and *Symphoricarpus* was suffering to about the same degree. The measurement of the shrubs in the low prairie on August 9 yielded the data given in table 32, p. 177.

The average gain for the mulched row was 64%, for the clipped 26%; the loss for the watered row was 7% and for the unaided 20%.

TABLE 31—*Growth of Rhus and Symphoricarpus in 1925*

Species	No. lvs.	Ave. length of leaf	Ave. width of leaf
Rhus glabra (high prairie)		*cm.*	*cm.*
Mulched	29	25.5	13.0
Clipped	23	25.0	12.0
Watered	16	20.0	11.5
Unaided	14	18.0	11.0
Symphoricarpus (low prairie)			
Mulched	147	4.5	2.7
Clipped	73	5.0	2.6
Watered	13	4.5	3.0
Unaided	22	4.5	2.7

TABLE 32—*Growth of Rhus and Symphoricarpus in 1926*

Species	% gain (+) or loss (−) during 3 yrs.	Ave. ht.	Ave. diam. base	Ave. spread of crown
Rhus glabra		*in.*	*mm.*	*in.*
Mulched	+78	19.9	12.3	20.6
Clipped	+38	14.7	8.8	14.8
Watered	—13	16.7	8.1	10.9
Unaided	—14	14.0	7.9	10.3
Symphoricarpus				
Mulched	+50	19.5	7.9	7.0
Clipped	+13	11.0	3.3	3.1
Watered	0	17.3	3.6	5.0
Unaided	—25	18.3	4.5	3.3

Summary—Two important facts are disclosed by the tables for survival and growth in 1925 and 1926. One is the striking loss in the unaided rows, which ranged from 67% to 100%, while the loss was even greater in the watered one. The other is that the survival of the three species was in close accord with the rate of growth and the normal requirements of each. The rapid-growing tall but intolerant *Rhus* excelled the other two in both mulched and clipped rows with an actual gain, but was a complete failure in the other two rows. The low and rather slow-growing *Symphoricarpus* gained less than *Rhus*, but also fared better when shade and water were critical, while *Corylus* naturally endured shade better than either, though falling far below the others in the mulched row.

COMPETITION BETWEEN HIGH PRAIRIE AND THICKET

Plan and methods—The ravines of the true prairie sometimes contain thickets of one or more species of shrubs that constitute the marginal chaparral of the deciduous forest. When extensive and continuous, these are probable relicts of a former forest border, but they may also arise from recent invasion due to birds, especially at times when the hold of the grass sod is weakened by flooding. A thicket of the latter type composed of *Rhus glabra* has existed for more than thirty years in the high prairie at Lincoln and hence offered a good opportunity for the study of the competitive relations involved. The thicket extended upward from the ravine over the lower and middle parts of a southwest slope. At the upper edge it was spreading into the prairie in spite of annual mowing, which appeared indeed to handicap the grasses to a greater extent; at the lower edge it was in contact with the consocies of *Spartina cynosuroides*, which it was also invading to some degree (Weaver and Thiel, 1917:35).

Physical factors—The thicket and a portion of the adjoining prairie were enclosed by a fence in the late spring of 1924. The enclosed area was charted twice each year by means of 5′ squares; photographs were made from time to time of the ecotone along a definite line and tristats taken of fixed areas in the invaded portion of the grassland. Three instrument stations were established in the following situations: (1) on the hillside in the prairie and about 2 rods from the *Rhus* thicket; (2) in the tension zone 1 meter from the thicket where the grasses were taller than usual because of the shade; (3) in the thicket itself where the grasses had been replaced by *Poa pratensis* or the ground was nearly bare owing to the deep shade. In addition, two series of phytometers were employed and two lots of tree seedlings were grown during the three years of the experiment. In order to conserve space, the tables of factor data are given only for the middle year, 1925, with occasional readings from the other two years, while the course of development of the two series of phytometers is recorded throughout.

TABLE 33—*Chresard in prairie, ecotone and Rhus thicket, 1925*

Date	0.0′–0.5′			0.5′–1′			1′–2′			2′–3′		
	Prairie	Ecotone	Rhus	Prairie	Ecotone	Rhus	Prairie	Ecotone	Rhus	Prairie	Ecotone	Rhus
	%	%	%	%	%	%	%	%	%	%	%	%
May 6	7.0	14.9	11.3	12.5	13.7	12.0	10.5	11.8	12.6	...	...	...
27	5.9	9.8	12.3	9.3	10.3	12.5	11.1	12.5	15.3	9.0	12.7	13.6
June 3	5.2	11.7	6.3	8.7	12.1	6.3	12.5	10.0	8.0	10.4	12.3	10.7
10	8.9	15.0	10.7	5.6	9.9	5.4	7.6	10.1	5.6	...	...	...
26	15.7	19.0	21.5	13.7	14.3	14.4	13.7	12.5	12.9	9.8	9.1	13.6
July 1	11.4	16.0	18.7	12.9	12.4	14.2	11.2	12.2	7.1	9.3	9.1	7.5
8	1.1	12.9	8.8	4.2	11.3	6.1	6.7	8.7	5.3	4.4	7.2	6.2
15	2.0	9.6	6.8	4.6	7.2	7.4	6.9	6.7	–0.6	6.6	8.2	–3.6
30	6.2	17.4	14.7	2.0	5.9	8.5	5.2	4.2	2.4	2.4	3.4	–3.7
Aug. 7	13.7	21.0	17.6	4.6	11.2	12.8	1.3	1.5	2.0	2.9	3.9	–0.8
14	6.9	14.0	19.1	1.1	11.0	13.7	–0.7	5.7	3.5	...	...	...
29	13.4	9.8	4.4	9.1	9.4	4.2	4.6	4.5	3.8	–0.1	3.6	4.6
Hygro. coef.	9.8	10.1	10.8	11.3	11.7	11.6	13.2	12.3	12.1	12.9	12.2	12.4

TABLE 34—*Average daily evaporation, 1925*

Date (ending)	High prairie	Rhus	Ecotone
	cc.	*cc.*	*cc.*
May 20	19.2	12.1	16.1
27	34.0	19.6	25.3
June 3	30.7	17.8	20.9
10	35.5	20.4	21.4
17	20.9	10.2	13.9
24	10.8	6.7	4.6
July 1	16.5	4.7	10.8
8	24.3	7.1	14.6
15	17.2	5.6	12.0
22	26.8	11.1	18.8
29	20.1	8.8	13.4
Aug. 7	16.3	7.3	10.7
14	12.4	4.8	8.4
22	11.8	4.3	8.0
29	20.9	7.3	14.8

TABLE 35—*Representative temperatures and humidities. (in shade 6" above ground, June 26, 1925)*

Station	Air temperature		Humidity	
	10:45 a.m.	*12:00 n.*	*11:15 a.m.*	*12:15 p.m.*
	° F.	° F.	%	%
Prairie ..	81.5	82.4	49	52
Ecotone ..	82.4	85.1	52	55
Rhus	80.6	80.6	54	57
Soil temperature				
	10:45 a.m		*12:00 noon*	
	1 in. °F	6 in. °F	1 in. °F	6 in. °F
Prairie ..	81.5	74.4	84.5	74.4
Ecotone ..	70.1	68.0	69.0	67.6
Rhus	68.0	66.2	68.0	66.7
Light intensity				
	10:15 a.m.		*12:00 noon*	
	3 in. below grass %	8–12 in. above grass %	Below grass %	Above grass %
Prairie ..	55.0	100.0	65.0	100.0
Ecotone ..	10.0	25.0	10.0	38.3
Rhus	1.4	5.0	2.0	9.0

Transpiration phytometers—The difference between the three habitats in terms of water-loss was determined by means of sealed phytometers for the week of July 17-25, 1925. The plants employed were seedlings of *Fraxinus lanceolata* established in the usual large containers of galvanized iron (Clements and Goldsmith, 1924), the average weight being 7.5 kg. The week was clear and hot, with the exception of two cloudy days. In accordance with the expectation, the transpiration was more than twice as great for the prairie as for the thicket, while the ecotone was almost exactly intermediate.

TABLE 36—*Transpiration from Fraxinus phytometers.*

No. of phyt.	Loss	Leaf area	Loss per sq. in.	Ave. loss per sq. in.
	cc.	*sq. in.*	*cc.*	*cc.*
	Rhus			
1	28.5	13.05	2.2	
3	24.0	9.15	2.6	2.50
5	10.5	4.85	2.2	
9	34.0	11.20	3.0	
	Ecotone			
2	14.5	4.65	3.1	
4	17.5	4.80	3.6	3.25
6	22.5	7.6	3.0	
14	30.5	9.3	3.3	
	Prairie			
7	37.5	6.40	5.8	
10	19.5	4.6	4.2	
11	45.5	6.65	6.8	5.42
12	65.5	12.30	5.3	
13	62.0	12.35	5.0	

Standard phytometers—For this series, plants of *Helianthus annuus* were grown in the usual containers out-of-doors and transferred to the several stations when the second pair of leaves was nearly full-grown. At this time the average leaf area was 38.7 sq. in. and the dry weight 0.64 gm. The containers were sunken in the soil at each station so that the top was only 2″–3″ above it; the phytometers were watered once during the period of measurement. Two phytometers were also installed in the greenhouse as controls.

The average transpiration in the prairie was about a third greater than for the ecotone and six times greater than for the thicket; the greenhouse loss was practically identical with that in the ecotone, a fact explained in large measure by the similar light intensity. The average increase in dry weight for prairie and ecotone corresponded closely with the transpiration and light intensity for these two communities; it was less than a third as much for the thicket as the prairie.

The growth measurements of the phytometers in the four batteries have been reduced to averages and given in the table below. In general, the agreement between greenhouse, prairie and ecotone was close, the latter leading slightly as to the stem and the prairie as to leaf. The stature was greatest in the thicket in direct correlation with the lowest light intensity, which also provides the explanation for a dry weight less than half that

TABLE 37—*Transpiration and dry weight of standard phytometers.*

Phyt. no.	Trans.	Ave.	Increase in area	Ave.	Increase in dry weight	Ave.	Light values at 11–12 a.m. just above plants
	cc.	*cc.*	*sq. cm.*	*sq. cm.*	*gm.*	*gm.*	%
	Prairie						
3	689		56.76		2.14		
5	913	801	93.8	73.63	2.93	2.44	88
4	804		70.34		2.24		
	Ecotone						
7	658		54.5		1.70		
6	638	611	78.58	66.46	1.65	1.89	61
							34 (p.m.)
8	540		66.3		2.33		
	Rhus						
2	127		56.36	56.36	0.76	0.76	9.5
	Greenhouse						
9	705		85.78		1.91		50 (about)
		643		72.75		1.62	
10	582		59.72		1.34		

of the battery in the prairie. The greater demand for water reduced the stature in the prairie below that for both ecotone and thicket, but the conditions for the growth of the leaves were somewhat better than for either of the other two situations.

TABLE 38—*Growth of standard phytometers.*

Criteria	Greenhouse	Prairie	Ecotone	Rhus
Height, cm........	21.5	21.2	25.1	30.7
Height to cotyls, cm.	3.3	2.5	3.1	2.1
Height to 1st pr. lvs. cm..........	10.8	11.2	15.2	11.5
Stem diam. 1″ above cotyls, mm.	5.25	5.67	7.0	6.5
No. fully grown lvs.	6	6.7	6.3	6
No. lvs. per plant..	10.5	13.3	11.3	12
Length blade largest pr. lvs. cm...	121.0	114.5	120.3	97
Leaf-area, one side sq. in..........	55.72	56.17	52.58	47.53
Dry wt. gm......	2.26	3.08	2.53	1.40

Tree phytometers, 1924—On April 24, seedlings of *Tilia americana* nd *Acer negundo* in 4″ pots were transplanted to this series of habitats, ix being placed in each. At the same time seeds of *Gleditsia triacanthus* ere planted in abundance and a month later those of *Acer saccharinum* lso. The first planting was made before *Rhus* was in leaf; the transplants were watered and protected until well established. The behavior of e several species during the three successive years is given in the following table and a summary of survival in the final one (p. 187).

TABLE 39—*Behavior of 1924 tree phytometers*

Date	Tilia americana		
	Prairie	Ecotone	Rhus
1924			
May 8....	5, good	6, good	6, good
May 22....	4, excellent	6, fine	6, fine
June 5....	4, all but 1 thriving	6, good, protected from wind by high grasses	6, fine,' 1/3 shade, warmer than prairie
July 3....	4, 3–5 lvs.	6, 2″–3″, as in high prairie, grass 15″ high, overlaps	6, 3–7 broad lvs., 3″–5″ high, lvs. 2x as broad as in prairie
Aug. 10....	3, 2″–2½″, 3 yellow lvs. each not thriving	5, some with 6 lvs., none over 4″, fair	6, best with 4 lvs. 3″–5″, lvs. very broad
Sept. 9....	1, remnant only, 1″ only	3, 2 good, 4″, grass level 17″, shaded but thriving	6, 5″, excellent growth
1925			
June 7....	2, poor, small lvs. 3″ high or less	0	6, all thriving, 5″–6″ high, lvs. large
July 1....	3, ½″, not good, badly shaded	....	6, 5″–7″, very thin lvs. much shaded
Aug. 10....	0	....	6, talles 6″, blue-grass competing strongly
Aug. 29....	0	....	6, 5″–7″, 4 lvs., good
1926			
June 2....	....	....	6, fine trees, 7″ tall, 4–6 lvs., shade dense overtopped by blue-grass
July 20....	....	....	6, 7″, very little growth not badly shaded
Sept. 1....	....	....	6, 4″ with 1 leaf to 7″ with 6 lvs., all thriving
	Acer negundo		
1924			
May 8....	6, good	5, good	5, good
May 22....	4 only, rest dead	5, good	5, good
June 5....	4	5, good	5, thriving, some 6″–8″, twice as tall as elsewhere
July 3....	4, 2″–6″, 3–7 pr. lvs., shaded by 12″ grass and yellow	5, 3″–6″, mostly 5 prs. lvs.	5, 5 much compounded prs. lvs., mostly 10″–11″ high
Aug. 10....	4, quite shaded, not thriving	5, good	4, badly insect-eaten. fair
Sept. 9....	2, remnants only, none over 6″	5, good, 5–7″	4, 8″–12″, excellent growth but somewhat insect-eaten
1925			
June 7....	2, poor, best 7″	4 only, quite good, 8″ or less	4, fine trees, 8″–12″
July 1....	2, 5″–7″, fair	4, 3″–10″, grasses 20″ high, shade dense	4, 6″–16″, quite shaded by grasses but thriving
Aug. 10....	1, 7″	2, about 9″, densely shaded	4, tallest 14″
Aug. 29....	1, 7″	2, 7″–9″, good but densely shaded	4, 4″–14″, good

TABLE 39—*Behavior of 1924 tree phytometers*—Continued

Date	Prairie	Ecotone	Rhus
		Acer negundo—*Cont.*	
1926			
June 2....	1, 7″	2, 8″	4, 1 cut back, the others 22″ tall, fine
July 20....	1 only	2, well protected by grass	3, tallest 23″
Sept. 1....	1, sprouting at base died back to 2½″	2, 5″–7″, a bunch of leaves at top only, spread of tops 1″–4″	3, 20″–24″, thriving
		Acer saccharinum	
1924			
June 5....	a few maples breaking ground	many maple seedlings	a few seedlings
July 3....	19, well spaced, 3–5 prs. lvs., 4″–6″, not badly shaded	20, 3–6 prs. lvs., 4″–7″ tall, badly shaded	18, tallest 8″, fair
Aug. 10....	18, 4″–7″, some dying	20, ave. 7″, few lvs. dead, grass level 16″	18, tallest 8″, fair
Sept. 9....	14, none over 7″, half dead	20, 7″–8″, good	17, 3″ 0″, excellent growth
1925			
June 7....	13, few leaves near top, mostly poor, none over 7″ high	19, 8″–12″, quite good	3 only, 3, 4, 8″ tall, good
July 1....	13, 5″–7″, leafy at top only, many poor, not badly shaded	18, 8″–12″, defoliated below but good, tops not much shaded	2, 5″–10″, thriving, all had been cut off by rodents
Aug. 10....	6, tallest 5″	18, leafy at top only, tallest 12″, considerable shade	2, tallest 10″, thriving
Aug. 29....	6, 4″–7″, 2 lvs. at top only	18, as above	2, 5″–11″, thriving
1926			
June 2....	1, remnant only	9, fair, 10″ or less	1, 14″, good
		9, fair, some shaded, others well lighted	1, 14″, good
July 20....	0		
Sept. 1....	0	9, 5″–11″, a few badly eaten lvs. at top	1, 15″, fine

Tree phytometers, 1925—Between April 7 and 22, ten seedlings each f *Acer saccharinum, A. negundo, Tilia americana,* and *Fraxinus lanceolata* vere transplanted to the three stations. Because of the very dry early pring it was necessary to water frequently in order to enable them to become stablished. Early in May, seeds of *Ulmus americana* and *Gleditsia tria-anthus* were likewise planted, those of the latter being scratched with a le to increase germination. The behavior is indicated in the following ables:

TABLE 40—*Behavior of 1925 tree phytometers.*

Date	Prairie	Ecotone	Rhus
		Acer saccharinum	
1925			
May 3....	3, established	3, established	3, established
July 1....	2, very leafy, about 2″	1, 2″, fair	....
Aug. 10....	2, 2″, poor, some leaves dead	0	2, about 3″, fair
Aug. 29....	2, 2″, poor	....	2, 3½ and 4″, 5 leaves, fair
1926			
June 2....	2 remnants	....	1, 5″, 4 leaves
July 20....	1 remnant	....	1, 5″
Sept. 1....	0	....	1, 4″, 4 leaves
		Acer negundo	
1925			
May 3.....	11, established	11, established	11, established
July 1.....	10, 3″–5″, rather yellow	6, 2″–6″, quite yellow, considerable shade	11, 6″–13″, very broad and thin leaves
Aug. 10.....	8, none over 5″, somewhat yellowish	5, one a mere remnant, tallest 6″	11, tallest 12″, leaves broad, thriving
Aug. 29.....	8, 1″–6″, average 4″	5, 2″–6″, good	11, 5″–13″, good
1926			
June 2.....	5, 5″–7″, yellow but fair	0	9, 5″–24″, fair
July 20.....	4, 6″–8″, shaded, yellow	2, well protected	9, tallest 30″
Sept. 1.....	4, 6″–8″, spread of tops not over 2½″	0	9, 13″–29″, thriving
		Tilia americana	
1925			
May 3.....	9, established	6, established	9, established
July 1.....	7, none over 2″, poor somewhat eaten	3, about 2″, leaves quite yellow, considerable shade	8, 2″–3″, broad leaves, thriving
Aug. 10.....	2 remnants	2, mere remnant to 2″	7, tallest 4″, doing well
Aug. 29.....	2 remnants, 1″	2, as before	7, 2″–5″, 2 or 5 leaves
1926			
June 2.....	0	0	1 remnant, 2 leaves
July 20.....	....	....	1 remnant, 2 leaves
Sept. 1.....	....	....	1 remnant, 1 leaf 3″

TABLE 40—*Behavior of 1925 tree phytometers*—Continued

Date	Prairie	Ecotone	Rhus
	Fraxinus lanceolata		
1925			
May 3.....	8, established	12, established	17, established
July 1.....	1, about 2″	8, 1″–5″, considerably shaded	14, 2½″–6″, leaves delicate, but plants thriving
Aug. 10.....	1, nearly 2″, good	6, from mere remnants to 5″, shade dense	11, tallest 6″, good
Aug. 29.....	1, about 2″, good	6, 1″–3″, badly shaded	11, 1″–6″, broad leaves good
1926			
June 2.....	1 remnant	3, 2 are remnants only	7, best 9″
July 20.....	0	3, 1 a remnant only	7, best 9″
Sept. 1.....	....	2, 3″ and 6″, both badly eaten	7, 4″–10″, thriving
	Ulmus americana		
1925			
July 1.....	24, 2″ or less, fair, none are much shaded	4, 1″, not much shaded	8, 2″, good
Aug. 10.....	18, 2″ or less	0	7, tallest 2″
Aug. 29.....	18, 2″ or less	....	7, 1″–2½″, fair
1926			
June 2.....	8, remnants, none over 5″	....	0
July 20.....	6, remnants, leaves drying, well protected by grasses	....	....
Sept. 1.....	6, remnants, 2″–3″, only a few green leaves at top	....	....
	Gleditsia triacanthus		
1925			
July 1.....	3, 3″–4″	4, 3″, not much shaded	12, 3″–5″, thriving
Aug. 10.....	2, 4″	4, 3″	10, 5″
Aug. 29.....	2, 2″–4½″	4, 2″–3½″, good	0, 4″–6″
1926			
June 2.....	0	4, 4″	8, 4″
July 20.....	....	4, 4″–5″	8, tallest 5″
Sept. 1.....	....	4, 4″, leaves sparse	4, none over 4″, fair

General course of development—On April 7, 1925, none of the transplants in the prairie had put forth leaves, but these were showing in the ecotone and 2–4 leaves were half grown on the seedlings in the protected and warmer site in the thicket of *Rhus*. The spring was very late and cold, and the sumac did not begin to leaf until May 17, casting no considerable shade before the first of June. On June 2 the sumac was found to have been cut back by trespassers and the tree plantlets were in full sunshine during the afternoon as a consequence. Those in the prairie and ecotone were well protected and considerably shaded by the grasses, which now formed a layer 10″–12″ high (fig. 16).

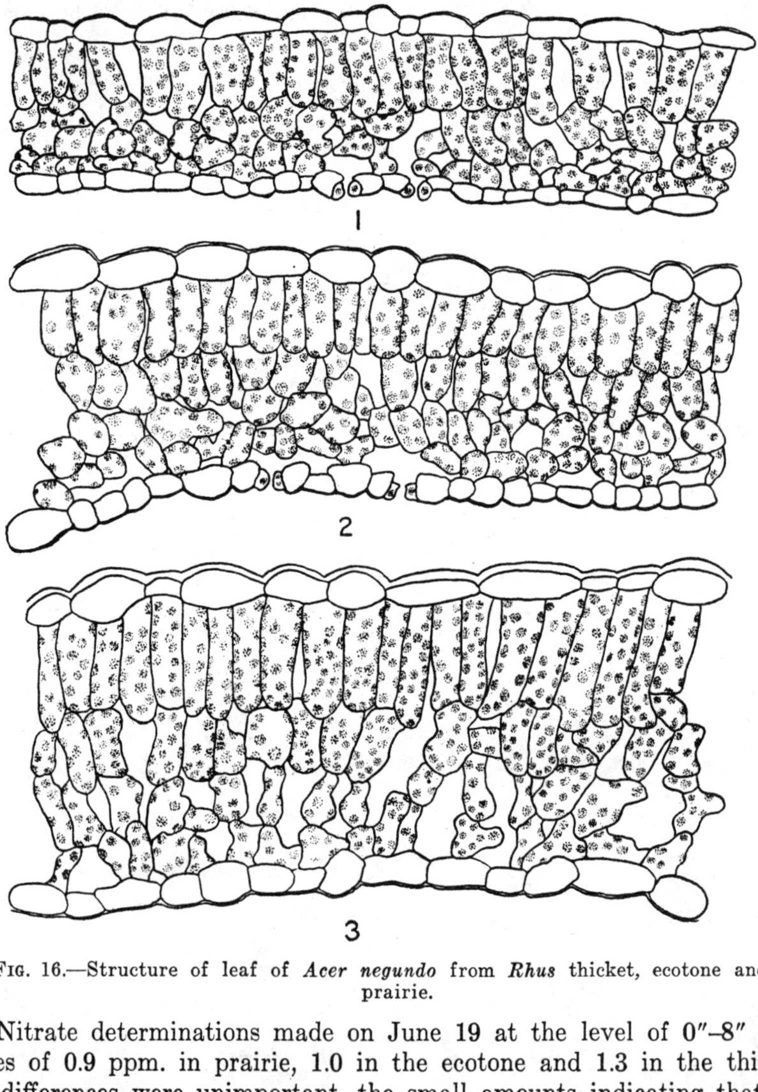

FIG. 16.—Structure of leaf of *Acer negundo* from *Rhus* thicket, ecotone and prairie.

Nitrate determinations made on June 19 at the level of 0″–8″ gav values of 0.9 ppm. in prairie, 1.0 in the ecotone and 1.3 in the thicke The differences were unimportant, the small amounts indicating that th nitrates were used as fast as they became available. On July 20 the soil in which the 1925 transplants were growing was bare and hard, owing to the cutting of the sumac on June second, which was resulting in a new crop of shoots. On July 30 the light values in the prairie at 3″ high were 21.3%, in the ecotone, 14% and in the thicket, 2%.

On August 6, 1926, the growth of grass was twice as good within the exclosure as on the outside, where it was mown the two preceding years. The level was 14″–17″ high, while the late-blooming *Andropogon* on the outside was but 5″–7″ tall and very dry and brown. Some of the transplants were dying and several prairie species were wilting. The sumac canopy was more open than usual and hence cast less shade. Conditions were much the same on September first, except that late rains had permitted the grasses to become greener and renew growth.

TABLE 41—*Summary of losses*

Species	% loss first year			% loss by 2nd year			% loss by 3rd year		
	Prairie	Ecotone	Rhus	Prairie	Ecotone	Rhus	Prairie	Ecotone	Rhus
1924 Transplants									
Acer s.	26	0	6	68	10	89	100	55	94
Acer n.	67	0	20	83	60	20	83	60	40
Tilia	67	0	20	83	60	20	83	60	40
Gleditsia	Only 3 seedlings grew; these were in the ecotone and died the second year.								
Average	53	0	15	78	43	43	89	58	58
1925 Transplants									
Acer s.	33	100	33	100	100	67	...	..	..
Acer n.	27	55	0	64	100	18	...	..	..
Tilia	78	67	22	100	100	89	...	..	..
Fraxinus	87	50	35	100	83	59	...	..	..
Ulmus	25	100	12	75	100	100	...	..	..
Gleditsia	33	0	17	100	0	67	...	..	..
Average	47	62	20	90	81	67	...	..	..

Thus, on an average, the first lot of seedlings did poorest in the prairie but equally well in the ecotone and *Rhus*. The second lot did poorest in the prairie and best in the *Rhus*. Averaging both series, the losses in prairie, ecotone, and *Rhus* were 90%, 70%, and 63% respectively. These figures show the decisive effect of the competition exerted by the grass dominants, as well as the aid rendered by a canopy of leaves.

Behavior of Rhus in the exclosure—The area was irregular in form and included approximately 100 unit areas 5′ square. The shoots of the sumac were abundant at the time of the first charting on June 10, 1924;

they were tallest and thickest near the thicket proper, but some extended well up the slope of the prairie toward the hill-top. On September 10 the general grass level was 13″ on the upper slopes and 21″ in the low prairie of the ravine. Nearly a half of the latter had not been cut in 1923 and now formed a tangled mass 2′ deep, the new grass in this area now being 40″ tall. The vigor of propagation in *Rhus glabra* was denoted by the fact that new sprouts came up under a layer of dead grass 1.5′ thick. Of the isolated stems that had appeared since mowing ceased, the total was 174, of which 70 were new in 1924. There had been a loss of 18 between June 10 and September 10, eight of which belonged to the current year. The average growth in height was 9″. In the low-prairie portion there were 3 new shoots with an average growth of 20″, while the old ones grew about 12″; there were no deaths here. In addition, there were 42 old unmown shoots in the plotted area, making a grand total of 216.

On June 8, 1925, the old shoots numbered 204, the decrease of 12 probably being caused by winter-killing, and 26 new ones. Many had been frozen back 6″–8″ and all the leaves had been frozen likewise and were just recovering. Eleven of the new shoots occurred in the *Spartina* community of the ravine, where water relations were most favorable. By August 17 the total had increased to 253, or 10%; of this number, 24 were new in 1925, while 9 dead plants were also found. The average growth for the season was but 3.7″, though the new shoots had grown 10″ before June 10 when they were badly frozen. Growth was best in the ravine, where the new shoots reached a maximum of 15″.

By September 2, 1926, the number had fallen to 156, a decrease of 28% from the original number in 1924. Of this total, only 13 were new shoots of 1926, while 14 stems had died since June second. The average height growth during this dry season was only 4.3″. These results, together with those obtained from the examination in 1927, showed conclusively that the community of *Rhus glabra* was not only making no progress up the slope into the high prairie, but was steadily losing ground. The rate of loss, however, was somewhat accelerated by two successive dry summers. Although an outlier was occasionally found several meters from the exclosure, such stragglers exerted no effect of any consequence upon the grasses. The stature decreased with decreasing holard up the hillside and hence the canopy of the thicket mass sloped to the prairie level like a roof. Beyond this the shrubs were never sufficiently dense and tall to produce any perceptible effect upon the growth of the prairie grasses. However, conditions were different toward the ravine, where *Rhus* was definitely invading the consocies of *Spartina* and gave promise of replacing it to a considerable degree. This outcome would appear inevitable if it were not for the action of recurrent flooding. As shown in the general discussion at the end of this chapter, the inability of one of the most aggressive members of the subclimax chaparral to advance into the high or true prairie is decisive evidence of the nature of the climatic control and of the corresponding climax.

COMPETITION BETWEEN DECIDUOUS FOREST AND PRAIRIE

Plan and methods—A series of experimental studies were organized in the ecotone between deciduous forest and grassland at the western edge of the climate of the subclimax prairie. These were located near Weeping Water, Nebraska, along the stream of that name, which has long furnished a center for the investigation of the margin of the deciduous forest climax. The rainfall averages about 30″, being 2″ higher than that of Lincoln in the true prairie; it is hardly adequate for even the marginal growth of the forest, except in the years with normal or excess precipitation. However, the river has cut a canyon more than a hundred feet deep in the Pennsylvanian limestone and this with its similarly cool moist laterals provided a refuge for the deciduous forest as it underwent shrinkage during the last major dry phase. Taking advantage of the shelter afforded by the rough topography, trees and shrubs constitute a belt of woodland a mile or so wide, with alternes of prairie only on the dry windy hill-tops and southwest slopes (plate 20).

The valley, bluffs and the ecoclines of the hills afford so much diversity with respect to water relations especially that an unusual series of communities is available for study. These reflect locally the structure of the climax deciduous forest to the eastward in general, but on a miniature scale and hence in an incomplete fashion. On the flood-plain of the stream is found an associes of *Ulmus americana, Juglans nigra,* and *Fraxinus lanceolata,* which is subclimax to the *Quercus-Hicoria* association. The lower slopes of the canyon walls are clothed with *Tilia americana* below and *Quercus rubra* above, while the higher slopes are covered with *Hicoria ovata* and *minima* and *Quercus macrocarpa,* which extend to the protected south slopes as well. The *Acer-Fagus* association of the deciduous forest climax is here represented only by *Tilia,* the other four species being representative of the *Quercus-Hicoria* association, which is more xerophytic in nature. The local grouping on the mesocline, *i.e.*, the northerly slope-exposures, is such that *Quercus rubra* is to be associated with *Tilia* rather than with its usual companions of the genera *Quercus* and *Hicoria*.. This is explained in part by the fact that *Tilia* is more xerophytic than *Acer* and *Fagus,* and *Quercus rubra* the most mesophytic of the oaks and hickories here represented, and partly by the telescoping of the dominant communities within narrow limits. The upper and drier margins of the woodland are fringed by an open border of the subclimax chaparral, which extends beneath the trees as a more or less interrupted layer. When the forest is cut, this layer at once takes possession as a subclimax stage that sooner or later yields to forest.

Four stations were selected to represent the major differences in habitat and vegetation from the flood-plain to the prairie. The first was located in an excellent stand of *Quercus rubra* with some *Tilia* and the relicts of *Quercus macrocarpa;* the second was installed in a pure consociation of the latter on a south slope, the third on the prairie alterne of the slope of this same hill and the fourth in an area of *Corylus ameri-*

cana that was invading the prairie. The permanent exclosures at these stations were installed on May 5, 1924, when they were equipped with the usual batteries of instruments.

Transect of Communities

Flood-plain associes—The flood-plain in this transect is nowhere more than one-third of a mile wide and sometimes much less, but it is better protected by the canyon walls than is the case upstream. The chief dominants are *Ulmus fulva, U. americana, Juglans nigra* and *Fraxinus lanceolata,* 25′–50′ tall and 10″–20″ in width. *Celtis occidentalis, Prunus serotina* and *Acer negundo* are also common and well-grown, while *Gymnocladus dioeca* and *Gleditsia triacanthus* are of less importance. The shrubby layer consists principally of *Symphoricarpus vulgaris, Ribes gracile, Rubus occidentalis* and *Sambucus canadensis,* with *Vitis vulpina, Parthenocissus quinquefolia, Celastrus scandens* and *Smilax hispida* as the characteristic vines. The soil of this community is not especially rich in humus, since it is periodically subject to overflow, the resultant lack of aeration excluding the climax dominants. Even a small rise above the flood level favors the latter and the flood-plain species often yield to them abruptly (cf. Weaver, Hanson and Aikman, 1925).

Quercus-Tilia community—This is dominated by *Quercus,* with *Tilia* more or less abundant on the lower slopes. Trees 50′ tall and with a trunk a foot or more in diameter are common. The humus is well developed and the forest floor closely resembles that of the deciduous climax generally. The annual rainfall is 2″–3″ greater than at the headwaters of the stream, the humidity is considerably higher and the holard much more adequate. *Ulmus fulva* is not infrequent toward the flood-plain, and other species of this associes occur here and there. The successional movement within the woodland is attested by attenuated or dead relicts of *Hicoria minima* and *Quercus macrocarpa,* now dominant on the upper and drier slopes. *Ostrya virginiana* is the characteristic member of the secondary layer, associated with *Staphylea trifolia, Xanthoxylum americanum, Euonymus atropurpureus* and others, but as a rule layers are not well developed (plate 20B).

Quercus-Hicoria association—This community consists almost entirely of an open stand of *Quercus macrocarpa,* with *Hicoria minima* scattered here and there, and an occasional individual of *Quercus rubra* or *Ulmus fulva* in lower moister spots. The oaks are only 30′–40′ tall and 6″ or less in diameter, but excellent specimens 13″ through are found in favored situations. The crowns are rather open and about half the forest is carpeted with *Poa pratensis,* the shade elsewhere being too deep for this grass. *Corylus americana, Symphoricarpus vulgaris, Rhus glabra* and *Cornus asperifolia* are the characteristic shrubs, although *Rhamnus lanceolata, Sambucus canadensis* and *Cornus stolonifera* also occurred. The first was found only near the forest margin or in dense thickets over more or less extensive cut-over areas. *Symphoricarpus* extended throughout, though it was often much interrupted, while *Cornus* made its best showing in open

Subclimax grassland in deciduous forest at Weeping Water, Nebraska.

A. General view of prairie and forest
B. Detail in the *Quercus-Tilia* community.

spots or along the forest edge, and *Rhus* was found chiefly as a fringe along the outside, owing to its intolerance of shade.

Chaparral associes—In the chaparral proper, *Corylus* exhibited a distinct preference for the more sheltered and moister areas, though it sometimes occurred on protected hill-tops. *Rhus glabra* was most often seen on exposed slopes and rocky crests, while *Symphoricarpus vulgaris* and *occidentalis* were more or less intermediate in position, the latter showing a decided xerophytic tendency. Associated with these species but especially with *Corylus*, were *Cornus asperifolia, Rhamnus lanceolata, Ribes gracile, Xanthoxylum americanum* and such lianes as *Celastrus scandens* and *Smilax hispida,* the whole mass often forming a dense community over large areas and sometimes indicating an artificial recession of the forest margin.

As the experimental results indicate, in this region the shrubs readily invade the subclimax of tall-grasses, chiefly by means of rhizomes and root-sprouts. Even such tall-grasses as *Andropogon* are soon outstripped by these sprouts in stature and gradually succumb to these woodland outposts, owing to lowered light values due in part to the accumulation of leaf litter. A thicket-like growth of bur-oak often follows in the wake of the chaparral and on moister slopes gives rise to woodland of this species. The approximate age of the larger trees was 60 years, which coincides with the period of general settlement of the region and the consequent disturbances due to fire and felling. However, the subclimax prairie is in full possession on the thinner soils of the warm dry slopes and will long remain in such situations in spite of the gradual encroachment of the chaparral along the margins of such alternes.

Physical Factors of the Stations

Soil profile—The prairie soil was typically brownish-black for the first 6″; the second 6″ layer was slightly paler and shaded into yellowish below, where it rested upon the decomposing limestone. The second foot or more consisted of a light-yellow sandy clay, well decomposed in the larger crevices but containing many small angular fragments of limestone at the margins of these in the smaller ones. The soil was moist to a depth of 2.5′ or more, and the roots of the grasses were remarkably developed to correspond. The soil of the *Corylus* area was very similar, though the black surface layer extends to a foot and a half or even 2′, where it merges into the subsoil derived directly from the decayed limestone, becoming light yellow at 2.5′. The rock fragments are smaller and less freqent than on the southerly slopes. The holard is much higher than in the grassland and there is a well-developed root level to the depth of 2′, the roots going much deeper where the soil permits.

In the bur-oak community, the soil is half bare and half covered with *Poa pratensis,* the duff not exceeding an inch in thickness. The black soil of the first foot is intermediate in color and texture between the preceding and that found in the red-oak areas. Below the first foot it passes rapidly into a clay similar to that found at a depth of 2′ under red-oak. Gravel and occasional small rocks occur at 1′–2′ and large rocks are found below this level. The soil in the red-oak community is black to 6″–8″ and grayish-black below this to 2′, where it becomes somewhat yellowish. The surface layer is very friable beneath a layer of duff 1″–2″ thick. The soil in general is so moist in spring that it can be fashioned into balls.

Water-content—The hygroscopic coefficients for the soil of the prairie, hazel and bur-oak stations were practically identical, ranging from 14% to 14.6%. The value for the red-oak community was much lower, 7.2%, owing chiefly perhaps to the larger amount of sand present. Determinations of the chresard during three seasons gave no very consistent differences between the four types. The forest soil was generally somewhat higher in the upper levels and lower in the deeper ones than that of the prairie, but not infrequently the values in the latter were the highest for any of the stations. This is largely explained by the drainage into the crevices where the soil samples could be taken and by the much larger interception exerted by the crowns of the trees. The lack of a uniformly higher holard in the red-oak community than in the grassland bears witness to the significance of insolation and of light intensity in the competition between woodland, chaparral, and prairie.

Determinations of other soil factors also yielded no important differences between stations. The nitrate nitrogen in the upper 6″ varied only from 1.3 to 1.8 ppm. on June 11, 1926, and was practically identical at 1.0 ppm. on July 16. All the soils were alkaline owing to their derivation from limestone, the respective values being as follows: prairie pH 7.6, hazel pH 7.5, bur-oak pH 7.5 and red-oak pH 7.1.

Temperature—The air temperatures were regularly higher in the prairie than in the bur-oak during the day-time, the average excess being 4°. Likewise, the temperatures in the bur-oak were higher than those in the red oak forest, the average excess being 4.3°. Soil temperatures from the surface to 6″ were much greater in prairie than in hazel, the difference ranging from 10°–20° for the surface and 7°–15° at a depth of one inch, while at 6″ it was but 2°. They were 1°–2° less in bur-oak than in hazel and averaged about a degree lower in red-oak than in the former. The temperatures of both air and soil were in general agreement with the screening action of the cover or canopy and the effect of the stand upon the wind movement.

Humidity and evaporation—As a rule the humidity of the hazel copse was about 5% higher than that of the prairie, while this differed from the open bur-oak woodland more than was expected. However, the differences in evaporation were much more uniform and significant. The average weekly loss from porous-cup atmometers during the summer of 1924 from May 3 to September 5 was 2.4 cc. greater for the prairie than for the hazel scrub, 2.2 cc. less for the bur-oak than for the latter, and 1.4 cc. more for bur-oak than red-oak. The difference between the latter and the prairie amounted to an excess of 7 cc. per week for the grassland. This was accord with the arrangement of the communities from valley floor to hilltop and on the various slope-exposures, and indicates that protection from insolation was probably the factor of chief importance.

Tree Transplants

Seeds of *Acer saccharinum, Ulmus americana, Fraxinus lanceolata, Acer negundo* and *Gleditsia triacanthus* were planted in the four stations in the spring of 1924 in order to test their growth and survival under the various conditions. Seeds of *Quercus macrocarpa, Q. rubra* and *Tilia americana* were also planted in the prairie in 1926, but none of them appeared above ground. The percentage of survival and the height at the end of 3 years are exhibited in the following table:

Table 42—*Summary of transplant growth and survival*

Species	Prairie	Copse	Bur-oak	Linden	Remarks
	%	%	%	%	
Maple	16 (9″–15″)	...	15 (7″)	25 (4″–10″)	Highest survival in linden, max. growth in prairie
Elm	0	0	100 (6″)	33 (3″–7″)	Highest survival and best plants in bur-oak
Ash	—	100 (3″)	0	51 (6″)	Highest survival in scrub, best trees in linden
Boxelder	52	0	79	—	Highest survival in bur-oak, best plants in prairie
Honey-locust ..	16 (5″)				
Average survival	21	33	49	36	

By July 10, 1927, many of the maples had died in the prairie, evilently most of them from winter-killing after the belated growth of a wet utumn following a dry summer. A few more had died recently, leaving ut 3 plants, 8″–19″ tall. Of the 16 boxelders, 12 remained, the tallest 0″–12″, but they were yellowish and in poor condition from drouth. All f the honey-locust were still growing, the tallest being 10″, but the leaves vere folded and the plants suffering from drouth. The ash seedlings were bundant in the hazel copse, but none were over 4″ in height as a consequence of shade and drouth.

The average survival is fairly significant in its increase from prairie hrough copse to bur-oak; it falls off again in the red-oak community o nearly the value in hazel and for the same reason, the denser shade. 'he effect of the shade in the copse is shown by the contrast in survival etween the tolerant ash and the intolerant boxelder, the respective percentages being 100 and 0. On the other hand, the ash made its best growth in linden, the habitat of which approaches its own most nearly. In general the highest survival was in some degree of shade, the best growth where the combined light and water relations were most favorable.

Natural Phytometers

A further test of the ability of trees to establish themselves in sub-climax prairie was made by means of seedlings that appeared naturally in the grassland. Thirty-two were selected as natural phytometers to determine the rate of growth and survival in competition with the tall-grasses. All of these were individuals of *Quercus macrocarpa,* with the exception of one each of *Quercus rubra, Ulmus americana* and *Corylus americana.* These were marked by means of permanent stakes and the growth recorded from year to year. The detailed results were tabulated each year for the entire number, but the following summary is based upon 20 individuals, 12 having been destroyed by fire in 1926.

The bur-oaks during 1924 made an average growth of 2″ (0.5″–4″ ir range), of 2″ (1″–3″) in 1925 and of 3.2″ (0.5″–7″) in 1926. During th three years but one plant died. The other species grew equally well, as di hundreds of unmarked bur-oaks. The greater growth during 1926 was evidently due to the more extensive root-systems and the increased leaf-area The exceptionally low mortality was even more significant than the rat of growth and was in striking contrast to the best results obtained b transplanting. In the burned area the oaks sprouted readily from the bas and made a vigorous growth. While the competition of the grasses appeare intense, it was rarely decisive during the period of study and seeme destined to become less so as the trees grew taller. The final examinatio and measurement on July 10, 1927, showed that practically all the tree were alive and making active growth. The clump of hazel that ha 2 shoots at the outset now comprised 6, with a maximum width of nearly 2 and was beginning to suppress the grasses by its shade.

Competition in Exclosures

First exclosure—Two areas in which shrubs and grasses were com peting were protected by fencing to prevent grazing, and the behavior o the competitors and their reactions followed in detail. The smaller exclo sure was 10′ square and comprised prairie chiefly, with the exception o a central mass of shrubs about 5′ wide. The area was charted in Ma 1924 and 1925 and in August of 1926. The shrubs were clearly encroach ing upon the grasses, spreading more or less uniformly in all directions during 1924 the increase was 15% and by the end of the third year this ha become 60%. In 1924 the original central stems attained a maximum 3′–4′, while those on the margin ranged from a third to a half foot; by th autumn of 1926 the central ones were 4′–5′ high and many of those on th periphery 2′. The total number of stems of *Corylus* rose from 75 to 10 they grew from 2″–11″ the first season and some to a height of 2′. Ma stems of *Corylus* and *Celastrus* were cut off near the ground by rodents in winter, and some of the oldest stems died during the drouth of 1926.

The other woody plants within the exclosure behaved in similar manner, but they were more affected by the grass on account of their being scattered. The shoots of *Corylus* increased only from 10–12 in the three years and their height from 1/3′–1′ to 2′–3′. *Celastrus* rose from 11–17 stems in the first two years and then fell off to 6 in the dry summer of 1926. *Rhus* increased from 19 to 22 shoots the first year, but lost all but 2 during the dry year. Of two seedlings of *Ulmus*, one died and the other grew but 3″ in the entire period, being far out of the habitat limits for this species. The decisive reaction exerted by the shrubs was upon shade, largely as a result of the canopy effect, but in part by the accumulation of their leaves and those of the adjacent woodland. In consequence, all of the native grasses disappeared beneath and about this outpost of *Corylus,* a thin stand of *Poa pratensis* alone being able to persist in its shade. A few seedlings of *Ulmus* and *Acer negundo* had also started to grow beneath the shrubs, but their persistence was very doubtful at best.

Second exclosure—This fenced area was 27′×40′ in size, and included two contiguous clumps of hazel, the larger 9′×17′, the smaller 11′×10′. These merged into one during the period of the experiment. The entire area was laid off in 3′×5′ units and was mapped at the same times as the preceding. Like this, it also was located on a hill-top and exposed to the full effect of wind and insolation.

During the first summer the hazel increased its territory 11.8% and the total gain was 42% by the end of 1926; this came almost wholly from the growth of rhizomes, though a few seedlings grew from seeds planted by rodents. As a rule the distance traveled by rhizomes did not exceed 12″, but in some cases it was over 2′. A large percentage of the increase was accounted for by the merging of the two groups, the oblique stems on the two fronts rapidly shading out the grasses and thus rendering the advance of the rhizomes much more rapid. Sometimes, new centers were established several feet from the parent areas, but their unaided progress against the grasses was necessarily slow. The general height of the mass of hazel increased from 4′–5′ in 1924 to 5′–6′ in 1926. The value of *Corylus* as a pioneer was shown by the presence in the group of an individual of *Ostrya virginiana* 10′ tall, one of *Morus rubra* 9′, *Rhamnus lanceolata* 7′ and *Fraxinus lanceolata* 8′.

In 1924, *Rhus glabra* was represented in the hazel mass by a few live stems and many dead ones; the poor growth and defoliated bases showed that it could not withstand the dense shade of the hazel, only those with tops above the hazel level persisting. *Symphoricarpus occidentalis* fared somewhat better; it was scattered thinly through the less dense portions of the hazel thicket, but was destined to disappear as the latter became thicker. Seedlings were found of *Parthenocissus quinquefolia, Smilax hispida, Rubus occidentalis, Ribes gracile* and *Morus rubra,* but none was abundant. A plantlet of bur-oak 5″ tall grew but a single inch in the three-year period. *Celastrus scandens* was abundant owing to its ability to compete successfully with the hazel for light, and reached lengths of 6′ or more as it clambered over the shrubs. With the growing density of the latter, the blue-grass that was originally found throughout much of the area practically disappeared, only a fringe remaining at the margin in 1926.

In the grassland around the main mass of shrubs were scattered many individuals of the latter. In 1924, *Rhus glabra* was represented by 79 shoots; the next summer this number increased to 93, but then fell off to 74 in 1926. These made an average growth of 10″ with a range of 4″–20″ during the first two years, but were set back by the drouth of 1926. *Symphoricarpus* comprised 47 shoots in 1924, increasing to 66 in 1925, but dropping to 44 the third year; it made a growth of 7″–8″ in 1924. During the same year, 145 isolated stems of *Corylus* grew from 6″–24″ in height, the number increasing to 188 the following season. Many of these were incorporated into the main masses in 1926, with the result that only 64 single shoots remained at this time. One plantlet of bur-oak grew from 5″–12″ and another from 17″–30″ in the three years; a third started but was winter-killed. No other species of trees occurred outside of the shade of the copse, but the presence of *Quercus macrocarpa* in the grassland was significant of the effect of the chaparral in preparing the area for the next woodland stage.

On July 10, 1927, it was found that the clump of hazel in the smaller exclosure had not merely made a vigorous growth from the old stems, but several new shoots 6″–15″ tall had spread among the grasses. In the large exclosure older shoots had grown 10″ and the new ones had attained the same stature; fruiting was abundant, even in the small scattered clumps. New stems appeared all around the mass, but they were especially numerous in the most vigorous portion, where some stalks 2′ from the parent stem reached a height of 20″. The bur-oak had grown 5″ and now had a spread of nearly 3′, thus casting considerable shade. *Morus* was also prospering, the larger plant now possessing a crown 3.5′ across. The prairie grasses were steadily decreasing in area and vigor, and the establishment of *Morus* and *Quercus* definitely foreshadowed the woodland stage.

Suppression of Andropogon—In the ecotone between chaparral and subclimax prairie the prairie-grasses gradually yield to *Poa pratensis* where they are shaded for a half-day or more by *Corylus* or other shrubs. This takes place under an average light intensity of 10% during the shaded portion of the day. When the light intensity approaches 3% as a consequence of the increasing density of the hazel, the blue-grass itself disappears, the accumulation of litter playing a part in this result. *Rhus* and *Symphoricarpus*, which usually form the vanguard, have vanished before this time, since they find their limit at 4%.

Several clumps of *Andropogon* were plotted and their fate followed in close detail. A bunch of *Andropogon scoparius* that grew near the edge of the hazel mass in 1924 was still sufficiently well lighted to make a good growth and bear flower stalks. By 1925 the mass had reached it, a large

Competition between *Corylus americana* and *Andropogon scoparius.*

A. Condition of *Andropogon* on August 21, 1925.
B. Condition on July 15, 1926.
C. Practical extinction of *Andropogon,* July 10, 1927.

branch extending well over the grass; however, it still bore inflorescences, except on the two shaded sides. The light value above the portion of the bunch too much shaded to produce flowers was 7%–15%, this diminution resulting from a single branch of the hazel. *Poa pratensis* now had complete possession of the 2′ interval between the grass and shrub.

By 1926 this bunch had become further overshaded and invaded, with the result that no flower stalks developed, and there were only a few spears of green present. A hazel shoot had arisen from a rhizome that extended beyond the grass and a young stalk of *Celastrus* had also appeared on the other side of the bunch. The light intensity at a height of 6″ had been reduced to 5% and at a foot to 4%–5.7%; on the side of the clump next the main mass of shrubs it was only 3.7%. The grass was now shaded throughout the morning and for part of the afternoon, and it seemed probable that it would die by the end of the year.

On July 10, 1927, only a few weak leaves were to be found and the steady advance of the hazel left no doubt of the immediate outcome. Two clumps of *Andropogon furcatus* were likewise traced in detail and found to behave in similar fashion. They first ceased flowering and then gradually died along the advancing front of the shrubs (plate 21).

Light as the Controlling Factor

It appears evident that the water relations at the margin of a thicket are secondary to the effect of light. It has already been seen that the holard is practically the same for grass and chaparral, and this is likewise true of the humidity and evaporation above the two. The modification of the latter is concomitant with the light effects and it is these that determine the demand for the water of the soil. As a consequence, the detailed light values were measured several times during the summers of 1925 and 1926 and starch tests were made at irregular intervals. These readings were taken at those places where the effects of competition were most visible and decisive.

June, 1925—The light intensity under the spreading branches of *Corylus* ranged from 6.7%–20% at 11 a.m. on June 9. The values above clumps of *Andropogon nutans* varied from 4.3%–13.3% until after 1 p.m. when they received full sunlight; further beneath the edge of the hazel where *A. scoparius* was dying, the light was full until 11 a.m., when it fell to 2.2%–3.3%. Where *Poa pratensis* was abundant in the shade of overhanging branches, the intensity was 10%–12%, but where the blue-grass was sparse and attenuated it dropped to 4.2%–5%. In spots where the blue-grass was nearly gone, the values were 3.3%–3.8%, and in completely bare areas they fell as low as 1.7%. Light intensities in the bur-oak woodland varied from a maximum of 20%–30% under an isolated tree, to 20%–25% where seedlings occurred in a cover of blue-grass and 12%–14% in the denser stands. These were in sharp contrast to values of 4.4%–5% in the red-oak and linden.

August, 1925—The shade under a single overhanging branch of hazel exposed to full sunshine gave a value of 8%–10%, while above the bunch of *Andropogon* that failed to flower in such a situation it was 7%–15%. Where the blue-grass was thick under the edge of the bushes, the intensity ranged from 4%–6%. In spots where this species was thin and dying, the value was 2%–3% and in bare places but 1.7%–2%. Slightly lower values were obtained a year later in the same situations.

June, 1926—The light values for *Andropogon furcatus* with a few poor culms were 5% at 6″ and 5.7% at a foot; just beyond the bunch toward the shrub mass where blue-grass grew sparsely it was 3.7%. A bunch of *A. scoparius* that was attenuated and dying was found in an intensity of 6%–6.7%. Other clumps shaded for half the day and still in fair condition yielded values of 7%–8%, while *A. furcatus* gave 8% also at a foot and 6% just above the leaf litter. The difference between 6% and 8% appears to be critical for *Andropogon,* though there is no question that it will gradually deteriorate at even the higher value. This is in accord with the fact that blue-grass replaced the prairie-grasses more or less rapidly at intensities of 7%–8%. In its turn the blue-grass began to disappear rapidly at values of 2%–3% and had vanished completely before the reduction reached 1%.

July, 1926—The situation as to the values for the grasses were essentially as in June, and especial attention was given to the values for the shrubs. The light intensity above seedlings of *Fraxinus* was 13.3%, which represented the strongest light they received during the day and in consequence little starch was present. The apparent minimum for *Rhus* was 4.3%, in which no starch was produced, by contrast with a large amount in full sun. Under a value of 3.1% the thin leaves of *Symphoricarpus* were turning yellow and were also without starch. The lower leaves of hazel had produced no starch and were turning yellow in a light intensity of 0.9%, but seedlings of *Celastrus* and *Ribes* still grew slightly at this value and yielded a light test for starch.

Summary of light values—The shade at the edge of the hazel copse ranged from 5%–20%, with an average of about 10%; this was too low for the prairie-grasses, which finally disappeared under overshading at 6%–8%. *Poa pratensis* thrived or at least maintained itself within a range of 5%–15% and an average of 7%, but decreased steadily from 5%–2%, the average being 3.3%. It was absent under values of 1%–3%, with an average of 2%. The most tolerant shrubs dropped out at a somewhat higher value, approximately 3%–4%.

Practically all the evidence from the several studies described points to the slow encroachment of thicket and woodland at the edge of the subclimax prairie, though under a rainfall of 30″ this is due in some degree to the modifying influence of stream and valley. To avoid repetition, this is considered in the following section, along with the relations between true prairie and chaparral.

GENERAL SUMMARY

The fascinating problem of the relation of the forest edge to grassland has been much debated on the basis of general observations, but has never before been attacked in a comprehensive and quantitative manner, or with the experimental procedure focused upon the essential functions of competition, reaction, and ecesis. It is evident that while the forest edge represents the major and climatic aspect of this competitive relation, the edaphic phase is equally susceptible of experimental study, whether in the form of prairie openings in climax forest or as outposts of woodland or chaparral in the grassland formation. In the present series of investigations, occasion has been found to study both of these, four of the installations having been made in the true prairie at Lincoln, and one with various subdivisions in the subclimax prairie at Weeping Water. In the former, both high and low prairie were taken into account, while the competition has dealt with shrubs as well as trees. Finally, the relative importance of climate and competition has been evaluated by means of four sets of conditions, which favored the woody plants in varying degrees. Mulching gave them the advantage both of factor and lack of competition, clipping the latter alone, watering compensated for the lack of this factor whether competitive or climatic, while the unaided row bore the brunt of both sets of conditions.

The detailed results of the several studies have already been summarized (pp. 169, 171, 177), and it remains to consider only their bearing upon the general problem. As is shown by its tall-grass cover and high holard, the low prairie is essentially subclimax in nature and the chances of survival of tree seedlings approach those of the regional subclimax at Weeping Water. In spite of this, the survival in the unaided rows at the end of three years was but 8%, while with watering it was but 21%; survival rose to 38% in the clipped and to 69% in the mulched, in close agreement with decreasing competition and the improvement of light and water relations. All this was in complete accord with the behavior of trees on the low prairie. They persist along the banks of streams or on the flood-plains where the grasses can not invade; they can be established by destroying the grass cover or by giving aid through watering, especially during the seedling and sapling period. Under the existing climatic and competitive relations, the chances are overwhelmingly against ecesis. Even when planting was done on the flood-plain in and about dense scrub, the survival of the more xeroid *Gleditsia* and *Acer negundo* was about 50%, while *Acer saccharinum, Tilia* and *Ulmus* vanished utterly in all three sites.

In the high prairie, the average survival of the several species was 11% at the end of the third year of the first series and 10% at the end of the second year of the other series, although the seedlings were watered and protected until established. In the more protected ecotone, as well as in the *Rhus* thicket, insolation was much less and survival correspondingly better, averaging 30% and 37% respectively.

The experimental results with the three genera of the marginal chaparral, *Symphoricarpus, Rhus* and *Corylus,* were in complete accord with their position and habitat relations in nature. As a species of the forest undergrowth and extending but little beyond the edge, *Corylus* was the least successful, earlier experiments having shown the inadvisability of even trying it in the high prairie. *Rhus glabra,* as the most vigorous grower of intermediate position, prefering lowland and ravine, fared the best of the three in mulched and clipped rows, but failed under the more rigorous conditions of the other two. *Symphoricarpus occidentalis* is the lowest and most xeroid of the three, and hence grows in drier situations further out in the prairie. As a consequence, its survival in the high prairie was 2–3 times greater than that of *Rhus,* while in the low prairie it was much less. The outcome for the three in the watered and unaided rows was so decisive as to leave no doubt of their inability to invade the true prairie under natural conditions. Indirectly, this confirms the conclusion drawn from the tree transplants, since the invasion by forest is all the more impossible when the vanguard of shrubs can not advance against the grasses, in spite of the use of rhizomes and a certain amount of care.

The investigation of the competition exerted by the native thicket of *Rhus glabra* in the high prairie and the protection afforded by it brought out several facts of significance. As was anticipated, the tree seedlings survived best in the thicket, less well in the ecotone and most poorly among the prairie grasses, a consequence indicated by both instruments and phytometers. Even more telling was the behavior of the thicket itself, which was greatly reduced in stature and density in contact with the high prairie. It had reached the limit of its upward invasion along the slope, and was steadily losing ground to the grasses. The fact that the sumac was tall only in the flooded bottom of the ravine afforded further proof that it was able to advance into the prairie grasses only during the wet phase of the cycle, and then only to a small extent.

The situation in the subclimax prairie at Weeping Water was naturally more favorable to shrubs and trees, partly as a result of greater rainfall and partly because of the canyon topography. The results obtained from unaided tree transplants and from natural seedling phytometers bear witness to this, at the same time that they demonstrate the severe pressure with heavy mortality that the trees experience. The low mortality of the natural seedlings of bur-oak by contrast with that of the transplants confirms the results of other studies to the effect that ecesis in nature usually takes place only under especially favorable conditions, and then often surpasses the best that planting can accomplish. The rate of survival rose from the prairie with a water deficit to the open bur-oak stand with intermediate conditions as to water and light, to fall off in the denser red-oak forest with a deficiency in light.

The slow but effective encroachment of the immediate forest edge is best exemplified by the behavior of *Corylus* in its competition with *Rhus* and *Symphoricarpus,* and with *Andropogon* and *Poa.* Its ability to grow much more vigorously than these in shade and the consequent cumulative reaction upon light intensity are the decisive features of the process. A much longer period of experiment is needed, however, to yield a complete understanding of the marginal relation of forest and prairie, and a fairly accurate evaluation of the control exerted by the climatic cycle. Between the true forest climate in which the shrubs and trees have only the competition of the grasses to overcome and the true prairie climate in which both climate and competition favor the grasses is a broad ecotone. In this, local differences in habitat will largely decide the fate of the competitors in accordance with life-form, but throughout the region also the wet and dry phases of the sunspot cycle will profoundly affect the outcome, especially where the competition is keenest. The slow advance of chaparral and forest will be hastened by the wet phase, retarded by the dry, or even converted into a retreat. It appears fairly certain that there can be no final victory for either, only periods of varying duration in which one or the other holds the ground won by the favor of the changing cycle.

6. COMPETITION IN CULTIVATED FIELDS

Plan and objects—The primary purpose of the series of experiments on the nature and outcome of competition in cultivated fields was to permit a closer correlation between studies in nature and those in the greenhouse. The former afforded natural conditions with little control of factors and processes, the latter permitted control in various degrees, while the field plots supplied conditions that approached those in nature but still made a certain amount of control possible. A secondary object was to study the functional and structural response under different densities and spacings, and thus to provide an ecological explanation of the effect of competition upon the growth and yield of cultivated crops.

The two crops selected were sunflowers and wheat, the one because of its vigor and rapid growth as well as its response to competition, the other as the most important representative of the grain crops of the Middle West. Furthermore, *Helianthus annuus* exemplified the regular spacing of a single-stemmed crop, and *Triticum sativum* that of the continuous stand found in all species that produce tillers or offshoots. The preparation of the plots and the planting of the seeds followed the best agricultural practice as closely as possible.

FIELD COMPETITION IN HELIANTHUS ANNUUS

SERIES OF 1924

Methods—The ordinary commercial seed was employed for this species, but only the largest seeds were selected for use. These were planted during two different years, 1924 and 1926, and provided two series for investigation, the second profiting in some measure from the experience gained with the first. The initial series of plots was planted on May 17, 1924, in a field adjoining the wheat plots. The planting was done in hills 64″, 32″ and 16″ apart respectively in the larger plots which were 86′×90′, but only 8″, 4″ and 2″ apart in the smaller ones, which decreased progressively with the rate of seeding to 12′×20′ in size. Two or three seeds were placed in each hill, but the plantlets were later thinned to a single one and the missing hills replanted to insure a uniform stand. After the preliminary preparation of the soil, no further cultivation was needed in the thicker plots with 2″ and 4″ intervals, but the 8″ and 16″ were hoed once to remove weeds, and the 32″ and 64″ several times as proved necessary to keep the plots clean.

Germination was delayed by the cold dry soil, but by June 13 the 8″ plants and those more widely spaced were about 4″ high and unfolding the third pair of leaves. It was evident that the 8″ had nearly reached the point where competition for light would begin; some of the 4″ were 9″ tall and some plants were becoming attenuated, though all still had a good color. However, competition began soon after germination among the 2″ and the cotyledons were now turning yellow; some plants had become suppressed at a height of 3″, by contrast with an average of 8″. The beginning of definite competition for light was marked by a greater growth in height and each plot exhibited in turn the greater stature, in the sequence of decreasing density. The 2″ plants were the tallest until after June 20.

At five different times during the growth of the crop a number of plants was selected from each plot, usually 10–30 of the 64″ and increasing progressively to 100 or more for the 2″, and the average values obtained for stem and leaf. With these as a guide, a representative plant was chosen from each density for photographing. Until the beginning of competition in any plot, the plants from it were also representative of those in the lesser densities and the latter in consequence were not measured.

Competition results on June 20—The first set of measurements of growth under the different degrees of competition were made approximately a month after planting, with the results shown in the following table:

TABLE 43—*Development of Helianthus on June 20*

Criteria	2″	4″	8″	16″
Ave. height, in.	15.8	11.9	12.2	9.7
Ave. diam. mm.	4.3	4.8	7.2	9.3
Ave. no. full-grown leaves	4.4	4.8	6.5	8.6
Ave. total no. green leaves	5.1	5.6	8.0	9.3
Ave. length largest leaf, cm.	5.9	8.8	13.3	17.2
Ave. width largest leaf, cm.	4.7	6.1	9.8	14.7
Ave. leaf-area both sides, sq. in.	15.0	33.6	100.4	150.1
Ave. dry wt. of tops, gm.	0.88	1.15	3.56	5.35

The effect of competition in the elongation of the stem is clearly evident, the 2″ being much the tallest and the 16″ the shortest, and the inverse effect upon the diameter revealed by the fact that the 16″ were practically twice as thick as the 2″ (plate 22A). At this time competition for light at least had barely begun among the 16″, as the leaves were just beginning to overlap between the rows. The 8″ plants were being differentiated into dominant and suppressed; some cotyledons were conspicuously yellow and a few lower leaves were losing their green color. Among the 4″ all the cotyledons were yellow or dry, the plants were also yellowish and the soil was completely shaded. The 2″ were still more yellowish and most of the lower leaves were deteriorating. The gradation from the densest to the least dense was especially marked in the diameter of stem, the number of full-grown and the total number of leaves, the 16″ being about twice as well developed as the 2″. The leaf dimensions were three to four times greater and the total area ten times, while the 4″ and 8″ were fairly intermediate. The leaf values were naturally reflected in the dry weights of the tops; these were a third greater for the 4″, four times larger for the 8″ and more than six times for the 16″.

Up to this time the differences in the chresard of the different plots were slight. The dense planting for the 2″ and 8″ was compensated by the smaller plants, so that the total demand for water was much the sam as for the more vigorous plants of the more open plots, where the loss b evaporation from the soil was much greater. The critical factor in the com petition was the light intensity, which was greatly reduced with the density

Competition results on July 2—At this time the 8″ plants were th tallest, having taken the lead from the 4″, which held first place for short time after June 20. In the 64″ plot there was still an interval o about 40″ between the leaves of two rows, but this was reduced to 6″ fo the 32″, where the crowns were 30″ across with all the leaves fresh an green. However, in the 16″ plot the crowns were about 25″ wide and man of the large leaves overlapped each other completely. As a result the firs pair of leaves was dead or yellow on nearly all the plants, the second pai frequently yellow and the third sometimes deteriorating. The crowns o the 8″ ranged from 24″ for the dominant individuals to 12″ for the sup pressed, and many more leaves were dead than with the 16″. The dominant of the 4″ bore a ratio of 2.5:1 to the suppressed and were but 9″ wide at th crown; as with the 2″, shading had produced such marked elongation tha the second pair of leaves was four times as high as in the 32″. The crow in the 2″ averaged only 6″ for dominants, which comprised half the total decumbent individuals were not infrequent, but as with the 4″, none ha entirely succumbed (plate 22B).

Table 44—*Development of Helianthus on July 2*

Criteria	2″	4″	8″	16″	32″
Ave. ht. in.	23.8	24.3	29.5	29.0	22.5
Ave. diam. mm.	4.5	5.7	9.8	16.4	19.5
Length of 2nd internode	24.3	25.9	18.2	8.4	6.1
Ave. no. full-grown green lvs.	5.4	6.4	9.5	13.6	16.4
Ave. no. half-grown green lvs.	1.0	1.0	2.5	3.4	3.7
Ave. no. dead lvs.	1.2	1.3	3.2	1.4	0.2
Ave. length largest leaf, cm.	6.7	9.1	13.2	22.6	24.9
Ave. width largest leaf, cm.	4.0	6.1	10.8	20.1	23.2
Ave. leaf-area, both sides, sq.in.	24.0	52.2	206.8	957.0	1201.6
Ave. dry wt. tops, gm.	1.1	2.4	9.0	31.3	35.8

Stem diameter was nearly twice as great in the 8″ as the 2″ and ove four times as great in the 32″. The latter had over thrice as many full grown leaves, which averaged nearly 4 times as long and over 5 times a wide. The actual number of dead leaves was greatest among the 8″, th lower ones being nearer the soil surface than among the 2″ or 4″ and mor shaded by the larger leaves above. Leaf areas were in the ratios (using whol numbers) of 1:2:8:40:50, and dry weights 1:2:8:29:32. The relativel small differences between the 16″ and 32″ were due to the short time th former had competed strongly. Apparently much of the leaf area of th denser plants was working poorly, light values being very low.

The light intensity was measured on July 5 at the level of the lowes pair of living leaves and the six exposures for each plot averaged, with th

A

B

Effect of density in field sunflowers, 1924.

A. Plant of 16″, 8″, 4″ and 2″ on June 20.
B. Plant of 32″, 16″, 8″, 4″ and 2″ on July 3.

following results. The value increased from 12% for the 2″ plot to 24% for the 4″, 31% for the 8″, 32% for the 16″ and 40% for the 32″. Since the lowest leaves of the latter plot were at 11″ and those of the first three at about 2′, the effective difference between the 2″ and 32″, for example, was much greater. As with light, the chresard rose regularly from the 2″ to the 32″ plot, the average for the three levels being twice as great for the latter as for the former. Showers were frequent until the middle of August and drouth did not complicate conditions at any time.

Competition results on July 17—By this date the plants in the 16″ plot were the tallest. The average lateral spread of the crowns from the least to the most dense was 44″, 38″, 25″ 14″, 8″ and 5.5″ respectively. In the 64″ plot there was a 20″ interval between the rows, but the 32″ overlapped 5″, the 16″ increased this to 9″, while it dropped to 6″ for the 8″ and fell even lower in the densely packed plot of 2″. There was little yellowing of the leaves among the 64″, though a few were dead; they were dead or yellow to a height of 24″ in the 32″, and to 28″ in the 16″, all below 20″ being dead in this plot. The leaves of the 8″ were dead to a level of 22″ and yellow to 36″, the 4″ to 26″ and 37″ respectively, and the 2″ dead to 36″ and yellow to the topmost leaves. There were no dead plants among he 8″ or 4″, the latter plot having about 1% of the plants decumbent and many suppressed at a height of 12″–18″. Among the 2″ over 65% were uppressed and 5% of these were already dying or dead (plates 23 and 24).

TABLE 45—*Development of Helianthus on July 17*

Criteria	2″	4″	8″	16″	32″	64″
Ave. ht. in.	29.8	36.5	45.3	53.4	49.6	48.4
Ave. diam. base stem, mm.	4.5	6.7	12.1	19.5	29.3	40.3
Ave. diam. top stem, mm.	2.8	4.0	6.8	13.0	19.3	21.4
Ave. no. full-grown green lvs.	3.2	4.6	8.0	12.5	23.5	31.3
Ave. no. half-grown lvs.	1.3	1.7	2.8	4.2	3.9	3.3
Ave. no. dead leaves	3.6	4.3	5.5	6.7	5.1	4.7
Ave. length largest leaf, cm.	6.3	8.7	13.5	20.7	29.6	35.1
Ave. width largest leaf, cm.	4.0	6.4	11.4	19.4	30.5	40.1
Ave. length petiole largest leaf cm.	2.1	3.9	8.5	15.3	22.9	25.6
Ave. leaf-area both sides, sq. in.	18.4	57.0	328.6	878.1	3426.0	5034.4
Ave. dry wt. tops, gm.	1.5	4.2	16.8	65.9	147.7	217.0

The stems of the 2″ had not increased in diameter since July 2 and ose of the 4″ had grown only 1 mm., but a marked gain had taken place all the others, the 64″ having doubled their width in the interval. In ct, the diameter of these exceptional plants at 6″ from the upper end was greater than the basal for any of the others, with the exception of the ″. While the four denser plots had decreased in the number of full-own leaves, the 32″ and especially the 64″ exhibited a marked increase. ead leaves became relatively more abundant as the density of planting creased, but the length of petiole and the size of leaf increased inversely with the density. The total leaf areas rose in the following sequence in the direction of decreasing density, 1:3:18:48:186:274, and the dry weights gave the ratio of 1:3:11:44:99:145. The two sequences correspond fairly

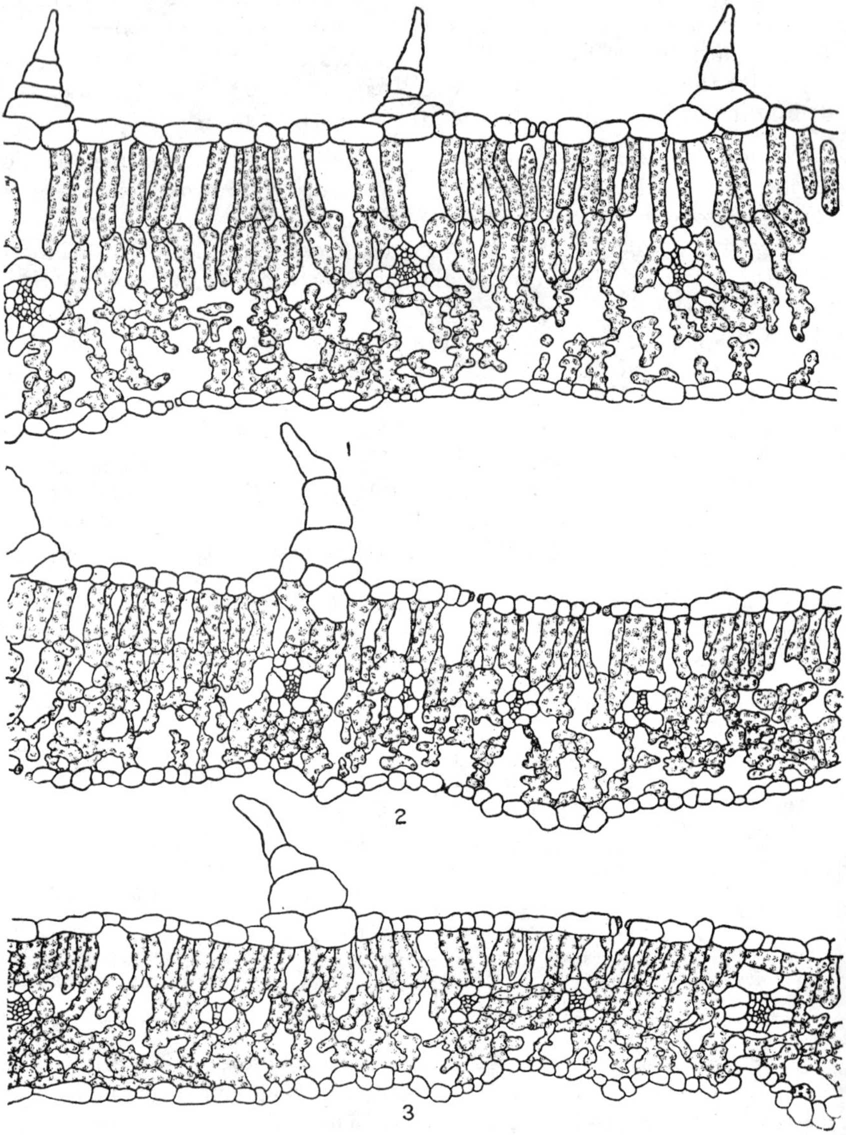

FIG. 18.—Structure of leaf of 32″, 8″, and 2″ field sunflowers, 1924.

well until the two lesser densities are reached, when the increased transpiration evidently takes a hand in disturbing the relation.

Root behavior—Four days later on July 21 and 22, an examination was made of the root systems of the 2″, 8″ and 32″ plots. In the 2 plants the taproots ended as a rule in the 5′ level, though a few reached a depth of 6′. The maximum lateral spread was 10″; none of the laterals penetrated more deeply than 2′ and their working level was entirely in the first 12″ (fig. 17). No branches of these laterals were more than 0.5″ long. In the case of the 8″, many of the taproots were more than 6′ long, the maximum being 7.3′. Their laterals spread to a distance of 2.3′ and were very abundant to a depth of 30″, the general working level extending to about 3′. As a consequence, the root system of a particular individual occupied a volume of soil tenfold or more greater than that of the 2″ plants. The roots of the 32″ ranged to a maximum depth of 7′–9′, and the laterals attained a spread of 3.5′. The soil was filled with a great mass of laterals especially to a depth of nearly 4′, making a working level four times as deep as for the 2″.

On July 29 determinations were made of the nitrate-content in four plots, with the following results:

TABLE 46—*Nitrate-content of sunflower plots*

Depth	4″	16″	32″	64″
0″–6″	0.68	0.77	1.18	1.99 ppm.
0″–12″	.71	.58	.55	.72
1′–2′	.56	.54	.40	.42

While the differences between the four plots were consistent in the first level, they were evidently too small to produce an effect in any way comparable with that of water and light, since all plants were using the nitrates as rapidly as produced.

Competition results on August 5—At this time the average spread of the crowns in the three denser plots was respectively 5″, 7″ and 10″ for dominant individuals and 3″, 4″ and 8″ for the suppressed. In the three less dense plots where no suppression occurred, the values were 18″, 31″, and 37″. While the plants of the 32″ plot were just touching the leaves, there was an interval of about 28″ between the rows of the 64″ (fig. 18). Dead leaves occurred up to 2″–3″ of the top in the 2″, 3″–6″ for the 4″, 12″–24″ for the 8″, 40″ for the 16″ and 57″ for the 32″ and 64″. In an average square meter of the 2″ plot, 60% of the plants were dominant, 10% suppressed and 30% dead. There was 10% dead among the 4″, but none in the other plots, though the 8″ showed about 8% of suppressed individuals. The flower heads were beginning to appear in all densities, but were most poorly developed in the denser.

The 32″ now exceeded in height the 16″, which were the tallest on July 17. The 2″ and 4″ had elongated about 10″, the 8″ slightly more, while the growth in height for the lesser densities was respectively 29″, 44″ and 38″. The increase in the diameter of stem in the 2″ and 4″ was due only partly to growth, the disappearance of the weaker plants accounting for much of this. The 8″ and 16″ had gained only about 1 mm., the 32″ and 64″ 4–5 mm. It is significant of the course of competition that

TABLE 47—*Development of Helianthus on August 5*

Criteria	2″	4″	8″	16″	32″	64″
Ave. ht. in....................	39.8	45.7	58.5	82.0	93.0	86.1
Ave. diam. base stem, mm......	5.7	7.2	13.1	20.9	33.2	45.0
Ave. diam. top stem, mm.......	2.8	3.5	6.5	8.8	12.0	14.8
Ave. no. full-grown green lvs...	3.6	4.0	8.6	18.0	23.4	28.3
Ave. no. half-grown green lvs...	0.8	1.4	2.9	3.2	3.6	4.3
Ave. no. dead lvs.............	9.1	9.7	11.9	16.0	22.0	15.6
Ave. length largest leaf, cm.....	4.7	5.9	10.0	16.4	28.9	38.3
Ave. width largest leaf, cm.....	3.0	4.4	9.0	15.8	30.2	41.1
Ave. length petiole largest leaf, cm.	1.4	2.2	6.3	13.7	21.9	26.9
Ave. leaf-area, both sides, sq. in.	12.8	20.0	180.0	810.0	3073.6	8429.4
Ave. dry wt. tops, gm.........	2.1	4.6	20.8	85.5	279.2	491.4

the top diameters had changed very little for the three denser plantings which were already greatly attenuated on July 17, but had decreased 4–7 mm. for the others, with increasing stature. The number of dead leaves was now much higher in all plots, the greatest increases occurring among the thinner plantings. The largest leaves were now shorter than any at the previous measurement except for the 64″, and the width was also less for all but this plot, as was the length of petiole likewise.

The leaf area reached the maximum among the 64″ at this time; the 32″ had decreased in area since July 17, while the decrease started in the 16″ on July 2 was progressive. The 2″ had also decreased consistently since July 2, but the 4″ and 8″ had attained their maximum on July 17. The dry weight of tops had increased in all plots respectively by the following percentages, 40, 10, 24, 30, 90 and 126; the gain among the 2″ was mostly apparent only, being due very largely to the death of subdominant individuals.

Competition results at the close of 1924—Following August 19, severe drouth began and lasted throughout the remainder of the growing season; a very small chresard occurred in any of the plots to a depth of 4′. As usual, the drouth accelerated blossoming in some measure, this beginning first among the 64″ on account of their excessive leaf-area. By September third nearly all the surviving 2″ were in bloom, about half the 4″ and 8″, only a few of the 16″ and 32″, while the ray flowers had fallen from the 64″ and the seeds were mostly mature. Six days earlier on August 28, 25 heads were measured in each plot, the maximum diameter to the tips of the rays being taken; similar measurements but without the rays were also made on September 15.

On September only the dominants among the 2″ were left; they were about 8″ apart, which permitted flowering, but the heads were so poorly developed that few seeds were produced. The remaining 4″ were also dominants that were about 10″ apart. The leaves of the 8″ were drying and hence overlapped less than before, as was likewise the case with the

Competition cultures of field sunflowers, July 17, 1924.

A. 32″ plot. B. 64″ plot.

TABLE 48—*Diameter of sunflower heads*

Plot	Diameter with rays	Diameter without rays
	cm.	*cm.*
2″	6.1	2.3
4″	7.7	4.5
8″	11.3	6.5
16″	16.4	9.5
32″	22.3	15.0
64″	26.6	20.0

16″. The interval between the rows of 32″ was about a foot, while the distance between the rows of 64″ had increased to more than 3′, as a result of the death of the leaves to a height of 4′.

TABLE 49—*Development of Helianthus on September 3*

Criteria	8″	16″	32″	64″
Ave. ht. in....................	73.8	105.3	114.8	120.1
Ave. diam. base stem, mm......	13.5	20.6	31.7	47.0
Ave. diam. top stem, mm.......	4.3	7.5	11.5	12.8
Ave. no. green lvs.............	8.8	20.4	24.4	25.9
Ave. no. dead lvs..............	22.6	22.5	31.6	33.2
Ave. leaf-area, both sides, sq. in.	65.2	553.9	1920.8	4450.5
Ave. dry wt. tops, gm.........	29.9	137.5	410.6	666.4

The 64″ now exceeded in height the 32″, which were the tallest a month earlier; both of these had gained slightly in the stem diameter at the base, but all four plots had fallen off in width at the top. Owing to the drouth also, there were no consistent changes in the number of leaves and the total leaf-area had been reduced in every case, to the extent of nearly 50% in the 64″ plot. However, there had been marked and consistent gain in the dry weight of tops, the percentages based on the weights for August 5 being respectively 44%, 60%, 47% and 35%. The greater gain of the 16″ plot probably corresponds to a better balance between leaf-area and water-supply.

As a measure of the total effect of competition on flowering and fruiting, ten heads were carefully selected for average size from each plot, air-dried and weighed, with the following results.

TABLE 50—*Values for sunflower head and seeds, 1924*

Plot	Ave. diam. heads, cm.	Ave. weight heads, gm.	Ave. no. seed per head	Ave. weight seed, gm.
2″	2.2	0.87	15	0.009
4″	4.5	5.15	89	0.011
8″	6.5	9.17	237	0.022
16″	9.5	28.91	507	0.030
32″	15.0	82.65	940	0.040
64″	20.0	122.90	1803	0.059

On account of the drouth, none of the heads attained normal size and weight and the seeds were only fairly developed. In spite of this, the results under the varying degrees of competition are strikingly consistent with respect to all four criteria. The highest ratio between the 2″ and 64″ is found in the average number of seed per head, where it is 1:120; this corresponds very closely with the ratio between average weight of heads, as might be expected. On the other hand, the ratio of 1:10 for the average diameter of heads is not far from that of 1:7 for the average weight of the seed, though this may be without significance.

SERIES OF 1926

Methods—Seeds of *Helianthus annuus* were planted on May 21 in five plots with the same density as in the first series, but because of the dry spring a full stand was secured with difficulty. After replanting the missing hills twice, it was finally necessary to fill the vacant ones by transplanting seedlings from pots. The season was one of serious drouth and the chresard was low throughout the summer. The dense plots showed signs of distress first and it finally became necessary to irrigate them freely from time to time. This was done by selecting representative areas in the 4″, 8″, 16″ and 32″ plots and making an embankment along the lower side to hold the water. The difference in treatment of the two areas of one plot is designated by the terms "watered" and "unwatered."

The severity of conditions is indicated by the fact that the 4″ and 8″ plots exhibited wilting several times during the first half of the season and even the 16″ began to give evidence of distress by the first of July. The chresard was determined each week for the three depths to 2′ or 3′ in the five plots; the amount was uniformly small and frequently below the hygroscopic coefficient for the 4″ and 8″ plots. Such a deficit occurred but twice in the 16″ and not at all in the 32″. The nitrate-content of the soil increased in the direction of less density; it was 1.3–2.9 ppm. for the 4″, about twice as much for the 8″ and double the latter amount for the 64″ plot. Water and nitrates were thus distinctly more favorable in the less dense plots and the growth in the latter was further promoted by much higher light intensities. Evaporation on the contrary was greater in the 8″ than in the 4″; between the first and second pair of leaves it was respectively 114 and 96 cc., while at the level of the tops it was 126 and 104 cc.

Effect of competition and drouth upon stomata—On June 25 determinations were made of the light intensity, stomatal behavior and starch content at intervals of a half-hour or hour from 8:40 a.m. to 4:15 p.m. in both the watered and unwatered portions of the 4″ plot. Three leaves were used in each reading, from as many different plants, and care was taken to employ only leaves that were still green. Lack of space renders it undesirable to give the details of the behavior through the day and these are summarized in the following statement (fig. 19).

In the case of the lowest pair of green leaves, the stomata were closed continuously or open very slightly, and for much of the day the leaves were partly wilted. These leaves contained no starch at any time, though the light intensity reached a maximum of 18%. The stomata of the upper pair just below the crown were mostly closed on account of the drouth and gave only a trace of starch until after they were watered at noon. After the leaves became turgid as a consequence, the stomata opened and the amount of starch at 4:15 p.m. ranged from slight to moderate under light values of 10%–30% approximately.

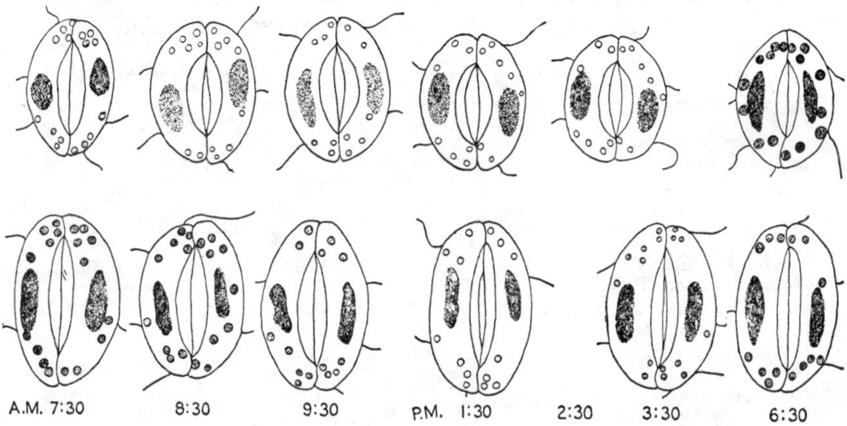

FIG. 19.—Stomatal behavior in lower epidermis of topmost and lowest leaf of 4″ field sunflowers, 1926.

On plants less wilted the stomatal openings were somewhat larger for both lower and upper leaves, but even in these the stomata were mostly closed until the plants were watered in the afternoon, and there was little starch present. The stomata opened after the leaves recovered, and the upper leaves gave a strong starch test and the lower a moderate one later in the afternoon. During this period the leaves of the plants in the more open plots were making starch in abundance, thus indicating that intense competition for water has the same effect as direct drouth in closing stomata and reducing the rate of starch production.

SUNFLOWER LEVEL PHYTOMETERS, FIRST SERIES

Method—The seeds were germinated out-doors in large containers under optimum conditions and the seedlings planted June 28 when they were 3″ tall and the first pair of leaves fully developed. At this time the plants had an average leaf-area of 13.5 sq. in. (both surfaces) and a dry weight of 0.21 gm. The containers were brought back to the original weight before fitting the corks into the opening about the stems and sealing the aperture with plasticene. Two phytometers were placed on the ground beneath the 4″ plants; two others were maintained at the general level in this plot where the leaves overlapped, and a third pair was kept 6″–8″ above this level. Two phytometers were likewise installed below the 8″ on the ground and two others at the level of the crowns. Atmometers were placed on the same levels as the phytometers in the two fields, and the usual determinations were made of the air factors. The evaporation for the 9 days of the experiment was 125, 207 and 233 cc. for the respective levels in the 4″ plot, and 228 and 230 cc. for those in the much more open and uniform 8″ plot. The rate increased rapidly from the ground upward where the plants were dense, and was also much less than in the more open stand (plate 25).

Growth and dry weight—The height, leaf-area and dry weight of the five pairs of phytometers in the two plots are shown in the following table:

TABLE 51—*Growth and dry weight of phytometers at different levels*

Criteria	4″ Plot			8″ Plot	
	Above Crowns	At Crowns	On Ground	At Crowns	On Ground
Ave. ht. cm.......	21.1	39.9	38.5	29.3	36.2
Ave. ht. to cotyledons, cm.......	3.3	4.2	3.8	2.9	2.3
Ave. ht. to 1st lvs. cm.	9.9	20.1	25.9	15.7	13.7
Ave. diam. stem, mm.	7.5	8	5.7	7	5.7
Ave. no. full-grown leaves	6	6	4.5	7	6
Ave. no. lvs. per plant	14	10	8	12	10
Ave. length largest blade, cm.......	11.1	10.8	9.2	10.3	9.9
Leaf-area 2 plants, both sides, sq. in.	108.8	116.2	60.2	109.0	77.2
Dry wt. 2 plants, gm.	4.34	4.64	1.79	4.11	2.35

In the case of both the 4″ and the 8″, the results were consistent throughout except for height, the discrepancy in this respect as well as the greater differences for the denser plot being due to the lower light intensity. As a rule, the values were higher for the phytometers at the crown level than for those above, evidently as a result of lessened transpiration without corresponding reduction in light.

FUNCTIONAL STUDIES IN COMPETITION PLOTS

Stomata and starch—On July 2, a study was made of the stomatal movement, light relation, and the amount of starch produced each hour during a fairly clear, warm day from 7 a.m. to 7 p.m. Owing to continued dry weather, the plants had been watered the previous day and were watered again during the experiment. The topmost mature leaves and the lowest green leaves of the 4″ and 8″ were used for observation. These plants were almost fully grown and approximately 3′ high.

During the day the stomata of the lowest leaf of the 4″ were very irregular in the degree of opening. A few of the stomata were slightly

Level phytometers in field sunflowers, 1926.

A. Method of installing phytometers.
B. Growth in the phytometer pairs at the three levels, top, middle and bottom, at end of test.

open at 7:30 a.m., reached a maximum at 11:30 a.m., and remained thus until 12:30 p.m., after which they began to close rather rapidly and were almost all closed by 2 p.m. They again began to open, but after 3 p.m. they gradually closed until complete closure was reached at 6:30 p.m. This leaf gave no starch test until about 10 a.m., when a slight amount was observed; during the remainder of the day starch was absent until 3 p.m., when a medium amount was found.

At 7:30 a.m. the stomata of the topmost leaf of the 4″ were slightly open, but within an hour and a half they were wide open and remained so until 3 p.m. They then gradually closed until complete closure was reached at 6:30 p.m. There was no starch present in the leaves at 7:30 a.m., but by 9:30 a.m. there was an abundance and this amount remained throughout the day. The stomata of the lower leaves were only open for a short time during the day because of the deficiency of the water supply. The chloroplasts were filled with starch while the stomata were closed, but starch was scarce when they were open. In the topmost leaves of the 4″, the amount of starch produced was closely related to stomatal movement and the per cent of light received. There was no starch present in the morning with a minimum amount of light and stomatal opening, but the starch increased rapidly as the light became greater and the stomata opened wider. A similar relation was found in the lowest leaf where the production of starch was delayed and decreased by the late opening of the stomata and the lower light intensity.

In the lowest green leaf of the 8″, a few stomata were slightly open at 9:30 a.m. They followed an irregular course of opening a little and closing throughout the day. The largest number were slightly open at about 2:30 p.m., after which most of them began to close and remained so for the rest of the day. There was no starch present until 10 a.m., when it began to accumulate as more stomata opened. Tests showed a medium amount of starch at 2 p.m.; after this, the amount became less until little was present in the leaves at 6 p.m.

The stomata of the topmost leaves of the 8″ were slightly open at 7:30 a.m. but wide open at 8 a.m. At 11:30 the leaves began to wilt and the stomata were found closed. After watering they slowly opened until they reached their maximum at 3 p.m., closing gradually until 6:30, when only a few were still slightly open. The starch-content of the leaf quickly increased from no starch at 8 a.m. to medium at 9:30 and abundant at 12 m. It remained thus until 3:30 p.m., but decreased slightly in amount as the stomata began to close. The stomatal movement of the topmost leaf of the 8″ was very irregular throughout the day due to wilting. An abundance of starch was present in the leaves after 12 m. The stomata of the lower leaves opened at 9:30 in the morning with a scarcity of starch, but closed at 10:30 as wilting began. The plants were watered at 11:30 and at 1:30 p.m. the stomata were open and little starch was present. They were closed at 4 p.m. and their chloroplasts exhibited an abundance of starch for the remainder of the day. The photosynthate for several types of leaf was determined at 3:30 p.m. with the following results:

TABLE 52—*Photosynthate for leaves of 4″ and 8″, July 2*

Plant and leaf	Area one side	Photosynthate per 100 sq. cm., gm.
	sq. in.	
4's Suppressed, top leaf........	5.80	0.030
4's Dominant lower leaf.......	5.10	0.046
4's Dominant top leaf.........	6.15	0.122
8's Dominant, 4″ below crown..	10.30	0.123
8's Dominant, top leaf........	16.45	0.127

Conduction—Two studies were made of the conduction of stems grown under three intensities of competition, namely, 4″, 8″ and 16″. Sections 5.5 cm. long were cut from the lowest internode and run in groups of four under an equal pressure of water, amounting to approximately 1/16 atmosphere. In the first series the 4″ with an average diameter of 8 mm. and leaf-area of 126 sq. in. gave a value of 3.8 cc. for the 30-minute period; the 8″ with averages of 12 mm. and 346 sq. in., 22.3 cc., and the 16″ with 18 mm. and 1,022 sq. in., 54 cc. The difference in conduction was due to differences in both diameter and leaf-area, as the correlation was practically the same for both, when the area of the cross-section of the stem was taken into account.

The second series was run on July 6, the conduction for a 30-minute period for the 4″ with an average diameter of 9 mm. being 11.9 cc., for the 8″ with a diameter of 11.2 mm., 25.8 cc., and for the 16″ with a diameter of 16.1 mm., 62.8 cc. With a stem but 1 mm. wider, the 4″ conducted more than 3 times as much water, while the values for the 8″ and 16″ were much the same as in the preceding, the increased value for the first probably being due to a relatively greater increase in leaf-area.

Root pressure—The root pressure of plants in the field was determined for two series, on July 9 and 15 respectively. Manometers were placed upon each of two plants in the 4″, 8″ and one in the 16″ plots; the stems were cut just below the cotyledons, except in the 16″, which was cut above them. All the stems were dry when cut and absorbed water quickly; the leaves of both 4″ and 8″ had begun to wilt by the middle of the forenoon. One of the 4″ gave a negative pressure ranging from —10 to —78 mm. of mercury; the other gave a positive pressure of +3 to +9 late in the afternoon. The 8″ showed a negative pressure of —9 to —38 mm. up to 11 a.m. and a positive pressure of +12 to +19 during the remainder of the day; the range in the 16″ was —5 to —25 mm. until eleven o'clock, followed by one of +30 to +60 mm. until 2 p.m., when it fell rapidly, reaching —41 at 5:30 p.m. The results on July 15 and 16 were less concordant, the turgid 8″ giving a pressure about a half higher than the 4″, but the 16″ denoting negative pressure practically throughout.

BEHAVIOR OF WATERED AND UNWATERED 16″ PLANTS

In order to throw light upon the behavior of individuals wilting under strong competition, two areas of the 16″ plot were set aside during a period of drouth for comparison. One of these was irrigated with sufficient thoroughness to wet the soil to 6″, while the other received no additional water. The experimental plants were selected with care in both areas and were labeled with tags to insure uniformity of material in so far as possible. The next day, July 23, the unwatered plants were slightly flaccid by 8 a.m., and wilting increased during the course of the day, though without the leaves becoming badly collapsed by night. The watered plants were completely turgid until noon, when increasing water loss caused them to be watered again to restore the optimum.

The stomata were stripped hourly from 5 a.m. to 6 p.m. of what proved to be a clear hot and windy day, and at frequent intervals tests of stomatal opening were also made by means of the cobalt-chloride method. Factors were measured throughout the period and leaf temperatures were also taken. Finally, material of leaves and stems was collected for the determination of the sap-content on the basis of dry weight. Upper, middle and lower leaves were cut at the base of the blade at 5 and 10 a.m., and 2:30 and 6 p.m., and at the same time sections of the stem 12–15 cm. long were cut out and the leaves removed.

Factor data—The holard for the first 6″ level was 19.4% for the watered group and 7.7% for the unwatered; for the next two levels it was 15.4% and 12.3%, and 16.6% and 15.9% respectively. The evaporation for the 23-hour period was 45.5 cc. for the watered and 54.6 for the unwatered area at 48″; at 30″ it was 31.5 and 39.2 cc. and at 12″ it was 36.4 and 42 cc. respectively. The semi-wilted plants permitted greater wind movement and stronger lighting within the group with the consequent increase in evaporation at all three levels. As in other cases, the minima occurred at the middle level, where the protection afforded by the foliage was greatest. In general the humidity ranged several per cent lower in the unwatered than in the watered area, while the leaf temperatures of the unwatered plants were 1°–4° C. higher and that of the lower leaf similarly higher than the upper. Tests by means of cobalt-chloride paper gave a rate about twice as rapid in the watered plants.

Sap-content—The variations in the sap-content during the day for the stems and leaves of the watered and unwatered plants were as follows, expressed in per cent of dry weight:

TABLE 53—*Sap-content of watered and unwatered 16″ plants*

Part	5 a.m.	10 a.m.	2:30 p.m.	6 p.m.
Stems, watered plants	538.51	538.36	659.11	650.06
Stems, unwatered plants	462.38	497.62	342.12	487.77
Top leaf, watered	251.43	296.92	274.16	312.73
Top leaf, not watered	246.66	228.40	231.30	240.37
Middle leaf, watered	324.68	354.73	329.06	383.16
Middle leaf, not watered	258.91	259.92	262.72	283.80
Lower leaf, watered	389.21	442.74	414.41	392.74
Lower leaf, not watered	360.55	277.66	313.26	301.47

As anticipated, there was a constant difference to the advantage of the watered plot, while the leaves of both increased in sap-content from the top to the lower. There was less regularity in behavior during the day, apart from the usual rise toward evening.

Behavior of stomata—The stomatal movement for both surfaces from 5 a.m. to 6 p.m. is indicated in the following table:

TABLE 54—*Stomatal movement in watered and unwatered 16" plants*

Time	Watered Series		Unwatered Series	
	Upper leaf	Lower leaf	Upper leaf	Lower leaf
5 a.m.	Closed	Closed	Closed	Closed
6	Closed	Closed	Closed	Closed
7	Open	Open	Open	Half of them open.
8	Open	Open	Almost closed on upper epidermis, a little wider open on lower epidermis, leaves flacid wilting	Closed
9	Open (more than half way)	One half slightly open	Closed	Closed
10	Open (more than half way)	Slightly open	Closed Whole plant wilting	Closed
11	Open (less than one half)	Closed	Closed	Closed
12 noon	Open (less than one half)	Closed	Closed to slightly open	Closed
1 p.m.	Closed above, slightly open below	Closed	Closed, few slightly open.	Closed
2	Closed	Closed	Closed	Closed
3	Very few slightly open, nearly all closed	Closed	Closed	Closed
4	Closed	Closed	Closed	Closed
5	Closed	Closed	Closed	Closed
6	Few very slightly open, rest closed	Closed	Closed	Closed

Starch-content—The amount of starch in leaves at the three levels at 6 a.m., 12 m. and 6 p.m. is given in table 55.

The correlation between stomatal movement and starch-content is fairly close throughout, as is seen especially by a comparison of the two extremes, the upper leaf of the watered group and the lower of the unwatered. The stomata of the former were open in some measure through most of the day and the starch-content ranged from fair to heavy, while the latter were closed practically all day and only a trace of starch was to be found at mid-day. The leaves of the unwatered series lagged far behind in the

TABLE 55—*Starch-content of watered and unwatered 16″ plants*

Leaf	6 a.m.	12 m.	6 p.m.
Upper leaf			
Watered	slight–fair	good–heavy	heavy throughout
Unwatered	trace	fair–good in spots	fair–good but with free areas
Middle leaf			
Watered	trace	fair–good	heavy throughout
Unwatered	none	fair	fair
Lower leaf			
Watered	none	light	heavy throughout except tip
Unwatered	none	trace near veins	none, wilted

amount of starch, and the latter decreased regularly and markedly from the upper to the middle and lower leaves. There was also the usual correlation between the starch in the chloroplasts and the degree of opening, the latter increasing as the starch diminished in the morning and decreasing as the starch increased to a maximum in late afternoon.

PLOT RESULTS

Development and behavior on July 27—Typical plants were selected from the five competition plots on this date and the following results secured:

TABLE 56—*Growth and conduction of competitors*

Plot	Leaf-area sq. in.	Dry weight, tops, gm.	Conducting power of stems		
			No. plants	Ave. cc.	Stem diam. mm.
4″	54	5.37	3	10	6.6– 7.6
8″	229	15.47	3	29	8.7–11.7
16″	800	42.91	1	40	22
32″	3,394	125.2	2	53	23.8–29
64″	5,010	200.0	2	61	31.0–35.5

The leaf-area, dry weight, conducting power and stem diameter all increased consistently and with some regularity as the density decreased. The correlation was most evident in the case of leaf-area and dry weight, the one increasing approximately fourfold, the other threefold to the 32″ plot, where the increase for the 64″ was about 1.5 times for each. Conduction in the 8″ was practically three times as great as in the 4″, the increase being closely correlated with that in dry weight, less with that of leaf-area and not at all with stem diameter. In the 16″, the latter was doubled, but conduction was augmented only a third, while in the two more open plots the correspondence was fairly close.

At this time the lower and upper epidermis were stripped from representative leaves in the several plots to permit the determination of the number of stomata per unit area, the average size of stomata, and the number of chloroplats per stoma (both guard-cells). The strips were taken at 7:30 p.m. when the stomata were closed and the starch-content high, potassium iodid-iodin being used to stain the chloroplasts. The latter were counted for each epidermis by means of a drawing machine. Each result in the table is an average of 50 counts for the chloroplasts and of 40 for the stomata.

TABLE 57—*Stomatal relations in the competition plots*

Criteria	4″	8″	16″	32″	64″
	Lower surface				
Ave. no. per sq. mm.	475	460	448	448	372
Ave. length, μ	17.9	19.0	22.0	25.1	27.1
Ave. diam. μ	13.0	13.4	14.0	16.4	17.6
Ave. no. chloroplasts per stoma	13.9	17.1	19.7	21.9	22.6
	Upper surface				
Ave. no. per sq. mm.	455	427	411	411	295
Ave. length, μ	18.2	19.2	24.3	26.4	26.5
Ave. diam. μ	13.2	13.8	16.6	16.2	15.8
Ave. no. chloroplasts per stoma	14.5	19.0	20.3	21.3	23.3

Practically all the values varied in the sequence of the plots for both surfaces. The number of stomata per unit area was highest in the 4″ and decreased from 475 to 372 in the lower epidermis, and from 455 to 295 in the upper, the minimum occurring in the 64″. The size of the stomata on the other hand varied inversely, the stomata being about a third longer in the 64″ than in the 4″ plants, the width corresponding in the lower but falling off slightly in the upper epidermis. This inverse relation between number and size is evidently the consequence of greater expansion and larger size of leaf in the more open plots where the competition was correspondingly reduced. This is also reflected in the conditions on the two surfaces, the number being greater on the lower surface to the extent of about 5% and the size slightly less.

As might be expected, the number of chloroplasts per stoma increases with the size of the latter, from approximately 14 in the 4″ plot to 23 in the 64″ one. Here again the larger stomata of the upper epidermis possess an average of 0.5 chloroplasts in excess of the lower. There is also some evidence that the chloroplasts are larger when the stomata are closed and the starch-content higher. Fifty measurements on stomata of the 32″ plants, taken on July 30 at 1:30 p.m. when the stomata were open, gave an average diameter of 2.32μ by contrast with an average of 3.34μ for 100 measurements made at 7:30 p.m. when closure was complete.

Sunflower Level Phytometers, Second Series

Plan and factors—A new series of level phytometers was started on August 2 at 12:30 p.m., soon after the various plots had come into flower. As usual, the 4″ blossomed first, followed by the 8″ and then the others in order, but at this time the 16″ were the most advanced. The experiment was continued for 9 days, until August 11; at the outset the phytometers were at the stage where the first pair of leaves was half grown. They were placed at the three heights shown in figure 20 and close to the competing individuals. The atmometers and anemometers were put in position on August 3 and continued to the close of the experiment. The usual readings of light, temperature and humidity were taken throughout the course.

The light regularly decreased from the upper to the middle and lower phytometer in correspondence with the shading from the top to the bottom of the stand in each plot. The general tendency was for temperature to decrease downward, while humidity was far from consistent, the differences for both being slight. Soil temperatures on the contrary were entirely consistent, falling from the 64″ to the 16″, the differences at 1″ ranging from 9°–16° F and at 6″ being 6°. The differences between the two depths in the 64″ was 7° and 14° at 10 and 11 a.m. respectively, while it was but 4° in the 16″ at both times. The wind movement was naturally much greater near the tops of the plots and decreased progressively downward.

Transpiration and evaporation—The number, arrangement and height of the various phytometers are shown in the accompanying figure, together with the respective values for transpiration and evaporation in the three plots, and the wind movement for the denser 16″ one (fig. 20).

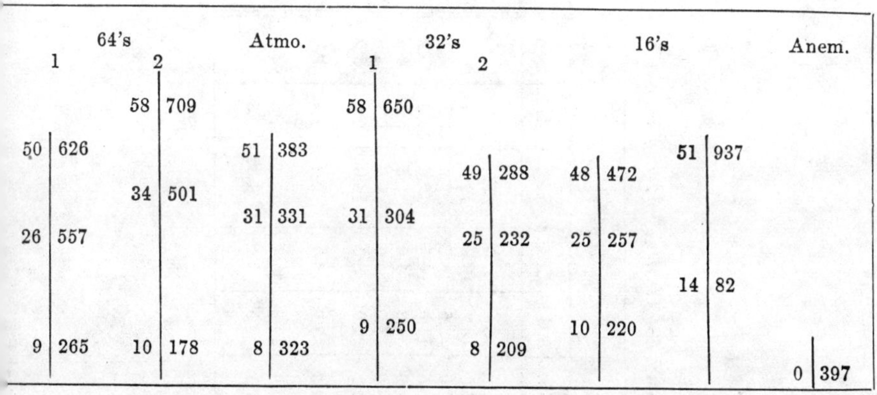

Fig. 20.—Position and water-loss of sunflower level phytometers, and atmometer. Height in inches and water-loss in cc.; wind in miles.

The decrease from the top to the bottom level was consistent and narked throughout. The transpiration at the upper level was more than wice as great as at the lower, the middle level being regularly intermediate, hough much nearer the value for the upper in the more open 64″ and for he lower in the denser 32″ and 16″. The evaporation from the atmometers fell off consistently in the same direction, the amount for the middle level being much nearer the lower one, but the difference between the two extremes was but a fifth or less of that for transpiration. Much of this effect was doubtless due to the heating action of the light absorbed by the leaves in contrast to the white porous cups. A phytometer placed in the soil in a bare area as a control gave a transpiration of 397 cc., in comparison with a maximum of 265 for the 64″, 250 for the 32″ and 220 for the 16″ at the lower level. The wind movement was measured only for the denser plot, the rate at approximately 4′ being more than five times that at about one foot.

TABLE 58—*Growth and dry weight of level phytometers, August 11*

Criteria	64″						32″			16″			Control
	Up.	Mid.	Low.	Up.	Mid.	Low.	Up.	Mid.	Low.	Up.	Mid.	Low.	Low
No.	10	13	3	12	2	8	5	7	9	14	11	4	
Ht. in.	17.0	18.2	19.3	19.5	21.0	22.6	15.0	24.1	19.6	14.0	31.7	19.3	12.5
Ht. to cotyls.	3.0	3.5	4.0	2.6	3.1	3.7	4.3	3.0	4.0	2.8	3.4	2.8	3.0
Ht. 1st lvs. in.	12.5	14.0	14.5	12.6	14.0	16.2	11.0	17.5	15.3	10.0	19.0	16.0	9.0
Diam. stem, mm.	4.0	4.2	4.0	5.5	5.1	4.5	6.0	5.0	3.0	5.0	5.0	4.0	5.0
No. full-grown lvs.	6.0	5.0	4.0	6.0	5.2	4.0	6.0	4.0	4.0	6.0	6.0	4.0	4.0
No. lvs. per plant	13.0	9.0	8.0	14.0	10.4	8.0	12.0	9.0	6.0	12.0	12.0	6.0	10.0
Length largest blades, cm.	9.3	8.2	6.9	9.9	7.9	6.8	10.6	8.4	6.4	9.1	7.5	8.2	9.3
Leaf-area both sides, sq. in.	69.8	42.0	30.5	78.8	51.6	24.0	88.3	44.7	15.7	64.1	42.2	26.4	51.8
Total dry wt. gm.	1.1	0.57	0.42	1.34	0.31	0.39	1.56	0.68	0.23	1.0	0.64	0.43	1.1

Growth and dry weight—In the preceding table are given the values r the growth of stem and leaf, and for the dry weight of the individual ıytometers in the three plots, as well as the control plant outside the mpetition groups (table 58).

In the most open plot, the total height of the stem, the height to the tyledons and to the first pair of leaves increased consistently from the per to the lower level as a consequence of reduced water-loss and decreased ght. In the two denser plots the effect of shade in elongating the stem as felt most strongly at the middle level, where the value was greater an at the others in five cases out of six. In all cases the diameter of the em decreased consistently from the upper to the lower level, this being rrelated with maximum light and hence being the reverse of stem length behavior. The total number of leaves and the number of full-grown ıes were likewise consistent as to diminishing values from upper to lower vels, but this relation was much more strikingly shown in the length blade of the largest leaves and in the total leaf-area. The latter in ırticular was 50% or more greater for the upper than the middle phytome-rs and two to four times in excess of that for the lower ones. With single exception, the dry weight followed the same order, the values for e upper being about twice as great for the middle, and two to four times much as for the lower.

In general, height values for the stem were slightly greater for the ″ plot than for the other two and slightly less for stem diameter, though ss regularly so. This was true also for the total number of leaves, but e effect of somewhat more shade caused the length and total area to be eatest in the 32″ plot, while still more shading led to a drop below the ″ in the case of the dense 16″ plot. In correlation with the leaf-area, e maximum dry weight occurred in the 32″ plot, with the 64″ next and e 16″ least, the control being practically the same as the latter. With one two exceptions the values for the control phytometer were less than r the upper phytometer in each of the three plots, the light relations ing similar but the bare ground causing the temperature, humidity and ind to be more extreme.

Comparative Development at End of Season, 1926

General measurements—On August 4 the diameter of the heads with ys was determined for the five plots with the following results:

Plot	4″	8″	16″	32″	64″
Cm.	9.3	13.1	21.6	27.2	27.0

hese values corresponded closely with those of the first competition series 1924, though drouth had reduced the difference between the three less nse plots.

The field measurements for the following data were secured on August 9. he average height of the 4″ plants was 26″ and the spread of the crown 8″; the green leaves were confined to the upper 4″–6″. Practically all ants exhibited some wilted leaves and on some all leaves were wilted.

Light was no longer an important factor, even the soil surface being but partly shaded. Nearly every plant bore a head, but on half the number this had dried without opening. The average height of the 8″ was 28″ and the spread 6.2″, while all the live leaves were within 10″–12″ of the top. With respect to wilted leaves and the size and condition of the heads, these plants were practically identical with the 4″. There were few subdominant individuals in either of these two plots, the competition being for water rather than light.

The average height of the 16″ was 55″ and the spread of crown 21″, the green leaves being absent only on the lower 15″. In general, the crowns overlapped about 5″, but in drier portions of the plot there was an interval of 2″–5″, owing to the semi-wilted condition of the plants. Every plant bore an open flower head, but most of them had lost the ray flowers early on account of the drouth. The 32″ plants averaged 63″ tall and the crowns 32″ wide, living leaves occurring to within 15″ of the ground. The leaves practically touched each other between rows; a few were wilted but most were in good condition. Each plant bore a head and the whol plot had reached the maximum of flowering. The average height in the 64″ plot was 64″ and the spread of crown 36″; on many plants green leave occurred to the base. The plants exhibited more wilting than in the 32″ plot owing to their greater exposure between rows, and in some even th upper leaves were not fully turgid.

Special measurements—The final values for the stem, leaves and th dry weight were determined on August 12, some of the plots having ceas their growth as indicated in the above description.

Table 59—*Development of Helianthus on August 12*

Criteria	4″	8″	16″	32″	64″
Ave. ht. in........	27.1	30.7	53.7	70.5	66.0
Ave. diam. base stem, mm.......	7.1	9.8	21.4	29.4	34.8
Ave. diam. top stem, mm.......	5.1	6.2	11.0	15.0	16.9
Ave. no. live lvs...	9.6	14.7	26.0	24.0	29.7
Ave. no. dead lvs...	10.7	11.1	15.3	11.0	13.8
Ave. length largest blade, cm.......	7.5	9.9	20.2	28.0	30.4
Ave. width largest blade, cm.......	5.9	8.8	21.0	28.1	32.3
Ave. length petiole largest lvs. cm...	5.0	5.0	12.7	18.4	20.3
Ave. leaf-area both sides, sq. in.....	67.27	192.6	1554.0	3485.2	5069.7
Ave. dry wt. tops without head, gm.	6.28	18.5	131.7	136.5	152.6
Ave. wt. head, gm.	0.8	2.0	15.1	27.1	26.8

With one or two exceptions there was a consistent increase in value from the 4″ to the 64″ plot; the latter was exceeded by the 32″ only i the average height and weight of head. The greatest difference between the extremes was naturally found in the leaf-areas and dry weight, the one being 75, the other 25 times greater in the 64″ plot. The sequence of values was essentially the same as in the plots of 1924, which was however more favorable for growth and yielded larger results.

FIELD COMPETITION IN TRITICUM SATIVUM

SERIES OF 1924

Installation—The field employed for the wheat competition was plowed 4″ deep and repeatedly harrowed until a good seed-bed was made. It was laid off in plots 16 rods long and 1 rod wide, four of which were used for planting. Marquis spring wheat was sown in these on April 5 to yield four different densities. The first plot was sown at the normal rate of 75 lbs. per acre, the second at one-half this rate, the third at twice and the fourth at four times the normal rate. In order to obtain an even distribution of the seed, each plot was divided into ten strips and half the seed sown in one direction and half in the other. The soil was in excellent condition at the time of sowing and by April 18 the wheat exhibited a fine even stand, while a week later it was mostly in the 3-leaf stage. On May 4 about half the plants bore a single tiller and some two.

Course of development—On May 8 the average height and number of tillers were determined for 60 representative individuals from each plot and the average dry weight of tops from 100 selected plants, with the following results:

TABLE 60—*Height, tillers and dry weight of wheat, May 8*

Plot	Ave. ht. in.	Ave. no. tillers	Dry wt. tops per 100 gm.
½N	3.5	3.6	16.21
N	4.0	2.9	17.43
2N	4.7	2.2	11.94
4N	5.5	1.0	7.54

The height decreased consistently from the least dense to the most dense and the number of tillers in the reverse direction. The dry weight was slightly greater for the N plot than for the ½N and fell off rapidly to less than half for the 4N. The plants in the two open plots were more spreading, in the others quite erect; the number of leaves was practically the same for each, as was the length, but the average width of 20 specimens was 15% less in the 4N than in the normal. On May 11 the dry weight of the roots in the upper cubic foot of soil was determined for the ½N and the 4N plots; the number of plants was 12 and 198 respectively. The corresponding weights were 0.375 and 1.775 gm., and the average weight per plant 0.0312 and 0.0164 gm., or about half as much in the densest plot.

The chresard was determined for each plot at weekly intervals from May 1 to July 11, at 3–5 levels as the roots grew in length. The values were considerably and regularly less in the denser plots, especially in the upper levels, the 4N frequently having less than half as much water available as the ½N. This is readily explained by the fact that the average of a number of counts gave 94 plants to the square meter in the ½N, 203 in the N, 360 in the 2N and 719 in the 4N plot. The unusually low temperature of the soil together with dry weather hindered growth in May and reduced nitrification to the point where nitrates were apparently used as fast as formed. As a result, the demands of the much larger number of individuals in the 2N and 4N plots caused them to turn distinctly yellowish by May 17, when those of the open plots were still quite green. The yellowing increased during the following week, until it became slightly noticeable even in the ½N. The stature was greater and the color greener along the edges of the 4N and 2N and to a slight extent the N plots, but the growth and color were practically uniform in the lesser competition of the ½N.

TABLE 61—*Height and condition of leaves in Triticum, May 22*

Plot	Ave. ht. in.	Total no. dead leaves	Total no. injured leaves
½N	5.8	4	30
N	7.0	14	44
2N	8.2	36	49
4N	8.7	46	59

These results were obtained from 25 plants from each plot, and are all the more striking in view of the fact that the more open plots exhibited 4–5 times as many leaves as the denser. Stature followed the rule in increasing with the degree of shading, while both shade and lack of water were instrumental in producing a loss of leaves tenfold greater in the 4N than in the ½N plot. The average light intensity on June 10 for the four plots was 80% for the ½N, 58 for the N, 37 for the 2N, and 24 for the 4N, while the humidity increased markedly in the same direction (fig. 21).

By this time competition in the 2N and 4N plots had become so intense as to overcome the effect of shade in producing elongation, and these plants were shorter than those in the two open plots, while the stem diameter fell off regularly to little more than half that of the ½N. The width and length of the fifth leaf decreased in the same sequence and at about the same rate. Differences in growth as measured by the number of tillers and the percentage of dead ones were much more striking. The ½N bore twice as many tillers as the N, this more than twice as many as the 2N, and the latter nearly twice the number for the 4N; the number of dead tillers was three times as great in the latter as in the ½N (plate 26A).

Competition cultures of field wheat.

A. Plants from the ½N, N(Normal), 2N and 4N plots on June 9.
B. Headed plants from the same plots on June 24.

The reduction in dry weight from ½N to 4N was nearly eight-fold and corresponded fairly well with that in leaf-area and in stem-area, these two naturally being in closer correspondence. The average number of roots decreased quite regularly to the 2N plot and then dropped less to the 4N, the latter being a fourth of the maximum.

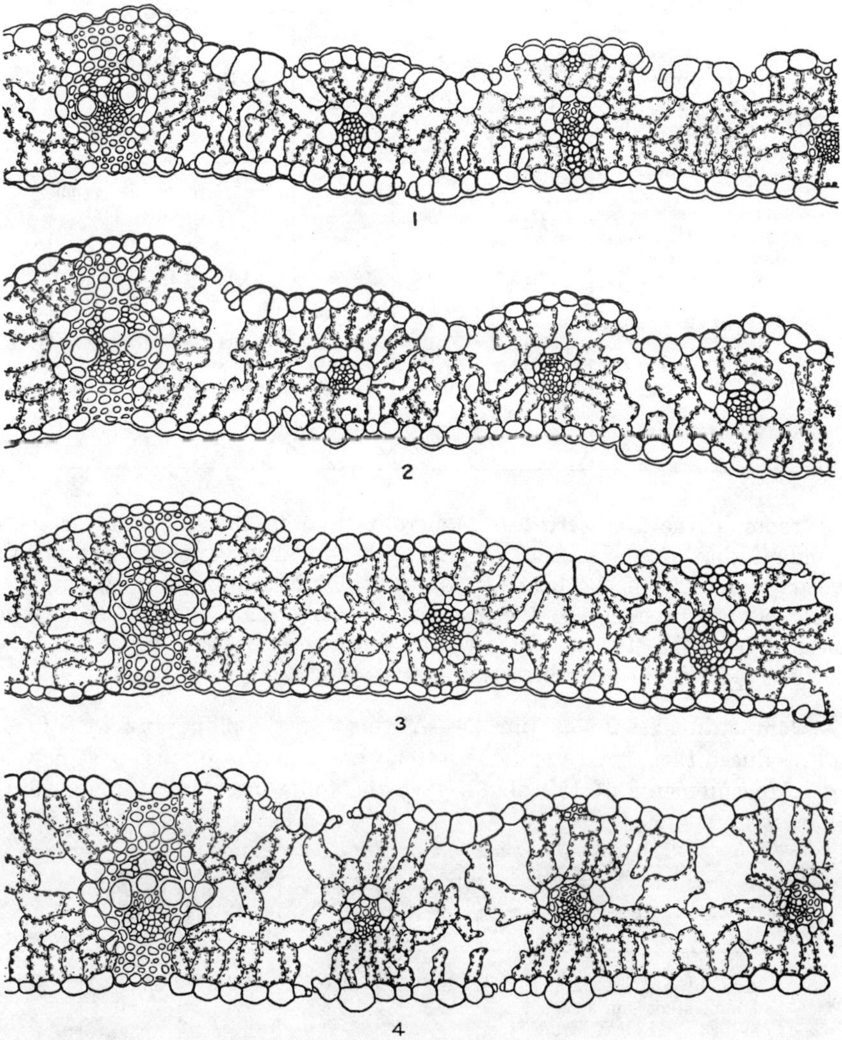

FIG. 21.—Structure of leaf of 4N, 2N, N and ½N field wheat, 1924.

Root relations—An examination was made of the root systems in the several plots on June 10. In the 4N plot the soil was densely crowded with roots to the depth of 34″, and roots were not infrequent to 4′, the maximum being 56″. In the ½N plot the more vigorous roots reached a depth of 5′, but the working level was much the same for all four plots.

TABLE 62—*Development of Triticum on June 9*

Criteria	½N	N	2N	4N	Remarks
Ave. ht. in........	18.0	18.0	17.0	15.0	
Diam. stem, mm...	3.5	3.0	2.6	1.9	Based on 100 plants from each plot
Width 5th leaf, mm.	11.7	10.0	8.6	6.1	
Length 5th leaf, mm.	182.1	148.6	128.0	87.7	
Ave. no. tillers....	9.2	4.1	1.6	0.5	Based on 300 plants from each plot
% dead tillers....	2.7	6.4	15.6	17.3	
Leaf-area, both sides, sq. in.....	926.86	358.48	144.8	69.08	10 typical plants
Dry wt. 100 plants, gm.	237.32	137.12	60.53	31.09	
Stem area, sq. in..	179.32	61.48	34.84	19.88	
Total area, sq. in..	1106.18	419.96	179.64	88.96	50 ½N to 150 4N used
Ave. no. roots	37.5	24.5	13.2	9.4	

The roots in the 2N were less numerous than in the 4N, but far more abundant than in the normal. Heading began about June 11 and within two days a third of the heads in the 4N plot had appeared; the number decreased progressively to the ½N, where only an occasional spike had emerged from the sheath. This was the usual response to a water deficit, the drier plot invariably blossoming first.

Competition results on June 24—At this time the plants in all the fields had produced their spikes and the latter were in flower or just past flowering. Measurements of 150 plants gave the following results (plate 26B):

TABLE 63—*Development of Triticum on June 24*

Criteria	½N	N	2N	4N
Ave. ht. in.................	30.25	28.0	25.0	22.0
Diam. stem, mm............	3.4	2.8	2.5	1.96
Width 2nd leaf from top, mm.	11.4	9.5	8.0	6.3
Length 2nd leaf, mm........	184.0	138.0	119.0	95.0
Ave. no. tillers............	8.9	3.8	1.3	0.6
% dead tillers	34.7	48.1	88.0	90.0
No. heads per plant........	3.9	1.8	1.0	0.9
Ave. leaf-area one side, sq. in..................	41.87	14.53	4.34	2.53
Ave. dry wt., gm..........	6.64	2.26	0.83	0.58

Competition cultures of field wheat.

A. Spike from each of the four plots: ½N, N, 2N and 4N.
B. Yield of grain from 1,000 heads taken from ½N, N, 2N and 4N.

On May 8 when the first measurements were made, the 4N plants, which had tillered least, were the tallest, and they remained in the lead until after May 22. Shortly after they were exceeded by the 2N, but before June 9 competition in this plot had increased to the point where height-growth fell behind that of the N and ½N. However, by harvest time, the latter as a consequence of the least competition were tallest. The original relation of the tillers, which decreased in number with the density, remained essentially the same throughout as to sequence, but the number more than doubled in the ½N plot between May 8 and June 9. It increased only slightly in the N, while the 2N and 4N showed a distinct drop, owing to the death of many tillers. The number for the ½N was about the same on June 24, but had now dropped in the N, as also in the 2N.

The dry weight rose in all the plots until June 24, but much faster in the N than in the 2N or 4N, while it was greatest by far in the ½N. When the root systems were examined on July 21, the average maximum penetration was 4.7′ in the 4N and 5′ in the N plot, the respective working levels being 35″ for the 4N and 38.5″ for the N. Not only was the number of roots greater on the normal plants, but they penetrated more deeply and had a somewhat larger spread. However, in comparison with the shoots, the roots of the 4N were much more extensive.

Competition differences at maturity—When the wheat was fully grown and in the "hard-dough" stage, 500 representative heads were taken from each plot, and the following data secured. These are contrasted with the weight of grain from 1,000 ripe spikes from each plot.

TABLE 64—*Growth of spikes in Triticum*

Plot	Length spike, cm.	Spikelet width, mm.	No. spikelets with grains	Weight of grain, gm.
½N	8.82	12.1	16.2	903
N	7.93	11.4	14.85	855
2N	6.8	10.3	11.8	681
4N	5.46	9.84	9.04	479

A consistent and fairly regular decrease occurred in all four criteria rom the ½N to the 4N plot. The weight of grain was a little more than ıalf as much for the 4N, and this was more nearly correlated with the numɔer of spikelets bearing seed, as would be expected.

The plants from 40 meter-quadrats were cut from each plot on July 5–16, and the weight of 3,000 grains from each was also determined, with he results indicated in table 65, p. 228.

This table exhibits in striking manner the essential contrast in the ehavior of wheat under different densities and corresponding degrees of ompetition. The height was greatest in the ½N and decreased consistently o the 4N; the number of heads was naturally largest in the latter and ecreased to a third as many in the ½N. The number of spikelets with

TABLE 65—*Yield of Triticum plots, 1924*

Plot	Ave. ht. in.	Ave. wt. per sq. m.	No. heads per sq. m.	Yield of grain, gm.	Bu. per acre	Wt. 3000 grains, gm.
½N	31	389	178	5000.5	19.07	85.62
N	30	398	216	5525.0	21.07	92.8
2N	29	438	293	6235.7	23.71	91.33
4N	24.5	417	534	5667.7	21.61	83.66

seed and the weight of the latter was greatest in the ½N (table 64), but this was not enough to compensate for the much smaller number of heads. As a consequence, the maximum yield of both shoots and grain occurred in the 2N, with the 4N next, but this must also be accounted for in large degree by very favorable weather after June 7. This minimized the severity of competition in the denser plots and correspondingly increased the advantage derived from higher number of plants. Although the heads were smaller in the 2N than in the normal, there was 36% more of them per unit area, and hence the total yield was larger. The number of individuals in the 4N plot was 82% greater than in the 2N, but the spikes were so much reduced in size that the total yield was less. The weight of a definite number of grains was highest in the normal, nearly as high in the 2N, and lowest in the 4N. The lower value for the ½N is to be explained partly by the better opportunity for vegetative growth at the expense of flowering and partly by the handicap of greater exposure during dry periods.

SERIES OF 1926

Installation—The ground was plowed and harrowed on March 26, and a few days later it was marked off in four tenth-acre plots 4×1 rods in extent, with a 4′ space between them. The seed was planted by the drill method at the same rates as for the first series, and the general treatment was likewise the same. Unfortunately, the season was one of serious drouth, and consequently the results of the competition were less valuable as a check upon the first series than for a contrast with this and for the purpose of following the course of competition under conditions of drouth. The physical factors were determined for the several plots through the season, as in all the competition studies made, but the mass of data was so great as to make it necessary to compress them into the following summaries.

HABITAT FACTORS

Water relations—In comparison with the normal, a deficit in precipitation occurred during each month of the growing season, the total deficiency for the three months being 5.46″ or more than 50%. The most serious lack was for April, the rainfall of which was 0.36″ or about an eighth of the normal for the month. The rainfall for May was a third below normal and that for June more than a third.

Although there was sufficient available water to permit good development early in the season, by May 25 the chresard of the upper 6″ was only 2.2% in the 4N and 3.5% in the ½N, while it was but 4.3% and 5% respectively for the next level. By June 7 it had been reduced below zero in the upper level of soil from the normal to the 4N plot and was only 2.1% for the ½N. The effect of the greater density was shown in the lower levels by the fact that the 2N and the 4N had distinctly lower values. On June 17 the chresard of the upper 6″ level of the 4N plot was fivefold greater than for the next two levels, apparently owing to the deterioration of the superficial roots, and a similar result was found in all the plots by June 24. The water-content was also greater at the deeper levels in the more open plots, but was still low in the two denser ones.

The high temperatures combined with low humidities produced rapid evaporation and correspondingly high transpiration. Readings made with cylindric porous-cups with the lower edge at the surface of the soil gave the following results during June. The losses for the 24-hour period were averaged to give the daily loss for each week.

Week Ending	*Ave. Daily Loss*
June 10	36.5 cc.
" 17	29.8 "
" 24	22.3 "
July 1	50.8 "

The average daily evaporation during a normal season in eastern Nebraska does not exceed 24 cc., according to a long series of determinations by Weaver (1919). The average loss during the four weeks' period exceeded this by 10.8 cc., while during the last week the evaporation was more than twice the normal daily loss.

Light intensity—The light values for the different plots were measured at noon on two days during June; the results are expressed in percentage of full light at meridian and are the average of five readings in each case. The level of the readings was 3″.

Plot	*June 7*	*June 17*
½N	39.1%	52.5%
N	37.0	50.7
2N	21.0	55.2
4N	11.7	60.9

On the first date the plants of the denser plots were still growing well, and the intensity fell rapidly in the direction of density, being less than a third of that for the ½N and N, and approximately a tenth of that of full sunlight. By June 17, the plants in the denser plots were dying in large number, and the light consequently increased from the N to the 2N and 4N plots. The normal plot was still in good condition and the light remained somewhat lower than in the ½N.

Nitrate-content—When the first samples were taken on May 6, the upper level contained nitrates as follows from the ½N to the 4N plot: 14.4, 16.2, 19.4 and 10.4 ppm. By May 15 the effects of competition had become marked in the two denser plots, the plants being yellowish-green in color and the nitrate-content falling to 4 ppm. in both and in the two 6″ levels. At this time the values for the 6″–12″ level in the ½N and N plots were nearly the same and practically five times greater than for the 2N and 4N. In late May and early June the nitrate-content of the open plots had fallen to about 3–4 ppm., while it was about half as high in the denser ones. From this time the values remained low in all plots, though on June 26 they were slightly higher in the denser, being 3.7 for the 2N and 3.5 for the 4N by comparison with 2.8 for the ½N and 2.2 for the N.

Phytometer Readings

In order to integrate the effects of water, light and temperature, several series of phytometers were employed at different periods during the growth of the wheat plots to determine the use of water as measured by the transpiration. These were all insert phytometers of the type in which the planting conforms to that in the plot concerned.

First series—Large galvanized-iron containers 1′ square and 2′ deep were sunk in representative areas of the N, 2N and 4N plots, 2 being placed in each. As the holes were made, care was taken to see that the soil from the different levels was not mixed and this was then placed in each container to simulate the original soil profile as nearly as possible. The containers were then planted with wheat, sown thickly to insure a good stand and to permit later thinning to the desired degree. As a basis for the latter, meter-quadrats were located in the N, 2N and 4N plots on May 6 and the number of individuals counted. Five quadrats were employed for each plot and the average number of plants utilized as the basis for thinning the phytometers. The averages were 185 for the ½N, 280 for the N, 470 in the 2N and 1,000 in the 4N. In accordance, the phytometers were thinned on this date to 25, 50 and 100 plants respectively for the three denser plots. At this time the plants in all were in excellent condition, with tillering just beginning; the height rose slightly with the density, as usual.

No further attention was given the phytometers until May 28, with the exception of two waterings to maintain the chresard at the approximate level of that in the plots. On this date one of the pair from each plot was carefully removed and weighed by means of a portable scale, after which they were all replaced for a five-day period. This was a typical summer period with one light shower and considerable wind movement thus favoring high transpiration. Many of the plants in the plots were already badly dried as a result of drouth, this being most marked in the 4N and decreasing to the ½N, which were drying at the leaf-tips merely.

On June 2 the phytometers were removed and weighed again; the one from the 4N had lost 11.2 lbs., the 2N 6.5 lbs., and the N 3.5 lbs. The first now contained 125 plants, some having appeared since thinning; these averaged 10.5″ tall and were little or not at all tillered. The 2N phytometer contained 56 plants, which averaged 9″ high and were fairly well provided with tillers; the normal phytometer comprised 30 plants 6.5″ in height and with abundant tillers. The respective water-loss in grams per plant was 40.6, 52.6 and 52.9, a small percentage of this coming from the soil surface. Thus, the transpiration per square foot of surface was nearly twice as great for the 4N as for the 2N, and nearly twice also for the latter as for the normal. However, the loss per plant was greatest in the latter, owing to the fact that the plants were better developed than in the other two plots.

Second series—The other phytometers of each pair were removed from the plots and weighed on June 9, after which they were replaced for a three-day period before the final weighing. This interval was very hot and dry, and the evaporation correspondingly high. The atmometers in the plots showed a daily loss of 35 cc. in the N plot and 27 cc. in the 4N, which was a third greater than the average daily loss for a normal season. The transpiration from the 4N phytometer was 5.5 lbs., from the 2N 6.25 lbs., and from the normal 8 lbs. The respective numbers and heights were 103 plants 15″ tall, 56 with an average of 13″, and 27 of the same height, while the losses in grams per plant were 24.2, 50.6 and 134.5.

This reversal of the results obtained in the first series was due largely to the open stand with few tillers found in the normal phytometer. The plants were much greener, less mature and bore few dead leaves, while those of the 4N plot were just the reverse, being dead also to a height of 5″. The topmost leaves were rolled and the others badly dried; the soil surface moreover was much better protected from direct evaporation by the thicker stand. The effect of these differences was greatly enhanced by the drouth itself.

Similar though less striking results were obtained with a series of small phytometers in containers 18″ deep and 5.5″ square. Each phytometer contained 8 plants 12″–16″ high and bore numerous tillers. The battery was exposed from June 15 to 18, with the following results: 4N 258 gm., 2N 350 gm., and normal 424 gm.

Development in Plots

Competition results on June 11—The data in table 66, p. 232, were based upon measurements of a hundred plants from each plot, and the leaf-area and dry weight were determined from ten average plants out of this number.

All of the data obtained follow the rule of decreasing value with increasing density, with the exception of height of stem. As has been repeatedly seen, this is regularly the reverse, but in the present case the effect of drouth had overcome that of elongation due to shade, and the 2N and 4N plants were shorter than those of the ½N plot. There was again a fairly good correspondence between leaf-area and dry weight, the extreme values for the former giving a ratio of 4:1, and for the latter 3.6:1.

Competition differences at maturity—By June 24 the wheat was drying rapidly in the 4N and 2N plots, the plants of which had grown little

TABLE 66—*Development of Triticum on June 11, 1926*

Criteria	½N	N	2N	4N
Ave. ht. in.	15.5	17.8	15.4	14.0
Ave. diam. base stem, mm...	2.6	2.5	2.0	1.8
Ave. no. roots per plant....	17.3	15.9	11.0	8.3
Ave. no. live tillers per plant	2.2	1.6	1.2	1.0
Ave. no dead tillers per plant	2.1	1.9	0.9	0.2
Ave. width top leaf, mm.....	10.1	10.3	8.5	7.5
Ave. length top leaf, cm.....	9.9	10.2	8.2	7.3
Ave. no. live lvs. per plant.	7.2	5.3	3.5	2.7
Ave. no. dead lvs. per plant.	10.9	9.8	6.7	4.8
Ave. leaf-area both sides, sq. in.	21.7	17.8	7.2	5.4
Ave. dry wt. gm...........	1.54	1.38	0.52	0.29

during the month. In the N plot the plants were still green and in fairly good condition, though some were beginning to dry. Soon after this date, the ½N plot started to ripen in the usual fashion instead of drying out, as had occurred in the other three plots. By July 1 the wheat was mature in all the plots, and 1,000 representative spikes were selected from each. These were measured for length and width, and were later air-dried and their weight determined.

TABLE 67—*Size and weight of Triticum spikes*

Plot	Length cm.	Width mm.	Weight 1,000 spikes, gm.
½N	7.67	11.0	399
N	7.37	9.6	342
2N	6.82	9.2	290
4N	6.56	8.6	254

The dimensions of the spikes decreased consistently from the ½N to the 4N plot, as did the weight also. The values for the former were not far from those of the first series of plots in 1924, but the weights were mostly less than half as great, owing to the dry season. The spikes in the ½N alone bore fully developed grains, those of all the others being much reduced in size and somewhat shriveled in appearance.

GENERAL SUMMARY

In the first series of sunflower cultures, the discussion has been devoted chiefly to tracing the course and effects of competition in detail at regular intervals through the season. To conserve space and avoid needless duplication, the treatment of the second series has been devoted especially to the results obtained with phytometers and the functions effects of competition in terms of stomatal behavior, conduction, root-pressure, etc. These are all in essential correlation with density, and thus serve to explain the consistent results obtained with respect to growth and structure.

The measurements given for the growth of shoots in the first series (tables 47 and 49) and in the second (table 59) are not only consistent but also fairly regular in their correlation with the density of the plots. As integrations of growth, the average leaf-area and dry weight are of primary significance in this connection. The correlation of diameter of heads with density was also consistent for both series, though the values were somewhat higher for the second, in response to a warmer drier season. The values for sunflower heads and seeds at the close of the season of 1924 exhibit a striking progression from the densest 2″ to the most open 64″ plot, and this table furnishes one of the best exemplifications of the rôle of competition in controlling reproductive growth.

The final values for wheat were practically consistent for the two series, as well as with those of sunflower, though the severe drouth of 1926 affected the more open ½N more than the N plot and caused it to fall behind in one or two instances. However, the values for the spikes in the two series were perfectly consistent, decreasing with increasing density with much regularity. With respect to yield per unit area, the necessary balance between size and number resulted in the greatest yield in the 2N and the greatest relative weight in the N plot.

7. THE RELATIVE IMPORTANCE OF THE FACTORS IN COMPETITION

General plan—The chief purpose of the following series of experiments was to throw light on the relative influence of the three major factors, water, nutrients and light, upon the course and outcome of competition. The general method employed was to vary one of the factors or the demand upon it at the same time that the others were maintained at an approximate optimum. For this reason, these experiments were carried on in the greenhouse, where methods of control were more practicable than in the field and the measurement of response simpler and more accurate. The species employed were *Helianthus annuus, Triticum sativum, Xanthium canadense* and *Andropogon nutans,* though a few collateral experiments were made with other native plants. All of these had been used in competition cultures or plots in either prairie or field, and some in both, with the consequence that it became possible to apply the control results directly to the interpretation of those obtained outdoors.

The series of experiments were essentially the same for the four major species, including competition for water, for nutrients, or for light, with the other two factors equalized or maintained at an optimum in so far as possible. Special series for a particular purpose were added in several instances and one of mixed cultures was employed for studying the effect of *Helianthus* and *Xanthium* upon each other. The details of the general method are discussed under the section on *Helianthus annuus.* and the modifications of this under the species concerned.

CONTROL EXPERIMENTS WITH HELIANTHUS ANNUUS

SERIES OF 1924

Plan and method—The containers employed were 2′ deep and 1′ square, and were made of galvanized iron. They were reinforced at the top with heavy iron wire over which the metal was turned, provided with handles and furnished with a half-inch opening near the bottom. This was for drainage and aeration, and was kept closed most of the time. The containers were filled with well-screened silt-loam to which one-third as much sand was added. The amount of soil-water above the hygroscopic coefficient averaged 12.5%. Seeds were selected for uniformity and after soaking for a few hours were planted a half-inch deep on May 28, and a gravel mulch applied to the soil. The number of seeds was respectively 4, 8, 16, 32 and 64 for the five containers, while a sixth one without plants was employed as a check. When the seedlings had reached the proper stage the numbers were reduced by thinning to 2, 4, 8, 16 and 32 in the series of containers. An additional series in wooden buckets was started on the same day, and on May 31 two further groups were constituted, one with a chresard sequence from 2.6%–16.5% and the other with a

Control competition of *Helianthus* in the greenhouse, 1924.

A. Water plus series with five densities: 32's, 16's, 8's, 4's and 2's.
B. Nutrients plus series. C. Light series.

progressive increase in the amount of fertilizer. However, in these each container was planted with 16 seeds and the seedlings later reduced to 8.

The series of experiments were as follows:

Competition primarily for water, 3 groups with 5 containers in each, one of the sets in wooden buckets.

Competition primarily for nutrients, chresard maintained, 5 containers.

Competition primarily for light, chresard and nutrients mantained, 5 containers.

Shaded series, chresard and nutrients maintained; 5 containers.

Special series, chresard or nutrients increasing progressively; 3 containers each (plate 28).

Competition Primarily for Water

Growth in first set—The method employed in this series was to maintain an optimum supply of available water, approximately 12.5% above the hygroscopic coefficient in the containers with 2 plants. The containers were weighed frequently and the amount lost from those with 2 plants was replaced, while the others received this amount also, in spite of their greater needs. As a consequence, the soil became progressively drier from the 2-plant containers to those with 4, 8, 16, and 32 plants. Two weeks after planting the effects of competition were already visible in the containers with 8 or more plants, being especially noticeable in the number of suppressed individuals (table 68, p. 236).

The number of suppressed plants increased strikingly with the number of plants per container, almost half of those in the 32's being suppressed by June 23. While the average number of leaves was about the same in all on June 12, it was largest in the 4's on June 23, in the 2's on June 30, and smallest in the 32's after June 18, the final number being less than half that of the 2's. The height of the stem followed the usual rule, being 12 cm. in the 32's at the outset and 9 cm. in the 2's. Six days later the 16's were tallest with a height of 34 cm., the 32's nearly as tall and the 2's the shortest at 24 cm. On June 23 the 4's were tallest, and a week later this was still true, though they were now almost equalled by the 8's, while the 32's were not much more than half as tall.

Final results for first set—One set of containers was terminated on June 23, when the plants were about a month old and approximately half a meter tall. At this time the chresard at the three usual depths through the series was as follows:

Table 69—*Chresard on June 23*

Cont. no.	Number plants	Depth 0′–6′	6′–12′	12′–24′
		%	%	%
6	2	11.5	11.6	11.1
7	4	8.6	9.5	10.1
3	8	5.2	5.9	6.4
4	16	4.0	4.2	5.3
5	32	3.0	3.5	3.1

TABLE 68—*Growth and suppression in water series, 1924*

Cont.	Cult.	June 12				June 18				June 23				June 30			
		Number dom.	Number sup.	Ave. no. lvs.	Ave ht. cm.	Number dom.	Number sup.	Ave. no. lvs.	Ave ht. cm.	Number dom.	Number sup.	Ave. no. lvs.	Ave ht. cm.	Number dom.	Number sup.	Ave. no. lvs.	Ave ht. cm.
1	2's	2	..	2.4	8.2	2	..	5.8	24	..	..	..	...	..	..	11	76.5
6		2	..	2.4	9.5	2	..	6.0	24	1	1	11	41.5	..	..	...	...
2	4's	4	..	2.6	11.0	2	..	6.2	25.5	..	..	..	...	..	..	10.6	77.7
7		4	..	2.4	9.5	2	..	6.0	28.7	..	1	8.4	50.2	..	..	...	...
3	8's	3	2	2.4	9.6	3	2	5.2	27.7	3	2	7.4	46.6	..	..	...	...
8		1	3	2.4	8.5	2	3	5.2	27.5	..	...	..	...	2	4	8.0	69.8
4	16's	4	3	2.4	10.5	6	6	5.4	31.7	7	6	6.3	47.4	..	..	...	...
9		3	3	2.6	12.6	3	5	5.8	36	..	..	..	...	3	7	7.2	66.2
5	32's	4	7	2.4	13.1	3	14	4.2	32.6	3	13	4.4	41.8	..	..	...	...
10		..	9	2.4	11.6	..	14	4.0	28.9	..	..	..	...	1	14	4.4	46.6

The difference between the values for the 2's and the 32's was great and fairly constant at all the depths, the latter being little more than a fourth. The results exhibited the usual increase of water-content with depth, except in the 2's and 32's at the second foot.

TABLE 70—*Development in Helianthus, first set, June 23*

Criteria	2's	4's	8's	16's	32's
Ave. ht. cm.......	41.5	50.2	46.6	47.4	41.8
Ave. diam. stem, mm.	8.0	8.0	5.2	5.2	4.2
Ave. length largest lvs. cm..........	13.2	13.5	11.0	9.2	7.8
Ave. width largest lvs. cm..........	9.7	10.5	7.8	5.9	4.9
Ave. no. full-grown lvs.	10.0	7.4	6.8	5.8	4.0
Ave. no. half-grown lvs.	2.0	2.0	1.2	1.0	0.9
Ave. ht. 1st pair lvs. cm..........	17.5	17.0	22.0	25.7	30.0
Ave. leaf-area both sides, sq. in.....	100.4	81.6	56.3	26.2	14.7
Ave. dry wt. tops, gm.	2.58	2.67	1.68	1.12	0.7

The plants in the 2's and 4's were about equal in most respects, the latter being a fifth taller and the former having a fifth larger leaf-area, while the dry weights corresponded closely. The 8's and the 16's were intermediate in every case between these and the 32's, the latter exhibiting the minimum for all criteria except the height of the first pair of leaves, the deepest shade producing the maximum elongation. The stem diameter was nearly twice as great in the 2's as the 32's, the number of leaves more than twice, the leaf-area 7 times and the dry weight 3.5 times greater.

Final results for second set—The plants of this set were removed on June 30, when they were a little more than a month old and ranged from a half to three-fourths of a meter tall. The individuals of the 16's and 32's were growing in an inadequate chresard as shown by their wilting during the day. The chresard at the various levels and the nitrate-content for the mixed sample are shown in table 71.

TABLE 71—*Chresard and nitrate-content on June 30*

Cont. no.	Number plants	Chresard Depth			Nitrate-content
		0′–6′	6′–12′	12′–24′	
		%	%	%	*ppm.*
1	2	8.4	11.5	11.9	64.9
2	4	5.0	4.6	5.5	58.3
8	8	4.7	5.1	4.6	47.7
9	16	3.7	3.7	2.9	38.6
10	32	1.4	3.0	2.6	33.6

As in the previous series the chresard decreased progressively with the density, but the correspondence with depth was less consistent. The values for the 32's were so low as to indicate severe competition for water throughout the 24 hours, while in the case of the next three containers the chresard was inadequate only during the high transpiration of the day. However, the close agreement between the final results for the 2's and 4's suggests that the chresard must have been considerably higher for the latter during much of the period of growth. The nitrate-content likewise decreased with the density, the value for the 32's being practically half that for the 2's. The respective chresards ranged from 1.4%–3% and from 8.4%–11.9%, with averages of 10.6% and 2.3%. The 2's had in consequence more than four times as much available water at their disposal, and this must have been far more important in deciding the outcome of competition than the difference in nitrate-content.

TABLE 72—*Development in Helianthus, second set, June 30*

Criteria	2's	4's	8's	16's	32's
Ave. ht. cm.......	76.5	77.7	69.8	66.2	46.0
Ave. diam. stem, mm.	9.8	8.6	6.8	5.6	4.0
Ave. length largest lvs. cm.........	15.3	15.4	11.2	9.8	7.8
Ave. width largest lvs. cm.........	11.7	11.7	8.2	6.4	4.8
Ave. no. full-grown lvs.	11.0	10.2	7.6	6.6	4.0
Ave. no. half-grown lvs.......	....	1.0	1.0	1.2	1.0
Ave. ht. 1st. pair lvs. cm..........	20.0	17.8	23.5	27.2	27.0
Ave. leaf-area both sides, sq. in.....	288.03	265.54	129.54	73.32	44.56
Ave. dry. wt. tops, gm.	5.245	5.325	2.787	1.861	0.995

The 2's and 4's were in even closer correspondence than in the first series, the latter having felt the effect of competition but little up to this time. This was not true of the 8's and 16's which in every case but that of the height of the first leaves fell off rapidly toward the minimum in the 32's. The stem diameter was now more than twice as great in the 2's than the 32's, the number of leaves nearly three times, the leaf-area about 7 times and the dry weight more than 5 times. In both series the relation of leaf-areas to dry weight was similar and corresponded much more nearly to the difference in chresard between the two extremes than to that of nitrate-content. While this was about twice as great for the 2's, the chresard in the upper soil level was nearly 7 times greater and the average almost 5 times greater.

Final results for third set—The cultures of sunflowers in wooden pails was terminated on June 19 when the plants were three weeks old, as the 32's gave indications of wilting at this time. The roots of the 4's had barely reached the bottom of the pail and there were but few in the lower half, while in the 8's they were well distributed over the bottom. This was true likewise of the 16's, but the soil was still moist; on the contrary the roots of the 32's were exceedingly numerous throughout and had reduced the chresard to a very low point.

TABLE 73—*Development of Helianthus, third set, June 19*

Criteria	4's	8's	16's	32's
Ave. ht. cm.	25.7	25.7	27.1	23.2
Ave. diam. stem, mm.	5.0	3.8	3.9	2.9
Ave. leaf-area both sides, sq. in.	65.4	41.6	33.4	19.6
Ave. dry wt. gm.	0.86	0.68	0.47	0.3

The results were essentially consistent and in good correspondence with the other sets; height of stem followed the rule in reaching a maximum in an intermediate density. The 32's were much the lowest in all the other values, exhibiting approximately a third of the leaf-area and dry weight of the 2's. The rate of decrease through the four cultures was fairly regular likewise.

COMPETITION PRIMARILY FOR NUTRIENTS

Growth and suppression—In all the cultures of this set, the chresard was maintained approximately at 12.5%, water being added to counterbalance the transpiration, and hence the competition was chiefly for nutrients. The number of dominant and suppressed plants, the height and the number of leaves were determined at three dates during the experiment and are recorded in the following table:

TABLE 74—*Growth and suppression, nutrient series, 1924*

Cont.	Cult.	June 12				June 18				July 1			
		Number		Ave. no. lvs.	Ave. ht. cm.	Number		Ave. no. lvs.	Ave. ht. cm.	Number		Ave. no. lvs.	Ave. ht. cm.
		dom.	sup.			dom.	sup.			dom.	sup.		
11	2's	2	0	2.4	8.8	2	0	5.9	24	2	0	11	77.0
12	4's	4	0	2.5	10.2	4	0	6.2	26.5	4	0	12	87.5
13	8's	8	0	2.5	8.6	6	2	5.6	24.1	6	2	10	75.4
14	16's	15	1	2.2	10.0	11	5	4.6	32.1	12	4	8	80.2
15	32's	27	5	2.5	13.4	24	8	4.8	38.0	23	9	7.6	71.4

It is evident from the results that the number of suppressed individuals increases with the density of planting, and likewise with the age of the culture, increasing demand and keener competition being the explanation in both cases. The most significant fact is that a little more than a fourth of the plants were suppressed in competition for nutrients when the water-content was maintained, while nearly a half were suppressed in competition for water. The greatest height growth occurred in the dense 32's at the outset, but the increasing competition caused these to lag behind the intermediate cultures, though the results were not as consistent as usual. The number of leaves was greatest in the 4's, closely similar in the 2's and 8's, and least in the 16's and 32's.

By contrast with the water-competition series, there were 2–3 more leaves per plant in the nutrient-competition series. The plants in the latter were 10–35 cm. taller than in the former and were also more uniform in height, both facts indicating that a reduced supply of nutrients is less serious than an inadequate water-content.

Final results for nutrient series—This experiment was concluded on July 1 when the plants were somewhat more than a month old. An examination of the soil showed that all the cultures contained an adequate supply of water, but that the nitrate-content had fallen from 35.2 ppm. for the 2's to 18.7 ppm. for the 32's.

TABLE 75—*Development in nutrient-competition series, July 1*

Criteria	2's	4's	8's	16's	32's
Ave. ht. cm.......	77	87.5	75.4	80.2	71.4
Ave. diam. stem, mm.	10	9.9	8.1	7.2	6.2
Ave. length largest lvs. cm.........	16.5	17.5	13.6	10.6	10.3
Ave. width largest lvs. cm........	13.0	12.9	9.6	7.5	6.3
Ave. no. full-grown lvs.	13	11.4	8.2	7.8	6.6
Ave. no. half-grown lvs.......	0	1.4	1.8	1.6	1.0
Ave. ht. 1st pair lvs. cm........	15.0	13.6	17.5	30.1	29.8
Ave. leaf-area both sides, sq. in.....	341.2	339.6	176.0	116.2	90.2
Ave. dry wt. tops, gm.	7.0	7.41	3.60	2.38	2.05

The results were essentially consistent, the values decreasing fairly regularly with the density, except in the case of stem height, which is a resultant of insolation and degree of shading. The values for stem diameter, width of largest leaf and number of leaves were approximately twice as great for the 2's as the 32's, the intermediates decreasing regularly toward the latter. The leaf-area and dry weight were more than thrice larger for the 2's and 4's than for the 32's, the 8's and 16's being intermediate. It is significant of the relative importance of water and nutrients in competition that the maximum dry weight with adequate water-content was one-half to thrice greater than in either series with inadequate chresard, and the minimum two to three times greater, while the average dry weight for plants of all cultures was 4.7 gm. by comparison with 3.2 and 1.6 gm.

Competition Primarily for Light

Growth and suppression—In this series the containers were weighed frequently to determine the water lost by transpiration and the corresponding amounts replaced by tap-water. In addition, Knop's nutrient solution was added at the same time to the denser cultures but less frequently to the more open ones. As a consequence both the water-content and the nitrate-content were maintained at an adequate level, as was demonstrated at the close of the experiment. At this time the holard was optimum in all, while the nitrate values from the 2's to the 32's were respectively 39.1, 42.2, 47.6, 38.4 and 40.5 ppm. Hence, it appears certain that little or no competition for water or nitrates took place in any of the containers.

As in the preceding series, the course of competition was followed by determining the number of dominant and suppressed individuals, the number of leaves and the height of stems on several occasions, as shown by the table below.

Table 76—*Growth and suppression in the light series, 1924*

Cont.	Cult.	June 12				June 18				July 1			
		Number		Ave. no. lvs.	Ave. ht. cm.	Number		Ave. no. lvs.	Ave. ht. cm.	Number		Ave. no. lvs.	Ave. ht. cm.
		dom.	sup.			dom.	sup.			dom.	sup.		
16	2's	2	0	3	12	2	0	6.7	33	2	0	13	86
17	4's	4	0	2.7	10.6	4	0	6.7	28.5	4	0	12.4	88
18	8's	7	1	2.5	9.6	8	0	6	32	7	1	9.4	88
19	16's	14	2	2.5	13.2	13	3	5.0	39.4	11	5	8.6	81.3
20	32's	24	8	2.5	14.4	17	15	4.2	37.2	12	20	7.2	64.6

The results are similar to those obtained in the series that precedes, except that the number of suppressed plants is much higher for the 32's and the height considerably less. On the other hand, the height in the other cultures is a little greater than in the preceding, due probably to the abundant water, while it is much greater than when the competition was chiefly for water, indicating that light is secondary to water-content in importance.

Final results for light series—This series was likewise terminated on July 1, when the plants were cut off at the soil level and carefully measured. An examination of the soil in each container showed that the water-content was abundant; the roots penetrated to the bottom of each and became more numerous from the 2's to the 32's, where they ran along the bottom for several inches. The roots were somewhat more branched than in the first series, where the low water-content reduced growth. The nitrate-content at the close was fairly uniform throughout.

TABLE 77—*Development in light-competition series, July 1*

Criteria	2's	4's	8's	16's	32's
Ave. ht. cm.......	86	88	80.2	81.3	66.4
Ave. diam. stem, mm.	11.5	9.7	7.8	7.7	6.1
Ave. length largest lvs. cm.........	19.0	14.7	11.6	11.4	9.3
Ave. width largest lvs. cm..........	15.0	12.1	8.0	7.9	6.2
Ave. no. full-grown lvs.	11.4	12.0	7.6	7.6	6.8
Ave. no. half-grown lvs.......	2.0	1.4	1.2	1.2	1.2
Ave. ht. 1st pair lvs. cm.	20.7	17.6	26.7	29.7	30.6
Ave. leaf-area both sides, sq. in.....	442.5	295.2	210.4	125.8	75.6
Ave. dry wt. tops, gm.	9.27	5.96	4.56	2.56	1.61

With two slight exceptions in the 4's, the values were greatest in the least dense 2's and decreased more or less regularly to the 32's. The sole exception to this occurred in the height of the first pair of leaves. This was a consequence of the earlier reduction of light in the densest culture but the final height fell behind owing to the increasing severity of competition. The leaf dimensions were twice as great in the 2's as the 32's and the number of full-grown leaves nearly twice as many also. The leaf-area and dry weight were more than five times greater for the 2's than for the 32's, and they were closely correlated throughout the sequence. The order was 100:67:48:29:17 for the former, and 100:64:49:28:17 for the latter.

The dry weight of the 2's was a third greater than in the preceding series where the water-content alone was maintained and approximately twice to four times greater than in the two sets where the water-content was inadequate, while for the 32's it was less in the one case and about half more in the other two. Furthermore, the average dry weight for all cultures was only a little higher than in the preceding series, again indicating the much greater importance of water.

SERIES WITH ARTIFICIAL SHADE

Growth and suppression—In this the water-content and nutrients were maintained as in the preceding series, but in addition the four less dense cultures were shaded artificially in such a manner as to receive no more light than the 32's. This was accomplished by employing artificial leaves cut out of cardboard and placed on stakes in such a way as to cast a shade approximating that in the densest culture. As the plants grew, these were raised as well as increased in size, this being done six times during the period of the experiment. Water and nutrient solution were likewise added at five intervals in order to maintain the supply of each at an optimum. The course of development in terms of suppression, number of leaves and height of stem is indicated in the following table.

TABLE 78—*Growth and suppression in shaded series, 1924*

Cont.	Cult.	June 12				June 18				June 29			
		Number		Ave. no. lvs.	Ave. ht. cm.	Number		Ave. no. lvs.	Ave. ht. cm.	Number		Ave. no. lvs.	Ave. ht. cm.
		dom.	sup.			dom.	sup.			dom.	sup.		
21	2's	2	0	2.5	9.7	0	0	5.0	33	0	..	8.5	82.5
22	4's	4	0	2.7	12.7	3	1	6.4	35.4	0	..	10.0	84.5
23	8's	2	2	2.5	11.0	2	3	5.0	30.3	0	..	8.0	83.2
24	16's	2	1	2.3	12.9	2	6	4.9	34.2	..	..	6.8	72.2
25	32's	1	7	2.3	13	2	14	4.7	35	..	..	6.8	71.5

Since the three primary factors were all more or less equalized, the differences were less than in the preceding series, the 4's being tallest and bearing the most leaves, and the 32's the smallest. Suppression was somewhat more marked, owing to the decreased light intensity.

Final results for shaded series—This experiment was discontinued on June 28, when the plants were a month old. The holard was adequate in all containers at this time and was doubtless such throughout the period. In the 2's and 4's a few roots had reached the bottom of the containers; in the 8's and 16's they were abundant throughout and in the 32's the bottom itself was well covered with roots.

TABLE 79—*Development in shaded series, June 28*

Criteria	2's	4's	8's	16's	32's
Ave. ht. cm.	82.5	84.5	83.2	72.2	71.5
Ave. diam. stem, mm.	9	8.6	8.8	6.0	5.9
Ave. length largest lvs. cm.	14	11.9	12.1	10.6	9.7
Ave. width largest lvs. cm.	11.2	8.7	8.6	7.3	6.5
Ave. no. full-grown lvs.	8.5	10.0	8.0	6.8	6.8
Ave. no. half-grown lvs.	2.0	1.0	1.5	1.1	1.4
Ave. ht. 1st pair lvs. cm.	25.0	32.8	28.2	31.3	29.6
Ave. leaf-area both sides, sq. in.	171.5	132.2	128.9	94.7	79.7
Ave. dry wt. tops, gm.	3.25	2.82	2.43	1.75	1.43

The extent to which the artificial leaves equalized conditions and in many cases reduced the values is evident. The height of stem was much the same for the 2's, 4's and 8's, and for the 16's and 32's, and there was no great or consistent difference in the height to the first pair of leaves. Comparable results were exhibited by the length of largest leaf and by the number of leaves. As to leaf-area and dry weight, the values for the 32's were similar to those in the preceding series, but they were greatly reduced in the three lower densities. Instead of being more than five times larger for the 2's, they were little more than twice. The agreement between the series as to the 32's and the marked reduction for the other cultures was the obvious result of the additional shading given the latter, and demonstrates the importance of light as a factor in competition. The correspondence between leaf-area and dry weight was nearly as close as in the preceding series, the respective sequences being 100:77:75:55:46, and 100:87:75:54:44.

Excess-Deficit Series

Growth—Six containers were employed in this series, three to provide a sequence in holard and three for nutrients. In the former the soil in one container was wet, in the second with normal water-content and in the third it was dry; in the latter the nutrients were present in excess, in normal amount, and deficient. Eight plants were grown in each container, but early in the experiment the +N plants met with an accident and were abandoned.

Table 80—*Growth in excess-deficit series, 1924*

Container	June 16		June 26		July 1	
	Ave. no. lvs.	Ave. ht. cm.	Ave. no. lvs.	Ave. ht. cm.	Ave. no. lvs.	Ave. ht. cm.
nW	4.5	15.9	7.4	53.3	10.6	76.5
−W	2.4	7.7	5.0	19.5	6.4	26.9
+W	4.2	18.1	8.8	64.1	11.4	81.0
nN	4.2	16.1	7.2	47.6	10.2	71.1
−N	4.0	15.4	6.0	42.8	7.8	59.7

The plants growing in dry soil were at first a half and then a third as tall as those in normal water-content, while those in wet soil were slightly taller than the latter. As to number of leaves, the +W culture was about 10% better than the normal and almost twice more than in the —W one. The plants with normal nutrient were about a fifth taller than the —N and the number of leaves a third greater, the differences in both series increasing with age.

Final results for excess-deficit series—This series was terminated on July 1, when the roots of the nW plants were found to reach the bottom of the container, while those of the —W were 8″ shorter. The latter were darker in color and more pubescent than the nW; they were uniform in size and after wilting became turgid again very readily. In the nutrient series the soil-water was ample and the roots reached the bottom in all.

TABLE 81—*Development in excess-deficit series, 1924*

Criteria	nW	—W	+W	nN	—N
Ave. ht. cm.	76.5	26.9	81.0	71.1	59.7
Ave. diam. stem, mm	9.6	3.3	9.5	9.1	6.1
Ave. length largest lvs. cm.	14.0	8.1	13.6	14.6	11.7
Ave. width largest lvs. cm.	11.2	4.0	10.9	11.4	7.7
Ave. no. full-grown lvs.	9.6	6.2	10.4	9.4	6.8
Ave. no. half-grown lvs.	2.0	0.6	2.0	1.6	1.8
Ave. ht. 1st pair lvs. cm.	21.1	14.4	23.5	23.3	26.2
Ave. leaf-area both sides, sq. in.	225.4	26.4	229.8	174.0	120.4
Ave. dry wt. tops, gm.	5.39	0.59	5.46	6.01	2.52

The most significant results are found in the differences between the nW and the —W plants, for which the respective chresards were 12.3% and 2.6%, giving a ratio of 5:1. The values for the +W were slightly larger as a rule than for the normal, but the small difference suggests that a lack of soil-air exerted an inhibiting action. The stem dimensions were little more than a third as great in the —W plants, and this was true also of leaf width. The discrepancy in dry weight was even more striking, being only 11% of that for the nW plants, while for the leaf-area it was but 12%, the two values being in close correlation. In the nutrient cultures, the dry weight of the nN plants was more than twice as great as in the —N and the leaf-area about one-half greater. The water-content was practically the same for both, but the former had 24 ppm. to 8.6 ppm. for the —N culture. The differences between the dry weight and leaf-area in the two sets are eloquent of the much greater significance of water-content in competition.

SUMMARY FOR COMPETITION SERIES, 1924

The best comparison of the effects of water, nutrient and light in relation to competition is afforded by the leaf-area and dry weight of the plants in the respective cultures. In all series these values rise more or less regularly from the 32's to the 2's, in accordance with decreasing density. The primary significance lies in the relative position of the three curves. The lowest values throughout were obtained when the chresard was inadequate, a deficit in nutrients being distinctly less serious. When water and nutrients were sufficient and competition was for light alone, the values were the highest in general, but lower in the 32's where the light effect was most marked. On the other hand, the effect of increasing the shade for all but the dense 32's was to produce a striking fall in values, especially noticeable in the less dense cultures. The curves and their position were in close correspondence for both leaf-area and dry weight, and confirmed the conclusion as to the essential connection between these two criteria.

SERIES OF 1925

Installation—The method and rate of planting were the same as for the preceding year, but the series was slightly modified as follows:

I. Competition for water and nutrients.
II. Competition for water, nutrients maintained (N+).
III. Competition for nutrients, water-content maintained (W+).
IV. Shaded series, water-content and nutrients maintained (WN+).
V. Light, water-content and nutrients maintained.

Two duplicate sets of series II and V were installed, but these contained only three cultures each, namely, 2's, 8's and 32's. A detailed record was kept of the amount of water and nutrient solution applied to all the containers of the various series at intervals of 4–5 days throughout the course of the experiment. The amounts were determined by weighing the containers, for example, 3.25 lbs. of water were added to all the containers in series I on June 25, because the 2-plant culture was this amount below the original weight. In the case of series III where the water supply was maintained, each container was given the amount of water necessary to restore the original weight. The much greater quantities needed by the denser cultures of the W+ series were striking in comparison with those required by the less dense ones. Moreover, the 8's, 16's and 32's in the W+ series used 50%–100% more water than in those series where the water supply was not maintained at an optimum. There was naturally much less water used by the 2's and 4's in the shaded series (IV) than in the unshaded (V), which served as a check on this one; the respective values for the 2's were 6.75 and 22.25 lbs., and for the 4's 12.5 and 20 lbs.

Water-content—The holard at the outset was 19.3% and the hygroscopic coefficient 6.3%. The holard was again determined for one set each of the duplicates at the several soil levels when they were removed on June 27 and 29, for the upper 2′ in the case of series I–III and for the three levels in series IV–V on July 2. In practically all cases the water-content decreased with the density and increased with the depth. Although wilting occurred frequently in the denser cultures toward the close, in none of the containers, even the 32's of the water-competition series, did the holard fall as low as the hygroscopic coefficient. The 16's and 32's of these series exhibited the effects of an inadequate water supply by poorer growth and more frequent wilting than in the other series, and this corresponded with a deficit of about 4%–5% in comparison with the 2's. It was evident that water could not be absorbed with sufficient rapidity to prevent wilting when the holard was 10% or less and the chresard about 3%. The effect of shade upon the water-content was striking, the values in the shaded series being sometimes twice as great as in the corresponding check (V). However, they were very uniform in the latter with respect to both density and depth, while they dropped rapidly from the 2's to the 8's in the former.

Nitrate-content—The original amount in each container was 25 ppm., and the control without plants gave 24.1 ppm. at the close of the series. As in earlier studies, the amount fell off with increasing density, the value for the 32's on July being one-half to a fifth that for the 2's. The lowest amounts were naturally found in the two nutrient-competition series, the minimum occurring in the third set, where the values were 10.3 ppm. for the 2's, 5 for the 4's and 2.2 for the 32's. On the other hand, the maximum was found in the shaded series, owing to the lessened use of water and the addition of nutrient solution. Contrasted with this was the behavior in series V, where all factors were at a fair optimum and the utilization of the nitrates greatest, as well as much more uniform.

Light—The light intensity decreased regularly from the 2's to the 32's. The differences between the several densities were smaller in the shaded series in view of the fact that the artificial leaves were continually adjusted to give values approaching those of the 32's. On June 19 the respective values from the least to the most dense were 7.5%, 6%, 5.6%, 4.2%, and 3.9%, and on June 27 they corresponded closely, the extremes being 6% and 3.6%. The efficiency of the artificial shading becomes more apparent when the light values are compared with those in the unshaded series. On June 25 the light near the top of the plants in the shaded series ranged from 9.4%–6.7% in 2's and 32's respectively in contrast to 27.1% and 6.3% in the unshaded one. The last values also bring out clearly the fourfold reduction of light with density, even near the top of the cultures. As the starch tests showed, this greatly reduced the rate of food-making, with a consequent reduction in the growth of all parts and in the dry weight (fig. 22).

Associated with the artificial shading went a marked reduction in the transpiration, especially in the 2's and 4's. The former lost 34 lbs. of water in the unshaded to 7.5 lbs. in the shaded series, and the respective losses for the 4's were 31 and 20.2 lbs.

Development on June 13—The first determinations on the several series were made on June 13 when the plants were 11 days old. At this time differences were already becoming apparent in the plants of the various densities as well as between those of the series, as the following table indicates. Throughout tables and discussion, competition for a factor is indicated by the minus sign, e.g., W—, N—, and the addition of water or nutrients by the plus sign, e.g., W+, WN+.

It is evident that competition was not yet intense in any of the series, but its effects were beginning to show in the first one where water and nutrients were at the minimum. The difference between the extremes was greatest in this set and least in the shaded series; the effect of shading was shown most clearly in the elongation of the hypocotyl, which was

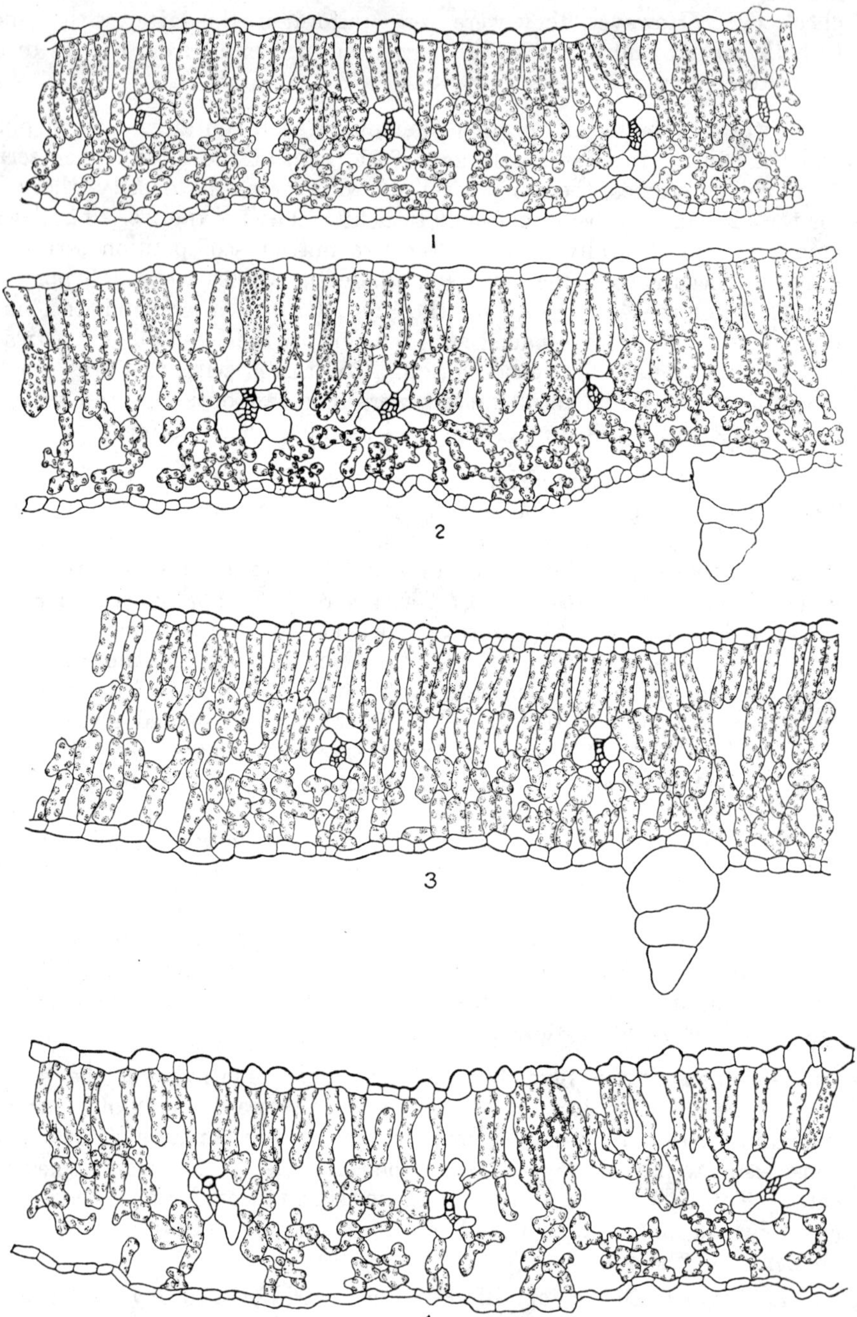

Fig. 22.—Structure of leaf of greenhouse sunflowers, 1925: (1) 8's, water and nutrients minus; (2) 8's, water and nutrients plus; (3) 32's, water aand nutrients minus; (4) 32's, water and nutrients plus.

distinctly greater than in the other two sets. Suppression had begun only in the densest culture of the WN— series, thus furnishing further evidence of the lack of serious competition at this stage.

TABLE 82—*Growth in control series, June 13*

Criteria	I. Competition for water & nutrients					IV. Shaded series, WN+					V. Unshaded series, WN+				
	2's	4's	8's	16's	32's	2's	4's	8's	16's	32's	2's	4's	8's	16's	32's
Ave. ht. cm. ..	7	6	7	8.4	11.8	9.5	9.9	9.6	9.8	11.7	10	7.4	9.2	9.9	10.8
Ht. cotyls, cm..	5	3.9	4.8	5.9	7.5	7.2	6.7	7.2	6.2	7.8	6.2	5.4	6.6	6.6	7.6
Diam. stem, mm.	3	3	3	3	2.8	3	3	3	3	3	3	3	3	3	3
No. leaves	2.3	2.1	2.2	2.2	2.1	2.2	2.2	2.2	2.2	2.2	2.5	2.3	2.2	2.1	2.1
Ave. size 1st pair leaves..	6 × 3.3	6.4 × 3.5	6.9 × 3.7	6.7 × 3.9	5.8 × 3.2	6.5 × 3.7	7.1 × 3.7	6.1 × 3.6	6.1 × 3.5	6.7 × 3.8	7.5 × 4.3	7.1 × 3.8	6.8 × 3.5	6.4 × 3.5	6.4 × 3.6
No. suppressed plants	0	0	0	0	1	0	0	0	0	0	0	0	0	0	0

Development on June 20—The second measurement in the series of control cultures of *Helianthus* were made on June 20, when all five sets were taken into account, with the results shown in table 83, p. 250.

In nearly all the containers the height of the plants had about trebled since June 13. In the 2's, the shaded series yielded the tallest plants with an average of 28.5 cm., in consequence of the elongation in reduced light intensity, and the unshaded series in which water and nutrients were at an optimum followed closely with a value of 27 cm. By contrast the minimum height of 20.5 cm. occurred in the first series where the competition was for both water and nutrients. Among the denser plantings the tallest plants as a rule were to be found in the series with the water supply maintained, but especially in the WN+. As in previous experiments, the stature was usually greatest in the 16's, then in the 32's and least in the 2's. When there was competition for water, nutrients or both, the stem diameter was greatest in the 4's, but in the two WN+ series, this was equalled or exceeded by the 2's. The minimum diameter occurred regularly in the 32's.

The leaves were largest in the W+ and the unshaded WN+ series, where conditions were the best; the relative value of water and nutrients is shown by the fact that the size in the N+ series was distinctly less and but little more than under competition for both water and nutrients. Leaf-size fell off with much regularity from the 2's to the 32's, the former being three times as large as in the W+ series. The lowest values occurred in the shaded series, in spite of the addition of water and nutrients; but they were also much more uniform, there being little difference between the 2's and 32's. This last effect was to be expected in view of the reduction of the intensity to that of the 32's by means of the artificial leaves, and the total result was significant of the control exerted by deep shade.

TABLE 83—*Growth in the five control series, June 20*

Criteria	I. Competition for water and nutrients					II. Competition for water, N+					III. Competition for nutrients, W+				
	2's	4's	8's	16's	32's	2's	4's	8's	16's	32's	2's	4's	8's	16's	32's
Ave. ht. cm.......	20.5	25.3	32.5	38.9	31.0	24.5	30.0	34.0	34.0	35.0	24.0	29.0	34.3	36.2	34.2
Length 1st internode, cm........	9.0	11.4	15.5	21.8	17.6	13	13	18	20	21.5	11.0	12.6	18.3	20.6	20.4
Diam. stem, cm...	5.5	6.0	6.0	5.5	4.0	5.5	6.5	5.8	5.5	5.0	5.7	6.0	5.5	5.5	5.0
No. leaves........	6.2	5.0	5.0	4.7	4.0	5	5.	5	5	4.5	4.8	5	5	4.5	4.0
Ave. size 2nd pr. lvs. cm.........	10.5 × 6.3	10 × 7.3	9.8 × 7.0	10 × 5.7	7.9 × 5.0	10 × 7	11.6 × 7.6	10.2 × 6.3	10.6 × 5.6	7.2 × 0.5	13 × 9.5	10.7 × 7.6	10 × 6.5	9.2 × 6.1	7.6 × 4.7
No. sup. plants...	0	0	0	0	6	0	0	0	2	6	0	0	0	1	6

Criteria	IV. Shaded series WN+					V. Unshaded series, WN+				
	2's	4's	8's	16's	32's	2's	4's	8's	16's	32's
Ave. ht. cm.......	28.5	30.7	34.0	35.3	35.0	27.0	25.0	32.1	37.9	36.7
Length 1st internode, cm........	17.0	13.5	17.4	19.1	20.5	9	11.5	16.3	21.5	22.4
Diam. stem, cm...	6	6	6	5.5	6.0	6.5	6.0	6.0	6.0	5.0
No. leaves........	4.5	5	5	4.5	4.2	6.3	6.2	6.0	5.0	5.0
Ave. size 2nd pr. lvs. cm.........	9.5 × 6.5	11.0 × 6.5	9.4 × 6.6	9.6 × 6.0	9.4 × 5.9	12.5 × 9.5	12.0 × 7.2	9.8 × 7.0	9.6 × 6.1	8.1 × 4.9
No. sup. plants...	0	0	0	2	5	0	0	0	1	10

Suppressed plants were present only in the 16's and 32's at this time, the number being greatest under competition for water, and least under that for water and nutrients where the growth in height was least and light values correspondingly higher.

Development on June 26—Six days after the previous examination, the plants in the 2's, 4's, 8's and 16's for all series had practically doubled in height, while the 32's had fallen off markedly in rate of increase, owing

TABLE 84—*Growth in the five control series, June 26*

Criteria	I. Competition for water and nutrients					II. Competition for water, N+					III. Competition for nutrients, W+				
	2's	4's	8's	16's	32's	2's	4's	8's	16's	32's	2's	4's	8's	16's	32's
Ave. ht. cm.	48.0	54	65	57.5	35.5	53.0	66.7	67.0	60.7	52.7	52	67.5	70	66.8	57.1
Length 1st inter-node, cm.	15	23.5	28.3	21.6	9.5	21	26	32.5	25.8	18.4	23.5	28	29	29.6	20.7
Diam. stem, cm.	7.5	8	7.8	6.8	4	9	9.1	7.3	6.5	5	9.5	9	7.5	7.1	4.8
No. leaves	9.5	7	7	5	3.5	8	7	7	6	4.5	9	8.3	8	7	4.7
Ave. size 2nd pr. lvs. cm.	23.5 × 9.5	14 × 10	10 × 8.7	8 × 6.6	5 × 2.3	14 × 12	13 × 10	11.5 × 7.6	9.9 × 6.3	6.5 × 4.3	15.5 × 14	13.3 × 10.5	12 × 8.5	8.5 × 5.8	7.1 × 4.2
No. sup. plants	0	0	0	3	16	0	0	0	3	13	0	0	0	3	8

Criteria	IV. Shaded series, WN+					V. Unshaded series, WN+				
	2's	4's	8's	16's	32's	2's	4's	8's	16's	32's
Ave. ht. cm.	47	58.7	59.2	60	57.3	57	60	63.3	69.2	55
Length 1st inter-node, cm.	18	25	23.8	22	20.5	15	25.7	27.1	27.6	19.5
Diam. stem, cm.	7.2	9	7.2	7	6.6	9.2	9	8.5	7.4	7.1
No. leaves	6.3	6.5	6	5	6	10	9	7.7	6.4	5.7
Ave. size 2nd pr. lvs. cm.	10 × 8.5	12.5 × 9	11 × 9	9 × 6	8.5 × 5.5	16.5 × 12	12.7 × 9.5	9.2 × 7.2	10.7 × 8	7.5 × 4.6
No. sup. plants	0	0	0	3	4	0	0	1	1	10

to the fact that the intensity of competition was also increasing more rapidly. As a consequence, the stature was now least in the two extremes, the 32's being usually a little shorter than the 2's, and generally greatest in either the intermediate 8's or 16's. The lowest values occurred under competition for both water and nutrients, where the range for 2's, 8's and 32's was 48, 65 and 35.5 cm., and the highest under competition for nutrients but with optimum water, where the heights were respectively 52, 70 and

57 cm. The stem diameter decreased regularly from the 2's or 4's to the 32's, being little more than a half as great in the latter under the severest competition, namely, that for both water and nutrients. The lowest values throughout occurred in this series, and the highest where the primary factors were near the optimum, *i.e.*, in the unshaded series with added water and nutrients.

The number of leaves was greatest in the 2's and least in the 32's, those of the latter being about half as many in three series, about a third as numerous in the water-nutrient competition, and essentially the same as for all the other cultures in the case of the shaded series with its equalizing effect. The size of leaf also decreased consistently with the density in accordance with the rule, the greatest difference between extremes occurring where competition for water was a factor. The poorest and most uniform development took place in the shaded series, but the series were in general less consistent than usual.

Suppression was greatest in series I and II where water was inadequate, and least in the shaded series with its uniform light. With a single exception, all the suppressed plants were found in the 16's and 32's, the number for the latter rising to 4–10 times greater than for the former.

Final results for the five control series, July 2—The experiment was terminated on this date because the 32's showed repeated wilting in the morning, in spite of recent watering, and the color of the other cultures testified to the general limiting effect of severe competition. The roots had also reached the bottom of the containers and other factors would soon enter to obscure the action of competition.

With respect to height of stem, the lowest general level and average growth and the lowest values for the densest plots, the 16's and 32's, where competition was most severe and hence most decisive, were found in series I, where the competition was for both water and nutrients. Here the average stature was 73 cm., in comparison with 77.4 cm. for the next lowest, the shaded series with water and nutrients plus. The highest general level and average stature occurred in series III with deficient nutrients and optimum water, where the range was from 85 to 101 and 72, the average being 88 cm. The unshaded series with water and nutrients plus was very close with a range of 86 to 95 and 69, and an average of 86.3 cm. In the intense competition of the two densest, the 16's and 32's, the average was the same, namely, 79 cm., the addition of nutrients affording no compensation by comparison with that of water alone, and actually decreasing both the maximum and the average for the whole series. On the other hand, the addition of nutrients in series II gave a definite increase over series I.

The minimum average stem diameter was 8.1 mm. for the shaded series and 8.2 for series I with inadequate water and nutrients; the maximum occurred in series V with all three factors at the optimum, where it was 9.1 mm. The most rapid decrease and the minimum value took place in series I, in which the 32's were only 4.5 mm. in diameter; series II with

Criteria	I. Competition for water and nutrients					II. Competition for water, N+					III. Competition for nutrients, W+				
	2's	4's	8's	16's	32's	2's	4's	8's	16's	32's	2's	4's	8's	16's	32's
Ave. ht. cm......	81.0	82.0	87.9	69.9	45.2	86.0	97.0	85.4	72.1	56.2	85.0	101.2	96.6	86.3	71.9
Ave. diam. stem, mm.	11.7	9.2	8.4	7.1	4.5	11.0	10.4	8.1	6.8	5.3	11.0	10.4	9.6	7.7	6.0
Width largest leaf, cm........	15.5	11.1	9.7	7.0	4.6	15.5	12.4	8.7	7.1	5.5	17.7	12.1	9.6	7.9	5.8
Length largest leaf, cm........	19.2	12.7	12.2	10.4	7.7	17.5	15.5	12.6	10.4	8.3	19.5	15.4	12.5	11.3	9.4
Ave. no. dead lvs.	0	0.2	1.0	1.6	0.8	0	1.7	1.5	1.8	1.6	1.0	0.7	1.6	1.3	1.2
Ave. no. full-grown leaves	12.0	8.0	6.7	4.9	4.0	10.5	8.0	7.2	5.2	4.1	10.0	10.0	7.6	7.0	5.1
Ave. no. half-grown lvs......	2.5	2.2	1.7	2.3	1.5	3.0	3.0	1.9	0.9	1.8	3.0	2.2	2.7	1.9	1.9
Ave. leaf-area one side, sq. in.....	178.73	88.83	60.05	33.68	15.65	205.83	125.13	61.08	35.43	18.10	230.80	129.53	70.58	62.95	25.03
Ave. dry wt. gm.	9.0	5.25	3.73	1.97	0.82	9.97	7.23	3.82	2.10	1.23	11.42	8.69	4.83	2.59	1.50

Criteria	IV. Shaded series, WN+					V. Unshaded Series, WN+				
	2's	4's	8's	16's	32's	2's	4's	8's	16's	32's
Ave. ht. cm......	77.5	81.8	78.2	80.2	71.5	86.0	95.2	91.1	89.1	69.2
Ave. diam. stem, mm.	9.5	9.5	7.5	7.4	6.5	11.5	10.0	9.6	7.9	6.5
Width largest leaf, cm........	9.0	10.1	8.6	7.4	6.8	18.2	11.6	9.7	7.9	5.8
Length largest leaf, cm........	12.0	13.0	10.8	10.8	10.3	19.2	14.5	12.6	10.8	9.0
Ave. no. dead lvs.	1.5	0.7	1.0	1.5	1.3	2.0	1.5	2.2	1.3	1.3
Ave. no. full-grown leaves	8.5	9.5	8.0	6.2	5.4	13.0	11.2	7.5	6.5	5.5
Ave. no. half-grown lvs......	3.5	2.0	2.1	2.1	2.0	3.5	2.0	2.4	1.5	2.7
Ave. leaf-area one side, sq. in.....	76.15	86.80	59.13	45.00	31.88	248.80	101.78	73.18	52.68	25.47
Ave. dry wt. gm.	3.63	4.06	2.88	2.12	1.70	15.03	6.95	4.49	2.69	1.57

water minus came next with a value of 5.3 mm., while the highest values were obtained in IV and V with both water and nutrients plus.

The results as to the width and length of the largest leaves were comparable to those for stem diameter. The lowest values and the smallest range occurred in the shaded series, the next lowest values in series I with the maximum competition. The largest leaves were found in series V with the factors optimum and the next largest in III with water optimum but nutrients deficient. Minimum values for the 16's and 32's obtained in series I among the four in which competition was active. The results as to number of full-grown leaves were less decisive, but the average was lowest in the two series with water minus and highest in V with all factors optimum.

The average leaf-area for all the cultures of a series was naturally distinctly lower in the shaded series with its low photosynthetic activity, the value being 60 sq. in. The next lowest were found in series I with an average of 75 sq. in. and II with 89 sq. in., both with inadequate water; the maximum value obtained in series III with water plus but nutrients minus, where it was 103 sq. in. This was nearly matched by series V with water and nutrients plus, where it was 100 sq. in. The general sequence of dry weights through the series was much the same; the general level and the maximum were far the lowest in the shaded series on account of the marked reduction of photosynthesis. The value for the 2's was little more than a third of that for the series with water deficient and a fifth for the unshaded series with optimum factors. On the other hand, it was distinctly better for the 16's and 32's than in the first two series with water minus. The range for the shaded series was 4.06–1.70 gm., for the unshaded 15.03–1.57 gm., and for the series with both water and nutrients minus 9–0.82 gm. The averages for I and II in which lack of water was controlling were 4.15 and 4.87 gm., while in III with water plus and nutrients minus it rose to 5.80 gm. It was 6.14 in the unshaded and 2.88 gm. in the shaded.

Summary of Competition Series, 1925

Quite apart from the regular and usually consistent decrease in all characters in relation to the density, the chief significance of the results from the five control series is found in the evidence furnished as to the relative importance of the three primary factors. In making comparisons, however, it must be kept in mind that the shaded series is not directly comparable with the others, as the major object was to demonstrate the effect of the degree of shade found in the 32's upon the other densities when not complicated by competition for water. The fundamental effect of a reduction in light intensity to this degree is revealed in practically all characters, but especially in the leaf-area and dry weight, where the values were a third to a fifth of those for the more favored series. However, with uniform shade and optimum water and nutrients, the greater demand for these two in the three higher densities brought about a consistent decrease in leaf-area and dry weight to less than a half of the maximum for the series.

The first three series were designed to furnish evidence as to the relative importance of water and nutrients, and the fifth to serve as a control by virtue of optimum conditions. Of the four, the lowest values were found regularly in series I with competition for water and nutrients, and the highest in V with both factors plus. The most significant evidence is afforded by the comparison of series II and III. With rare exceptions the values are higher when water is plus than when nutrients are adequate, and this is true throughout for the most important criteria, leaf-area and dry weight. An additional supply of nutrients brought about an increase of 13.5 sq. in. in the former over the average in series I, but the addition of water made an increase more than twice as great, namely, 27.6 sq. in. Strictly comparable results occurred with dry weights; the respective values for WN—, N+, and W+ were 4.15, 4.87 and 5.80 gm. The increase in series II was 0.72 gm. and in III 1.65 gm., the latter being also more than twice as great.

The importance of the two factors combined is revealed by a comparison of the data in series I and V, and particularly in leaf-area and dry weight again. The maxima for the two were respectively 179 sq. in. and 9 gm., and 249 sq. in. and 15 gm., while the corresponding averages were 75 and 100 sq. in. for leaf-area and 4.15 and 6.14 gm. for dry weight. When the averages are compared for series V with WN+ and III with W+, it is seen that water accounts for by far the largest share of the difference between the first series with competition for water and nutrients, and the last with these two factors maintained. When water alone was added, the leaf-area was 103 sq. in. and the dry weight 5.80 gm., while with both water and nutrients supplied the respective values were 100 sq. in. and 6.14 gm. In short, water alone was a little more effective in leaf growth, probably owing to a lower concentration of salts in the holard. The increase in dry weight over series I was practically 2 gm., of which approximately 1.65 gm. could be assigned to water and 0.34 gm. to nutrients.

It is interesting and probably significant as well of the general relation of density, to note that the 8's, which were four times as dense as the 2's and a fourth as dense as the 32's, yielded values in leaf-area and dry weight that were regularly between a third and a fourth of those for the 2's and three to four times as great as for the 32's. The only exception was furnished by the shaded series, in which the light was artificially equalized.

CONTROL EXPERIMENTS WITH XANTHIUM CANADENSE, 1925

Plan and method—The five series of competition experiments were similar to those of *Helianthus annuus* in all respects and the general treatment was also the same throughout, except that but three densities were employed, namely, 2's, 8's and 32's. As before, the use of water in the shaded and unshaded series was strikingly different, the 2's and 8's using respectively 7 and 19 lbs. in the one, and 9.25 and 25.5 lbs. in the other. In the series where the water-content was maintained, the increase in the absorption of water was not proportional to the density. Thus in series III, the 8's received about three times as much water as the 2's, but the 32's only about 50% more than the 8's. As in the series with *Helianthus, Triticum,* etc., the plants that received the nutrient solution always showed distress and wilting long before those without it. It was evident that under conditions of rapid water-loss the concentration of solutes in the soil-water tended to become too great, and this may have been true of the osmotic density of the cell-sap (plate 29).

The holard at the time of planting was 16.7% and the nitrate-content 45.6 ppm.; the hygroscopic coefficient of the soil was 6.3%. On August 6 when the experiment was concluded, the water values in series I with competition for water and nutrients were 8.3% for the 8's and 7.8% for the 32's, while series II with competition for water alone gave 8.7% and 8.4% for the respective values. In both of these the water-content had been reduced to within 1%–3% of the echard or non-available water. Readings of the light values in the three densities of the shaded series showed that these were essentially uniform, the maximum difference being not more than 3%–4%.

Development on July 28—The first measurements were made 18 days after planting, with the results given in the following table:

TABLE 86—*Growth in the Xanthium control series, July 28*

Criteria	I. Competition for water and nutrients			II. Competition for water, N+			III. Competition for nutrients, W+		
	2's	8's	32's	2's	8's	32's	2's	8's	32's
Ave. ht. cm.	11	15	14.5	9	12	12.7	8.5	13	21
Ave. length 1st internode, cm.	1.2	2.8	5.2	0.8	2.4	7.1	1.1	2.4	6.7
Ave. diam. stem, mm.	5.0	4.5	3.1	4.7	4.2	3.3	4	4.8	4.3
Ave. no. full-grown leaves.	5.5	5.5	4.5	5	5	4.5	5	5.5	4.5
Ave. size largest leaf, cm.	11 × 10	9.5 × 9	7.3 × 4.7	10 × 5.8	10 × 8.5	7 × 5.1	9.5 × 7	10 × 9	8.3 × 7
Ave. no. suppressed plants.	0	0	10	0	0	4	0	0	2

Criteria	IV. Shaded series, WN+			V. Unshaded series WN+		
	2's	8's	32's	2's	8's	32's
Ave. ht. cm.	16	17	20	9.5	12	13.7
Ave. length 1st internode, cm.	3.0	3.8	5.5	0.9	2.1	4.7
Ave. diam. stem, mm.	4.7	4.5	4.0	4.8	4.4	3.6
Ave. no. full-grown leaves.	5	5	4.5	5	5	4.5
Ave. size largest leaf, cm.	9.2 × 8	9 × 7.5	8.6 × 6.5	10 × 8.5	9.5 × 8	8 × 6
Ave. no. suppressed plants.	0	0	4	0	0	12

With rare exceptions, the sequence of values corresponded with the density in all the series, thus conforming to the universal rule; height and length of the first internode increased with the density, while all other

Control competition of *Xanthium* in the greenhouse, 1926.

A. Competition cultures with insert phytometers partly removed.
B. Water-series: (1) water and nutrients minus; (2) water minus; (3) water plus; (4) water and nutrients plus.
C. Series with the light intensity equalized (one plant removed from container 1).

dimensions decreased in this direction. Suppression occurred only in the 32's; it was greatest in the two extremes, viz. 10 for series I with inadequate water and nutrients, and 12 for V with both these factors plus. However, this contradiction was apparent rather than real, since the higher concentration of the holard doubtless hindered absorption. With respect to stem diameter and leaf size, the differences were fairly consistent for the 8's and 32's, the 2's experiencing so little competition that they responded much less to factor differences. The stem diameter was the same for both series with competition for water, being the lowest here; it was highest in III with water plus, next in the shaded series, and scarcely higher in the unshaded with water and nutrients plus than in the first two, again owing to osmotic relations in all probability. The respective averages for stem diameter of the 8's and 32's were 3.8, 3.8, 4.5, 4.2 and 4 mm.

TABLE 87—*Growth in the Xanthium control series, August 6*

Criteria	I Competition for water and nutrients			II. Competition for water, N+			III. Competition for nutrients, W+		
	2's	8's	32's	2's	8's	32's	2's	8's	32's
Ave. ht. cm.	23.0	27.9	21.6	20.5	27.1	23.3	19	33.6	36.5
Ave. diam. stem, mm.	8.7	6.6	3.9	8	6.4	3.9	8	8.2	5.3
Ave. width largest leaf, cm.	16.7	11.7	5.8	16.2	11.3	6.1	15.2	15.0	9.0
Ave. length largest leaf, cm.	16.5	11.7	7.3	15.5	11.6	7.3	14.5	14.7	9.5
Ave. no. dead leaves.	0	0.75	0.4	0	0	0.09	0	0	0
Ave. no. full-grown leaves.	8	5.9	4.1	7.5	6.6	4.7	6.5	7.5	6.1
Ave. no. half-grown leaves.	13.5	1.7	1.0	14.0	1.9	1.2	9.5	3.7	1.2
Ave. no. branches more than 1 in. long.	7.0	0.12	0	6.5	0.87	0.3	6.5	2.4	0
No. suppressed plants.	0	0	15	0	0	9	0	0	5
	Av. ht. 19 cm.			Av. ht. 20 cm.			Av. ht. 26 cm.		
Ave. leaf-area one side sq. in.	151.35	60.63	18.30	163.60	64.25	22.38	119.82	79.71	30.45
Ave. dry wt. gm.	5.23	2.94	0.82	4.82	2.71	0.82	4.00	3.61	1.58

Criteria	IV. Shaded series, WN+			V. Unshaded series WN+		
	2's	8's	32's	2's	8's	32's
Ave. ht. cm.	32	35.9	33.3	21	30.4	27.2
Ave. diam. stem, mm.	7	7.1	5.1	8.2	8.0	4.8
Ave. width largest leaf, cm.	13.5	12.1	8.4	16.2	15.1	7.9
Ave. length largest leaf, cm.	13.0	12.0	9.6	15.5	14.6	9.3
Ave. no. dead leaves.	0	0	0.03	0	0	0.06
Ave. no. full-grown leaves.	8	7	5.2	9	7.5	5.0
Ave. no. half-grown leaves.	2	1.7	1.1	10	4.2	1.4
Ave. no. branches more than 1 in. long.	3.0	0.12	0	7	3.6	0
No. suppressed plants.	0	0	4	0	0	13
	Av. ht. 18 cm.			Av. ht. 13 cm.		
Ave. leaf-area one side, sq. in.	86.15	45.20	27.59	147.50	103.04	30.01
Ave. dry wt. gm.	2.76	2.10	1.47	4.9	3.55	1.12

Final results for Xanthium control series, August 6—The behavior with respect to density was practically the same as in the earlier measurement, though increasing competition had now reduced stem height in the 32's below that of the 8's. Suppression was again present only in the 32's, being greatest in series I, next in V, and least in III and IV; the height of the suppressed plants was greatest in the series with water plus, being 26 cm. as against 19 and 20 cm. for the two with water minus.

The growth of 2's had been subjected to little competition throughout and the values are in consequence of little significance, while those for the denser cultures are fairly consistent with the treatment of each series. The average stem diameter for these two densities was 4.2 and 4.1 mm. for the two series with water minus, 6.7 for III with water plus, 6.1 for the shaded series and 6.4 mm. for the unshaded. Leaf dimensions followed practically the same course; they were least for the two series with competition for water, somewhat higher for the shaded series, especially the 32's, and highest in the series with water added and the unshaded one. The shaded series yielded the lowest value for leaf-area, followed by I and II with water lacking; the values were the same in III and V for the 32's, but much higher in the latter for the 8's. The dry weights were lower for the 32's in the water-minus series than in the shaded, but the reverse was true for the 8's the highest values occurred in the water-plus series, though for the 8's they were nearly the same for this and the unshaded one.

Starch tests on Series II and III showed a moderate amount present in the 2's and 8's of the water-minus by contrast with a large amount in the water-plus. The greatest difference was found in the 32's, as was expected little starch was present in the one and much in the other, thus indicating the limiting effect of water upon photosynthesis. This seems to have been exerted partly at least through the agency of the stomata in connection with the absorption of carbon dioxid, since the stomata of the 32's were almost closed with water lacking, and wide open with this factor optimum.

CONTROL EXPERIMENTS WITH TRITICUM SATIVUM

SERIES OF 1924

Installation—The type of container, kind of soil and general treatment were essentially as in the preceding series with *Helianthus* and *Xanthium* The holard at the outset ranged from 15.2% to 17.3%, and the nitrate content was 36.2 ppm. On May 1 Marquis spring wheat was planted at the rate of 2, 8, 16, 32 and 64 seeds in the five containers of each series, in such manner that the intervals were as uniform as possible, namely, 5″, 4″, 3″ 2″, and 1.5″ respectively. To insure a full stand two seeds were planted in each spot at a uniform depth of an inch, and the surplus seedlings were later removed.

The first two sets, as well as an additional one in wooden pails, were employed to determine the effects of competition for water. As in previou experiments, the 2's were repeatedly brought back to the original weigh by adding an amount of water equal to that lost, and the other containe received only as much as the 2's. A large container without plants wa used to determine the evaporation through the gravel mulch and thus serv as a check. Of the other two series, one was devoted to ascertaining th effects of competition for nutrients and light, the holard being maintained at the optimum, and the other those of competition for light, water and nutrients being kept at an approximate optimum.This was accomplished by adding Knop's solution from time to time.

Final results for W— series in wooden pails—The plants were grown until May 29, when they had attained a height of about 30 cm. and the roots had reached the bottom of the pails, about 30 cm. deep. During this period of nearly a month only 3 lbs. of water had been added to each container, and wilting was consequently beginning in the denser cultures. The effect of varying density on growth of wheat under competition primarily for water is shown in the following table:

TABLE 88—*Growth of Triticum in first W— series*

Criteria	2's	8's	16's	32's	64's
Ave. ht. cm......	23	26	30	32	30
Ave. no. lvs......	8.00	7.65	5.93	5.66	4.77
Ave. width 4th leaf, mm.......	7.3	6.9	6.6	6.3	5.5
Ave. no. tillers...	2.5	1.3	0.5	0.41	0.11
Ave. dry wt. roots, gm.	0.061	0.029	0.020	0.014	0.014

The values decreased consistently as the density increased, with the usual exception of height which responded to the reduced light intensity, but fell off again in the 64's as the shade became too intense. The number of leaves was nearly twice as great for the 2's as the 64's, and the tillers many times more numerous. The dry weight of the roots was more than twice as great in the 2's as in the 8's, thrice greater than in the 16's, and nearly five times greater than in the 32's and 64's.

Final results for the second W— series—The plants were removed and the usual measurements taken on June 12, when the 64's showed distinct signs of wilting (table 89).

In accordance with the rule, the height of stem increased with the density of planting, and the width decreased in the same direction. The length of leaf and practically all other values decreased from the 2's to the 64's, the number of tillers being ten times greater in the former. The total number of dead leaves was naturally in the inverse relation, there being none for the 2's and an average of nearly 3 for the 64's, by contrast with 6 and 4 full-grown living leaves respectively. The average leaf-area fell off rapidly from the 2's; it was little more than half as great for the 8's, less than a fourth for the 16's, less than a tenth for the 32's and about one-fortieth for the 64's.

Final results for the third W— series—By June 16 the plants in the 64-container and a few of the 32's were beginning to head. The individuals of both cultures but especially the denser, wilted more or less regularly by

TABLE 89—*Growth of Triticum in second W— series*

Criteria	2's	8's	16's	32's	64's
Ave. ht. cm.......	41.3	41.3	42.5	43.8	45
Ave. diam. stem, mm.	3.5	3.1	3.03	2.99	2.5
Length 5th leaf, cm.	29.5	25.9	24.4	23.8	21.1
Width 5th leaf, mm.	11.0	9.9	10.03	9.7	8.7
Ave. no. tillers....	5	4	2.5	1	0.5
Ave. ht. tillers, cm.	27	26	29	26	22
Total no. dead lvs.	0	5	5	20	164.0
Ave. no. full-grown live leaves......	6	7.4	6.7	6.0	4.0
Ave. no. half-grown live lvs...	0	1	3	6	15
Live lvs. per tiller	3.2	2.5	2.7	2.1	1.5
Ave. dry wt. tops, gm.	0.63	0.66	0.52	0.37	0.35
Ave. dry wt. roots, gm.	0.087	0.05	0.028	0.021	0.016
Total area live lvs. sq. in...........	116.96	249.23	219.81	136.48	90.04
Ave. leaf-area, sq. in..........	58.48	31.15	13.74	4.27	1.41

day but recovered at night. The plants were removed for measurement on June 28, when the leaves of these two containers remained wilted or permanently rolled in the morning at a humidity of 80%. At this time there were practically no rolled leaves on the 2's, some of the upper leaves of the 8's and 16's were drooping, some leaves were rolled or wilted in the 32's and practically all in the 64's.

Determinations of the holard showed that it fell off consistently for the three soil-levels from the 2's to the 64's, with the exception of the upper 6″ level where it was nearly 1% higher for the 64's than the 32's, owing to the deeper shade. On the other hand, it was 0.5%–1.5% lower at the other two levels. The decrease is well represented by the values for the middle layer, which were respectively 14.8%, 9.5%, 7.9%, 6.9% and 6.4%. Since the hygroscopic coefficient for the soil was 7.2%, it is evident that the denser cultures were suffering for water. This was shown also by the fact that the roots extended to the bottom of the container in all these and were abundant throughout. The nitrate-content was practically the same for the two open and for the three denser cultures, viz., 77, 74, 59, 58 and 57 ppm.

The stature was now greatest in the 16's, the 32's having fallen far behind as a consequence of increasing shade. The diameter of the stem and the leaf dimensions were larger in the 2's or 8's and dropped consistently to the 64's. The average number of living leaves decreased very regularly from the 2's, the 8's, 16's and 32's each having practically half as many as the preceding. The 2's had not yet produced heads and those of the 8's were not mature, so that these values are of little importance at this time.

TABLE 90—*Final results for water-competition in Triticum*

Criteria	2's	8's	16's	32's	64's
Ave. ht. cm.......	31	39.8	46.7	45	32
Ave. diam. stem, mm.	3.0	3.3	2.9	2.64	2.08
Ave. length 1st. leaf from top, cm.	25.2	24.2	21.9	18.4	14.6
Ave. width 1st. leaf from top, cm....	13.0	12.5	11.8	10.9	9.2
Ave. length 2nd leaf from top, cm.	27.2	28.0	25.0	22.8	19.5
Ave. width 2nd leaf from top, cm....	11.0	10.2	9.7	9.2	7.6
Ave. no. live lvs...	24	10	5.5	2.7	2.34
Ave. no. heads....	not out	0.76	0.88	0.92	0.77
Ave. length of heads, cm.......	......	4.5	6.5	5.9	3.82
Ave. no. live tillers	5.0	2.1	1.1	0.03	0.03
Ave. dry wt. tops, gm.	1.422	1.038	0.838	0.655	0.436

The dry weights decreased consistently and fairly uniformly with increasing density, the values for the 32's being about half and for the 64's about a third of those for the 2's.

Final results in series II and III—The plants in these series were cut on July 7, at a date earlier than planned owing to the necessity of minimizing the effects of infection by a mildew. The latter had begun to affect the growth and this probably furnishes the explanation of occasional inconsistencies in the results.

TABLE 91—*Final results for Triticum II and III*

Criteria	II. Competition for nutrients and light, W+					III. Competition for light, WN+				
	2's	8's	16's	32's	64's	2's	8's	16's	32's	64's
Ave. ht. in..........	15.5	19.0	22.0	23.0	21.5	14.0	20.5	22.0	23.0	24.0
Ave. diam. stem. mm.	3.2	2.9	2.6	2.4	1.9	3.5	2.8	2.7	2.5	2.4
Ave. length upper leaf, cm...........	25.5	27.3	24.6	22.6	19.7	24.0	28.6	24.5	23.0	21.1
Ave. width upper leaf, cm.	12.5	12.2	11.8	12.0	10.9	11.5	13.2	12.6	11.5	12.1
Ave. length 2nd leaf, cm.	27.5	26.8	26.0	26.1	22.7	24.0	29.3	26.4	24.7	23.9
Ave. width 2nd leaf, cm.	11.2	10.3	10.2	10.1	8.9	10.5	10.6	10.8	10.0	9.7
Ave. no. heads.......	1.0	1.4	1.4	1.13	0.94	1.0	2.2	1.5	0.96	1.0
Ave. length heads, cm.	4.35	5.8	6.4	6.3	5.9	5.1	6.8	6.4	6.8	6.8
Ave. width spikelets, mm.	5.75	7.0	7.2	7.2	7.8	5.0	7.0	8.0	7.6	8.8
Ave. no. developed spikelets	9.0	10.7	10.8	10.4	9.6	9.0	11.9	11.4	14.0	10.5
Total live leaves.....	26	88	118	131	186	35	103	125	130	191
Ave. no. live leaves..	13.0	11.0	7.9	4.2	2.9	17.5	11.4	7.2	4.6	3.0
Ave. no. live tillers..	3.5	4.0	2.3	0.7	0.1	8.0	4.0	3.2	1.3	0.7
Ave. leaf-area one side, sq. in........	24.12	30.32	16.86	8.77	5.68	27.3	29.38	20.6	10.38	7.08
Ave. dry wt. gm......	1.46	1.78	1.35	1.04	0.72	1.46	2.02	1.42	1.00	0.97

In general, growth was better in series III with competition for light alone than in series II with competition for both nutrients and light. This was most striking in the 64's where the demand was naturally greatest; in every case the values were higher in the first instance. On the contrary, the 2's, with the smallest demands and abundant light, were lower in about half the criteria when nutrients were added, probably owing to the further concentration of the soil solution produced by the greater insolation. In such cases, the beneficial effect of added nutrients was regularly felt in the 8's and tended to increase with more or less consistency to the 64's.

With respect to the average number of spikelets, live leaves and tillers, and the leaf-area and dry weight, there was a consistent decrease in both series from the 8's to the 64's, the 2's again being regularly lower than the 8's. The greatest effect was observed in the average number of live tillers, the difference between the extremes being sixfold or greater. This is to be explained by the fact that the tiller not only reflects the suppression of the parent but also a secondary suppression acting upon itself directly. As to leaf-area, the reduction from the maximum in the 8's to the minimum in the 64's was five times for series II and approximately four times for series III. In the case of dry weight, the difference between the two series was negligible, the minimum in the 64's being respectively 40% and 43% of the maximum in the 8's.

SERIES OF 1925

Installation—The methods employed were in all respects similar to those for the series of 1924. The holard of the containers varied from 20.7%–24.1% and the nitrate-content was 62.7 ppm. Planting was done on May 1 in such a manner as to yield 4, 8, 16, 32, and 64 individuals in each series, and the usual check container was provided.

The series utilized were as follows:

I. Competition for water and nutrients.
II. Competition for water, nutrients maintained.
III. Competition for nutrients, water maintained.
IV. Competition primarily for light, water and nutrients maintained.
II*a*. Duplicate of II, but with 4's, 16's and 64's only.
III*a*. Duplicate of III, 4's, 16's and 64's.
IV*a*. Duplicate of IV, 4's, 16's and 64's.

The amounts of water given the containers were determined by weighing at intervals of a few days to ascertain the water lost. In the W+ series the original weight was maintained by adding the amount lost during the interval between two weighings. In the W— series the amount of water lost by the 4's was added to each container in the set. At the close of the experiment, the amount below the original weight was twice as great as for series I and II with water minus as for III and IV with water plus (plates 30 and 31).

Control competition of *Triticum* in greenhouse, 1925.

A. Water and nutrients plus. B. Water and nutrients minus.

Physical Factors

Holard—The holard for the three duplicate sets was determined when the plants were about half-grown, with the usual result. The soil-water decreased with the density of planting, but increased with the depth. In the water-minus series, the holard was practically uniform at all three depths for the 64's, ranging from a half to a third of that for the 4's. In the series with water and nutrients plus, however, the values increased with density and were essentially uniform as to depth. This was due to the death of the lower leaves and the generally poorer condition in the two denser cultures, as a result of which there was a falling off in the demand for water.

In the four major series, I and II with water minus showed the usual striking decrease of holard with density, the values at the close of the experiment being about half as great in the 64's as in the 4's. The amount above the echard was about a third as great for the first as for the second. Where the water supply was maintained artificially, the values were essentially uniform for all densities in each set, ranging from 10%–14% above the echard. The efficiency of the gravel mulch in reducing evaporation was well demonstrated by the check container, in which the holard at the close was approximately the same as at the outset.

Nitrate-content—In the duplicate sets the nitrates at the middle of the growing period decreased regularly with the density; in the N+ set the respective values were 91, 78 and 74 ppm. and in the N— they were 79, 65 and 62 ppm. By the close of the period, the amounts had decreased greatly, the WN— set ranging from 160 in the 4's to 84 in the 64's. The content of the control container had increased from 63 to 106 ppm. In series III with competition for nutrients, the maximum fell at the 16's, namely, 153 ppm., and the minimum at both the 4's and 64's, respectively 110 and 112 ppm.

Light intensity—In order to reduce the high intensity and consequent insolation in mid-day, a cheese-cloth curtain was hung just below the glass of the greenhouse, to be drawn when occasion demanded. The respective values with the curtains drawn and open were as follows on June 5:

Table 92—*Light values with and without curtains*

Below	Curtains drawn Ave. %	Curtains open Ave. %
4's	11.9	19.4
8's	10.8	19.0
16's	7.7	11.2
32's	4.1	6.6
64's	3.9	5.5
Above plants	16.3	42.5

The use of the curtains reduced the general intensity above the cultures to nearly a third of that for the greenhouse generally, and also decreased the extremes somewhat. The light intensity within the 4's was thrice as great as that within the 64's when the curtains were drawn and nearly four times as great when they were open. The effects of this reduction were seen not only in photosynthesis and growth, but were also indicated by the greater number of dead leaves in the denser cultures. About June 15 the light values increased somewhat in the latter owing to the death of lower leaves, while at the same time it decreased in the more open on account of the abundant tillers and augmented number of leaves. This was exemplified on June 19, when the respective readings with curtains open were 15%, 9.5%, 7.6%, 3.5%, and 6.7%. On June 25 the values at a height of 10″ under the wheat plants were as follows:

TABLE 93—*Light values in series I and III*

Container	I. Competition for water & nutrients	III. Competition for nutrients, W+
	%	%
4's	35	17
8's	20	13.3
16's	12.5	11.2
32's	21.7	9.7
64's	14.2	5.0

The much lower intensities in series III were due to the more abundant tillering and the greater number of leaves. In consequence, the competition for light was much more intense in this, while the poorer growth in I with water and nutrients minus correspondingly reduced the competition for light.

GROWTH IN CONTROL SERIES

Growth on June 22—Measurements of nine characters were made in all series at weekly intervals from June 1 to June 29, but as all these were in close accord, the table for June 22 has been chosen as representative, though the values are naturally larger than for the three preceding ones.

As a rule, growth was better in the two series, III and IV, in which water was at an optimum, while as to I and II with water minus, it was regularly better in I where nutrients were not supplied. This was the probable consequence of the greater concentration in II where nutrients were added. This was not well exemplified by stem height, owing to the disturbing effect of light, but was evident in practically all other characters. As to the average number of tillers, series II was less than I only in the 16's and 32's, while IV made a distinct increase, especially in the denser containers, and III was higher still. The range for I with water and nutrients minus was 14 for the 4's to 2.5 for the 64's, the average being 7.8; for III with water plus and nutrients minus, it was 20 and 4.4 respectively, with an average of 9.6. For the average number of live leaves, the ranges were

Criteria	I. Competition for water and nutrients					II. Competition for water, N+				
	2's	8's	16's	32's	64's	2's	8's	16's	32's	64's
Ave. ht. cm......	35	48	50	48	42	35	42	50	48	40
Ave. no. tillers....	14	11	6.7	5	2.5	14	11	5.5	4	2.5
Ave. no. live lvs...	40	32	15	7.5	3.5	36	24	11	7	4.5
Ave. length top mature leaf, cm.	25	20	18	16	11	22	21.5	17	13	9
Ave. width top mature leaf, mm.	12	10	10	10	10	11	10	9.5	10	9
Ave. diam stalk at 20 cm., mm...	3	3	2.5	2.5	2.5	3	3	2.5	2.5	2.5
Ave. no. dead lvs..	9	12	11.5	11	10	15	9.5	12.5	8	10
Ave. spread of plant, cm.......	7	5.5	3.5	2	1	6.5	5	4	2.5	1
Stage of flowering	sheaths ½ swollen	spikes emerging	6 spikes out	20 spikes out	all spikes out	sheaths swollen, 1 showing	spikes emerging	8 spikes out	28 spikes out	most spikes out, stamens dry

Criteria	III. Competition for nutrients, W+					IV. Water and nutrients maintained				
	2's	8's	16's	32's	64's	2's	8's	16's	32's	64's
Ave. ht. cm.......	38	45	45	50	50	35	44	46	48	46
Ave. no. tillers...	20	10.5	7.25	6	4.4	15	10	7	6	5
Ave. no. live lvs...	56	29	21	17	8	43	30	16.7	11	6
Ave. length top mature leaf, cm.	21	21	20	20	19	22	22	18	18	18
Ave. width top mature leaf, mm.	11	10.5	10.5	11	11	11	11	10.5	10.5	10.5
Ave. diam. stalk at 20 cm., mm......	3	3	3	3	3	3	3	3	3	2.5
Ave. no. dead lvs..	10	9	10	10	9	6	8	10	10	9
Ave. spread of plant, cm.	9	7	5	3	1.5	8	6	5	3.5	1.5
Stage of flowering	3 spikes out	4 spikes out	2 spikes out	10 spikes out	13 spikes out	1 spike out	2 spikes out	5 spikes out	20 spikes out	14 spikes out

36–4.5 with water minus and nutrients plus, 40–3.5 with both factors minus, 43–6 with both maintained, and 56–8 with water plus but nutrients minus; the respective averages for all the cultures of each series were 16, 19, 21 and 26.

The average diameter of the stem was greatest in series III, though the differences were too small to be decisive. The average spread of plant was highest in III with water plus and next in IV with both factors optimum.

The development of the spikes was best and most uniform in series III, but nearly as good in IV; it was considerably poorer in I and II, and about equal in both.

TABLE 95—*Final results in Triticum control series*

Criteria	I. Competition for water and nutrients					II. Competition for water, N+				
	4's	8's	16's	32's	64's	4's	8's	16's	32's	64's
Ave. ht. cm.	49.0	59.5	56.6	53.6	43.3	Discarded	63.1	58.5	51.6	41.4
Ave. diam. stem, mm.	2.6	2.5	2.1	1.9	1.8		2.1	1.8	1.9	1.9
Length of top leaf, cm.	23.8	20.9	19.8	16.5	10.7		21.1	18.1	13.6	10.5
Width of top leaf, mm.	11.7	11.0	10.5	9.9	8.8		10.6	10.1	9.3	8.3
Length of 2nd leaf, cm.	26.2	22.3	23.0	23.6	18.6		23.3	23.1	22.9	19.5
Width of 2nd leaf, mm.	9.0	8.7	7.8	7.7	7.7		8.8	8.4	7.7	7.5
Ave. no. live leaves.	19.7	10.1	5.3	2.3	1.3		12.0	5.1	2.3	2.0
Ave. no. dead leaves.	51.7	43.2	26.9	19.2	11.1		30.4	19.0	14.6	11.1
Ave. no. live tillers.	11.4	6.9	3.8	1.7	1.0		6.9	3.1	1.5	1.0
Ave. no. dead tillers.	1.3	4.2	3.5	3.4	5.5		1.4	2.3	2.2	2.2
Ave. no. heads	6.0	6.4	3.3	1.7	0.9		5.4	2.7	1.3	0.9
Ave. no. heads in boot	1.7	0.3	0.2	0.06	0.03		0.57	0.07	0.06	0.03
Ave. length spikes, cm.	4.6	5.5	5.5	5.7	5.6		5.7	5.6	6.1	5.6
Ave. width spikes, mm.	7.2	7.1	8.1	8.4	8.2		8.4	8.4	9.3	8.6
Ave. no. spiklets per spike	8.7	9.9	9.7	9.0	8.1		9.3	8.6	9.0	7.8
Ave. leaf-area one side sq. in.	38.77	23.35	10.57	3.43	1.69		25.10	9.09	4.05	2.82
Ave. dry wt. gm.	4.59	4.44	2.18	1.25	0.67		3.66	2.02	1.19	0.67

Criteria	III. Competition for nutrients, W+					IV. Water and nutrients maintained				
	4's	8's	16's	32's	64's	4's	8's	16's	32's	64's
Ave. ht. cm.	65.4	62.2	58.4	61.0	53.5	55.7	64.3	59.4	59.4	53.9
Ave. diam. stem, mm.	2.2	2.2	2.2	2.0	1.8	2.1	2.2	2.0	2.0	1.8
Length of top leaf, cm.	23.0	27.6	22.2	19.9	19.3	20.8	20.9	20.5	18.7	19.4
Width of top leaf, mm.	10.5	10.3	10.5	11.1	10.1	10.2	10.5	10.5	10.2	10.0
Length of 2nd leaf, cm.	22.9	22.0	23.3	26.8	24.1	20.5	23.6	23.8	25.9	23.8
Width of 2nd leaf, mm.	8.9	8.3	7.1	8.4	7.0	8.3	8.5	8.1	7.5	7.4
Ave. no. live leaves.	27.7	23.2	7.1	4.5	1.8	25.0	19.5	5.7	4.5	1.7
Ave. no. dead leaves.	36.8	26.6	22.3	19.8	16.8	40.5	28.0	28.6	19.8	15.8
Ave. no. live tillers.	18.8	11.8	4.6	3.3	1.6	14.7	11.7	4.7	2.9	1.5
Ave. no. dead tillers.	0.7	0.2	2.2	2.6	2.2	1.7	1.7	2.2	3.0	2.2
Ave. no. heads	8.0	5.5	2.9	2.3	1.5	5.7	4.6	3.6	2.1	1.3
Ave. no. heads in boot	2.25	0.75	0.23	0.4	0.5	2.75	1.6	0.57	0.16	0.10
Ave. length spikes, cm.	5.8	5.7	5.0	5.4	5.5	5.6	5.6	5.4	5.6	5.2
Ave. width spikes, mm.	8.7	8.3	8.3	8.4	8.0	8.4	8.0	8.6	9.1	8.2
Ave. no. spikelets per spike	9.7	9.9	9.2	9.4	9.0	9.2	9.5	9.5	9.1	9.1
Ave. leaf-area one side, sq. in.	47.30	36.30	12.02	9.78	3.74	45.65	34.60	11.56	7.79	3.09
Ave. dry wt. gm.	4.90	4.41	2.24	1.78	1.02	4.92	4.56	2.56	1.71	1.04

Control competition of *Triticum* in greenhouse, 1925.

A. Water minus, nutrients plus. B. Water plus, nutrients minus.
C. Water-series, 1926: (1) 64's, water plus; (2) 16's, water plus; (3) 4's, water plus; (4) 16's, water minus; (5) 64's, water minus.

With respect to the effect of density on competition and resulting development, the measurements of the several series were in close accord with the rule, exemplified in practically all the cases in which the density has been varied.

Final results, July 7—The experiment was concluded on July 7, when a comprehensive list of measurements and counts was made, as shown in table 95.

The stature was greatest in series III and IV in which water was added, the effect of the latter being not merely to increase growth in general but especially the mechanical stretching of the stem. It was somewhat less in series IV in which nutrients were also supplied, owing to the increased concentration, and much less for I and II with water lacking. The maximum stem diameters were found in the 4's and 8's of series I, but the general level was slightly higher in III. The leaf dimensions also maintained a higher level in series III, though the differences were not very consistent either as to series or density. With respect to the number of live leaves, the W+ series were far in the lead, the two with competition for water being about equal; the respective averages for III, IV and I were 12.8, 11.2 and 7.8. The relation with respect to the number of live tillers was much the same, the averages being 8, 7 and 5 in the same order.

The number of spikes was greatest for series III with water plus, and nearly equal in series I and IV, the respective averages being 4, 3.6 and 3.5, while much the same relation obtained as to their dimensions and the number of spikelets per spike.

The values for the average leaf-area and dry weight are much the most significant, chiefly because of their nature as integrators, but partly also because of their consistence as to series and degree of density. With respect to leaf-area, the highest values were found in series III, where they ranged from 47 for the 4's to 3.7 for the 64's, and the lowest in I with both water and nutrients minus, with a range of 39 to 1.7 sq. in.; series II was a trifle higher, while series IV was but little below III. The averages for series III, IV and I were respectively 22, 20.5 and 15.5 sq. in. In the case of dry weight, the values were slightly higher for series IV with water and nutrients both plus than for III with water alone plus, but part or all of this may have been due to the weight of the additional salts absorbed. There was again little difference between I and II, which with one exception were considerably below the values for III for the most part. The average dry weights for the whole series were 2.62, 2.87 and 2.96 for I, III and IV respectively (fig. 23).

As a rule, the results in the control series for *Triticum* were similar to those of *Helianthus*. Growth was regularly better when water was supplied to compensate than when nutrients were added, and it was usually better also than when both factors were increased. It was so nearly the same under competition for both water and nutrients, and for water alone with nutrients plus as to indicate that nutrients were of quite secondary importance as compared with water in the case of the fertile soils used.

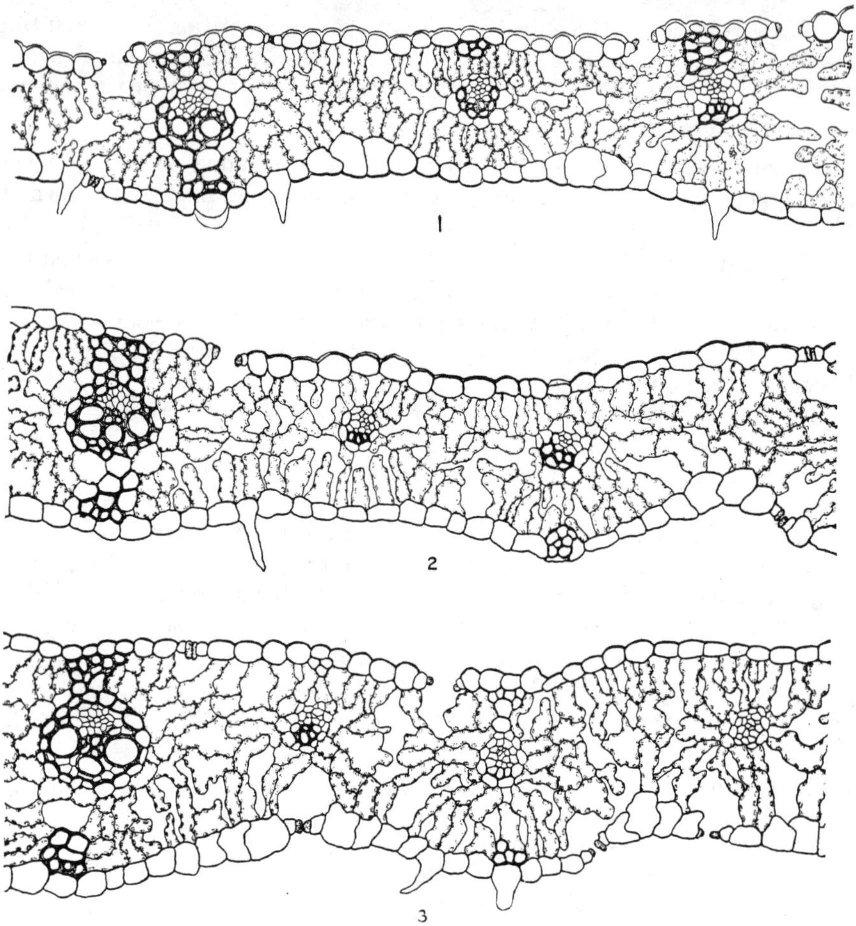

FIG. 23—Structure of leaf of greenhouse wheat, 1925; (1) 64's, nutrients plus, water minus; (2) 64's, water plus; (3) 4's, water plus.

CONTROL EXPERIMENTS WITH ANDROPOGON NUTANS

Installation—The methods employed with this native species were essentially the same as those for *Triticum,* except as to rate of sowing. The seeds were planted thickly on May 9 and were thinned on June 2 to give three densities, viz.: "thins" with 150 individuals per container, "mediums" with 400 and "thicks" with 1,100. The holard at the time of sowing was 19%–20% and the nitrate-content 53 ppm., the hygroscopic coefficient being 6.5%. Each set of containers was duplicated, the series being as follows:

I. Competition for water and nutrients.
II. Competition for water, nutrients maintained.
III. Competition for nutrients, water maintained.
IV. Competition primarily for light, water and nutrients maintained.

Physical factors—The holard was in general agreement with the values obtained in previous series, decreasing with density and increasing with depth, while it was higher as a rule in the sets with the water supply maintained. The nitrate-content was regularly higher in the "thins" when nutrients were not maintained and in the "thicks" when nutrients were added; it was also higher in the series with nutrients added, as a rule. It ranged regularly from 75 to 116 ppm. and hence had relatively little effect. After the middle of June, the light intensity at 3″–6″ in the "thicks" ranged from 1%–2%, in the "mediums" from 2%–3% and in the "thins" from 5%–8%. The differences between the series were less significant, but the light intensity for the "thicks" with N+W— was twice as great as with W+N—.

Growth in the duplicate series—The course of development in the duplicate series was followed by means of measurements and counts made on June 12, 15, 22, 29 and July 7. Differences between the series were late in appearing and were most significant in the final determination on July 16, but growth in the three densities was well differentiated by the end of the first month. The results obtained on June 12 were as follows:

TABLE 96—*Growth in the three densities, June 12*

Criteria	Thins	Mediums	Thicks
Ave. ht. cm.	16.5	20.5	23
Max. ht. cm.	29.0	32.5	38
Ave. diam. stem, mm.	1.3	0.9	0.5
Ave. width largest lvs. mm.	4.5	3.7	2.5
Ave. no. live leaves	4.0	3.5	3.0
No. suppressed plants	0	0	10%

The values were consistent throughout, height and suppression increasing with the density, and the others diminishing in the same direction. The largest differences were in stem diameter and the width of leaf, the values for the "thins" being approximately twice as great as for the "thicks." Suppressed plants occurred at this time only in the latter, these individuals being about 3.5″ tall with three leaves, while the dominants in the same container were 9″–15″ with 4 leaves. By June 22 about half of the "thicks" were suppressed and a third of the "mediums," in contrast to none in the "thins." The sequence of the other values was practically the same as earlier, except that maximum height was now greatest in the "thins" and the average height little less than for the other two containers.

The average dry weight on June 15 in the water and nutrient minus series was .04, .03 and .02 gm. from the thin to the dense culture; on June 29 for the water and nutrient plus series it was respectively .11, .08 and .03 gm.

Final results for duplicate series—The comparison of results is drawn from series II*a* and III*a*, since nutrients were added in the one and water in the other. The number of plants utilized was 50 for each of the three densities, with the exception of 100 of the "thicks" in the first series.

TABLE 97—*Final results in duplicate series of Andropogon*

Criteria	IIa. Competition for water, N+			IIIa. Competition for nutrients, W+		
	Thin	Med.	Thick	Thin	Med.	Thick
Ave. ht. cm............	58.8	49.7	39.8	63.5	60.0	48.5
Ave. diam. stem, mm.....	2.1	1.4	1.0	2.0	1.3	1.1
Ave. length top mature leaf, cm.............	44.9	38.7	31.7	47.4	47.4	39.3
Ave. width same leaf, mm.	5.3	3.5	2.7	5.4	3.8	2.6
Ave. no. live leaves......	5.9	3.5	1.9	7.4	6.0	3.0
Ave. no. live tillers......	3.1	2.1	1.1	3.9	2.3	1.6
Ave. dry weight, gm.....	0.27	0.12	0.05	0.65	0.16	0.10

With respect to density, the values were consistent throughout, the reduction of light intensity in the "mediums" and "thicks" having reached the point where elongation was more than offset by diminished manufacture of materials. As a consequence, all values decreased in the direction of increasing density. The series with water plus regularly yielded larger values, which were most marked in the case of height, number of live leaves and tillers, and dry weight. The average dry weight was more than twice as great when water was plus than when it was minus in the case of the "thins" and twice as great in the "thicks," with a smaller gain in the "mediums." Thus, as in previous series with other species, the addition of water gave decisively better growth than that of nutrients, confirming the conclusion that in fertile grassland soils water is the paramount factor in competition.

Final results for regular series—The general course of development in this series was parallel to that in the duplicate sets. The experiment was terminated on July 21, with the results given in table 98.

As a native perennial grass, *Andropogon nutans* grew more slowly than the vigorous annuals, *Helianthus, Xanthium* and *Triticum;* it is likewise late in development, flowering usually occurring in autumn. As a consequence, the differences between the series were often small and in such cases little consistent. The characters that yielded the best differences were height, leaf dimensions and dry weight; these were likewise entirely consistent in their relation to increasing density. The stem height was greatest for all densities in series III with water plus; the difference was slight in the "thins," but amounted to nearly a third in the intense competition of the "thicks." Series I, II and IV were much alike, except for the stature of the "thicks" which increased in this order. Leaf dimensions behaved in essentially the same fashion, series III giving the highest values and the "thicks" increasing in height from I to II and IV.

The largest differences as usual were found in the dry weights; these were not only strikingly different for the several densities, but also consider-

TABLE 98—*Final results for regular series of Andropogon*

Criteria	I. Competition for water and nutrients.			II. Competition for water, N+			III. Competition for nutrients, W+			IV. Water and nutrients maintained		
	thin	med.	thick	thin	med.	thick	thin	med.	thick	thin	med.	thick
Ave. ht. cm..........	71.1	59.6	37.7	69.0	59.9	46.0	72.8	65.9	48.2	68.5	60.5	48.6
Ave. diam stem, mm....	2.1	1.5	1.1	2.2	1.6	1.2	2.1	1.7	1.2	2.1	1.5	1.3
Ave. length top mature leaf, cm..........	54.7	47.3	29.6	52.6	47.3	37.1	54.4	51.0	38.3	51.5	42.8	38.4
Ave. width same leaf, mm.	6.0	4.0	2.6	5.8	4.2	2.9	6.1	4.3	3.2	5.6	4.1	3.2
Ave. no. live leaves......	7.9	4.8	3.2	7.8	5.2	3.5	7.6	5.1	3.6	8.6	4.3	3.8
Ave. no. live tillers......	4.7	2.8	1.8	4.5	3.1	1.9	4.4	2.9	2.0	5.5	2.5	2.0
Ave. dry wt. gm.	0.534	0.231	0.085	0.544	0.257	0.115	0.847	0.237	0.122	0.855	0.205	0.122
			30% were dead			26% were dead			29% were dead			21% were dead

bly so for the two major types of competition, namely, with water deficient nd with water optimum. The "thins" of series III and IV gave a value nore than 50% higher than those of series I and II with water minus. The values for the "mediums" corresponded closely in all four series, but he value for the "thicks" with added water was again about 50% larger han in series I with competition for water and nutrients (fig. 24).

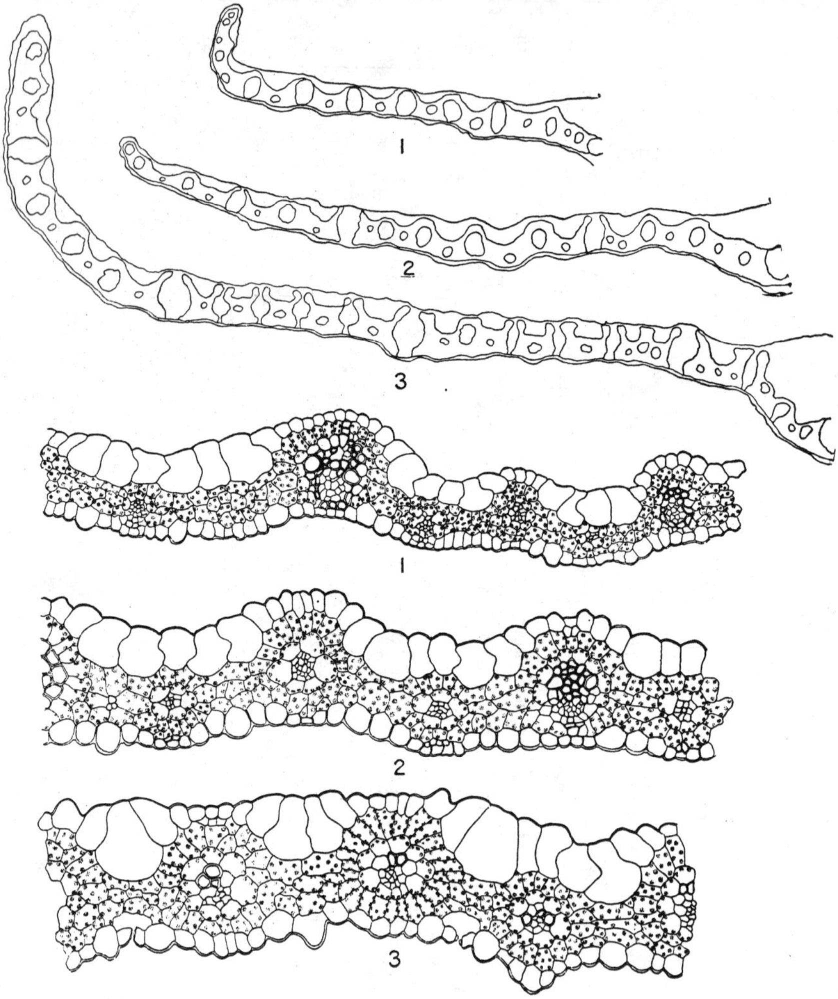

FIG. 24.—Outline and section of leaf of *Andropogon nutans:* (1) thicks; (2) mediums; (3) thins.

COMPETITION IN UNDERPLANTED CULTURES

Plan and method—In these series the individuals of one species were permitted to grow to a height of several inches before seeds of another were planted beneath them. The primary purpose was to throw light upon the course of competition when a species invades an established community and consequently upon the nature of the relation between dominant and subordinate. In addition to determining the effect of the former upon the latter, an essential point was to ascertain whether subordinates exerted any considerable effect upon the dominant. A preliminary series with *Helianthus* and *Xanthium,* and with *Kuhnia* and *Oenothera* was instituted in 1924. The original plantings were made on May 31, and the respective cultures were underplanted on June 19 when *Helianthus* was 14″ and *Xanthium* 6″ high. On July 11 the underplanted individuals of the latter were suppressed to the point of being greatly dwarfed, while those of the former were much attenuated and bent over.

On July 14 the upper layer was cut off to determine the effect upon the lower one of subordinate plants. At this time the underplanted cockle-burs were 8″ and the sunflowers 18″ tall; all were badly suppressed and suffering from drouth, and it had been necessary to add abundant water twice a day to prevent the upper layer from wilting. By August 9 the cockle-burs had attained a stature of 22″ and were essentially normal in appearance, though they had grown more slowly than those in the control. Though the sunflowers were now 33″ tall, they had fared less well, the stems being weak and sprawling on the ground.

The degree of control by the dominant layer was indicated not only by the marked suppression of the lower one, but reciprocally by the rapid growth and recovery of the subordinates after the dominants had been removed. Somewhat unexpectedly the sunflower succeeded less well as a subordinate in spite of its much greater stature, owing to the pronounced attenuation of the stem.

Reciprocal Underplanting of Helianthus and Xanthium

Installation—Four large containers were filled with the usual well-mixed soil consisting of 1/3 sand and 2/3 loam; the holard was 16.5%, the hygroscopic coefficient about 6.5% and the nitrate-content 45.6 ppm. On July 10, 1925, one container was sown with seeds of *Xanthium*, the other with those of *Helianthus;* after the seedlings appeared they were thinned to 8 per container. The cockle-bur culture was interplanted with sunflower seed on July 25, when the individuals were 5″ tall with 4 full-grown leaves and shaded the ground to a considerable degree. At the same time, the sunflowers, which were 8″ tall with 3 pairs of full-grown leaves, were interplanted with cockle-bur seed. An additional container of each was also planted at this time to serve as a control for the growth of the underplanted individuals. The four containers were placed near each other to insure essentially uniform conditions and were watered now and then to maintain an optimum holard.

Course of development—On August 3, the 8 underplanted cockle-burs averaged 3″ tall with a first pair of leaves 0.75″ long and 0.25″ wide, by contrast with the control plants which were 1″ high with leaves 1.5″×0.75″. The dominant sunflowers of this container were 22″ tall with 6 pairs of leaves, all of which were still green, and in consequence it was necessary to water this culture frequently to prevent the underplanted cockle-burs from wilting. The root system of the latter was much less extensive and efficient than that of the sunflower, and both the youth of the seedlings and the handicap of competition further increased the difficulty of absorption. The effect of reduced light intensity was also becoming marked, the consequent elongation rendering the underplanted seedlings twice as tall as those of the control container.

The underplanted sunflowers were much attenuated, being 5″ tall to for the control; the hypocotyl was 4″ for the former and 1.5″ for the latt The diameter of the stems was a third greater in the controls, and t leaves nearly twice as wide and long. The dominant cockle-burs of th container averaged 12″ tall, with 3–4 pairs of full-grown leaves; all t latter were still green and branches were starting in the leaf-axils. As consequence, the shade beneath was very dense and the sunflowers we correspondingly suppressed and etiolated. However, wilting was less com mon in the lower layer than in the reciprocal culture, since the dema made by the dominants for water was less. On the other hand, the su flowers exhibited a much more striking response to low light intensiti than did the cockle-burs.

Final results for the underplanted series—The experiment was term nated on August 7, and the usual counts and measurements made, as show in the following table:

TABLE 99—*Growth in underplanted cultures of Xanthium and Helianthus, August*

Criteria	Xanthium		Helianthus	
	Control	Under-planted	Control	Under-planted
Ave. ht. cm.	5.1	5.6	14.0	21.6
Ave. diam. stem, mm.	1.9	1.9	3.5	2.2
Ave. width largest leaf, cm.	3.2	1.7	4.4	2.8
Ave. length largest leaf, cm.	6.9	4.0	7.6	5.7
Ave. no. lvs.	2.9	2.0	4.1	2.1
Length root system, cm.	57.5	24.7	57.5	27.5
Ave. leaf-area 1 side, sq. in.	4.98	1.40	12.37	3.17
Ave. dry wt. gm.	0.18	0.05	0.33	0.12

The cockle-burs of the dominant layer were 16 cm. tall with abo 7 full-grown leaves reaching a maximum of 5.5″×5.5″, while the su flowers were 30 cm. tall with 14 full-grown leaves and an area of 6″×5 By comparison with the controls, the growth of the stem in *Xanthium* w little affected, while in *Helianthus* it was elongated 50% with a corr sponding decrease in diameter. The reduction in leaf dimensions was mu the same for both species, but in leaf number it was more marked in th sunflower. The decrease in leaf-area was approximately the same in bot the value being little more than a fourth as great in the underplante individuals, and they agreed nearly as well with respect to dry weigh underplanting reducing this to a little more than a third in *Xanthium* an a little less than a third in *Helianthus*.

The difference between the controls and the underplanted individua is explained in large measure by the much longer and more extensive branched root system of the former. The roots of the two species corr sponded closely in length under the two conditions, but the greater eff iency of the sunflower in absorption is attested by the fact that those of he dominant layer were twice as tall and leafy as the dominant cockle-urs. On the other hand, the sunflowers were distinctly less tolerant of hade, as shown by the very much greater elongation and attenuation of he stem. This was confirmed by the results of starch-tests; all the leaves f the control contained an abundance of starch, the two leaves of the nderplanted cockle-burs a moderate amount, and those of the underplanted unflowers only a small amount. As was shown repeatedly by competition ultures in the prairie, the advantage of early germination or rapid growth, nd especially of both combined was usually decisive, the occupant prac-ically always maintaining itself against the invader in the absence of dis-urbance or other change in conditions.

Factor Control in Underplanted Cultures of Helianthus

Installation—In connection with studies of function under competition arried out in 1926, a special method was developed for eliminating water r light as a factor. This was accomplished in the case of water-content y underplanting in a cylinder which prevented root competition with the lants of the culture; with light it was secured by tying back the tops of he upper layer. In addition, competition for both factors was practically liminated by tying back the tops of plants growing above those in the ylinder, while in the absence of both devices the usual competition for water nd light took place. The four containers thus offered the following series f conditions:

1. Insert cylinder, tops tied back; no competition.
2. Insert cylinders, shoots erect, competition for light.
3. No cylinder, tops tied back; competition for water.
4. No cylinder, shoots erect; competition for both light and water.

First series—On May 18, when the sunflower 8's were 14″ tall with pairs of leaves, the cultures were underplanted with the same variety. he containers were restored to the original optimum holard and kept at his point by adding water every day or two in accordance with the trans-iration. A cylinder 3.5″ across had been sunk to the bottom of two of the ontainers at the time of filling them, and after underplanting these were ealed with a gravel-mulch. The latter practically eliminated evaporation rom the surface, while the cylinder prevented root competition.

So effective were the cylinders in eliminating competition for water hat the plants in them were turgid at all times, while those outside in the ulture proper were often wilted in spite of frequent watering. The roots f plants in the older upper layer occupied the soil so completely that he demand for water exceeded the supply at nearly all times. The plants ere cut at the ground line on June 3, with the results indicated in the ollowing table (plate 32A):

TABLE 100—*Growth of phytometers in underplanting, first series*

Criteria	Container			
	1	2	3	4
	No competition	Competition for light	Competition for water	Competition for light and water
Ave. ht. in.	13.0	11.5	7.0	7.5
Ave. length roots, in.	16.0	11.0	10.5	7.5
Ave. length largest leaf cm.	9.5	5.5	4.6	3.8
Ave. width largest leaf, cm.	7.0	2.8	2.0	1.9
Ave. leaf-area both sides sq. in.	24.0	5.8	4.7	3.1
Ave. dry wt. tops, gm.	0.63	0.16	0.16	0.11

Second series—The installation and treatment were essentially the same as before, the underplanting being done on June 22. On June 28 the light intensity under the upper layer was 18%; at this time the tops of the stems were tied back in containers 1 and 3. The light value above the underplanted sunflowers was 23% on June 30, but by July 7 it had dropped to 10%. The nitrate values were respectively 15.4, 11.2, 6.5 and 4.4 from container 1–4. The plants were removed on July 7 and the following results secured:

TABLE 101—*Growth of phytometers in underplanting, second series*

Criteria	Container			
	1	2	3	4
	No competition	Competition for light	Competition for water	Competition for light and water
Ave. ht. cm.	18.2	21.5	18.4	16.5
Ave. diam. stem., mm.	3.8	2.8	2.7	2.2
Ave. no. leaves	7.5	6.5	5.5	4.5
Ave. length largest blades, cm.	5.9	4.8	6.0	4.0
Ave. width largest blades, cm.	3.3	2.3	2.6	2.2
Ave. leaf-area both sides, sq. in.	75.8	48.3	49.0	18.9
Ave. dry wt. gm.	1.08	0.65	0.81	0.39

In all features but one, the plants grown without competition made a much larger growth than those with competition for either light or water, the single exception being the usual elongation of the stem in reduced light. With respect to all the measurements in the second series except height the plants in container 1 were 50% larger than those in container 4 with competition for both factors. The results for the two most important

A. Insert phytometers of *Helianthus,* from 8's with water plus and minus, and 32's with water plus and minus.

B. Plants from the second underplanting series of *Helianthus:* (1) no competition; (2) competition for light; (3) for water; (4) for light and water.

criteria, leaf-area and dry weight, were much more decisive; the former was four times and the latter two and a half times greater without competition than with it, in the case of the second series. The first series yielded even greater differences, the value for leaf-area being nearly eight times and for dry weight six times greater when competition was absent (plate 32B).

With respect to the relative importance of light and water, the two series are not in accord, but this is explained by the much lower light values in the case of the second. In the first series the values for the plants grown under competition for light are consistently higher, except for dry weight. This is explained by the fact that the stretching of stem and leaf are effected by water, while the elaboration of material is dependent primarily upon light.

DIRECT RESPONSE TO PRIMARY FACTORS IN COMPETITION

Object and method—A series of experiments was carried out for the purpose of ascertaining the direct response of the several species to varying amounts of the three primary factors. It was hoped that this would be an aid in interpreting the results obtained in competition cultures, as well as throw further light upon the relative importance of the factors. A small number of plants were grown at uniform intervals in the usual containers and subjected to three different amounts of light, water or nutrients, under the general control afforded by greenhouse conditions.

Light Series

Installation—Six large containers were filled with the usual soil mixture with an original holard of 19.3%; the hygroscopic coefficient was 6.3% and the nitrate-content 25 ppm. On June 2, three of the containers were planted with seeds of *Helianthus* and three with those of *Triticum,* in such a way as to yield 8 individuals equally spaced in each. Two shade-tents were constructed to give light intensities of 1% and 10% of full sunshine outside the greenhouse, the general value within the latter being approximately 50%. A container of each species was placed in each of the three light values, and the water-content was maintained as near the original amount as possible by replacing the amounts lost through transpiration.

Comparative development—The usual counts and measurements were made on June 12, 19, and 27, with the results indicated in table 102, p. 278.

On June 12 the differences between the plants growing under the three light intensities were already marked, especially in height, erectness and leaf development. Those in the shade-tent with a value of 1% were much etiolated, the sunflowers being thrice as tall as those in the light of the greenhouse, and those in 10% light twice as tall. The plants in 1% light possessed no rigidity whatever, bending over until they touched the ground or some support; the 10% plants were more erect, but were still weak and declined. The dense shade was unfavorable to the formation of chlorophyll and the plants were much paler than the others. All these differences were much more pronounced in the sunflower than in the wheat. At this time

TABLE 102—*Development under different light intensities*

Criteria	June 12						June 19						June 27			
	Helianthus			Triticum			Helianthus			Triticum			Helianthus		Triticum	
	50 %	10 %	1 %	50 %	10 %	1 %	50 %	10 %	1 %	50 %	10 %	1 %	50 %	10 %	50 %	10 %
Ave. ht., cm.	5	10	15	11	14	15	16	18	16	22	25	15	47	26	25	20
Ave. ht. cotyls, cm.	4	10	15	..	..	..	5	11	15	..	..	..	..	..	..	..
Ave. no. leaves	2	2	1	2	2	1.3	4	2.5	1	3	3	2	8	5	6	4
Ave. size leaves, cm.	3.5 × 1.6	2.5 × 0.7	1.2 × 0.3	4mm. wide	3	3	8.9 × 4.9	6.7 × 3.5	2.5 × 0.6	4	3.5	3	11.0 × 7.5	7 × 4	..	..
Ave. diam. stem, mm.	2.7	2.5	2.5	1.3	1	1	4.5	3	2.5	2	1.8	1	7.5	5.1	..	..

the plants were only ten days old, but it was already evident that those in 1% light would not long survive (figs. 25, 26).

On June 19 the 1% plants were almost dead; no growth had occurred in the wheat and very little in the sunflowers, the leaves being very pale-green or yellowish. By June 25 all these plants were dead; none of the sunflower roots had penetrated more than 2.5″ below the surface and the wheat bore only the initial roots 4″ long. Evidently very little food had been translocated of the small amount made, none of the shoots developing

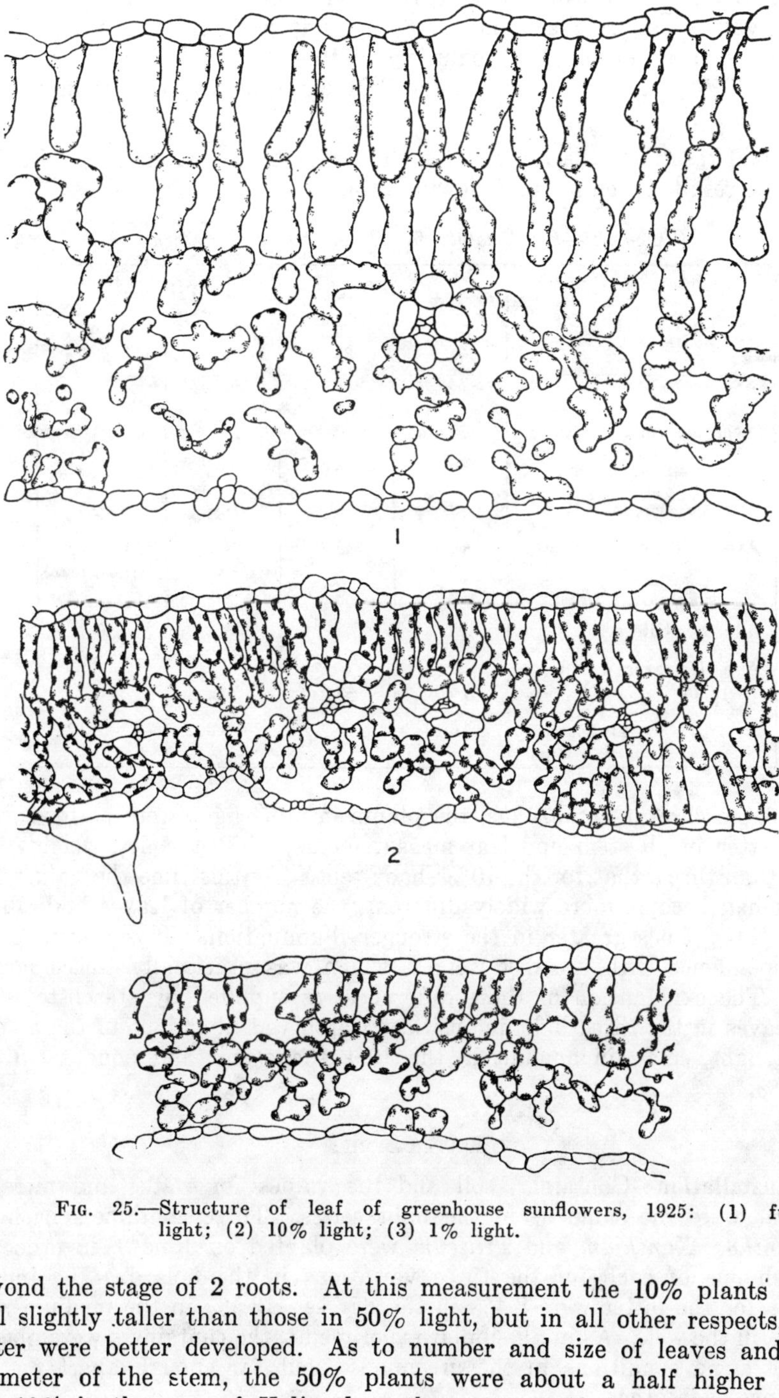

FIG. 25.—Structure of leaf of greenhouse sunflowers, 1925: (1) full light; (2) 10% light; (3) 1% light.

beyond the stage of 2 roots. At this measurement the 10% plants were still slightly taller than those in 50% light, but in all other respects, the latter were better developed. As to number and size of leaves and the diameter of the stem, the 50% plants were about a half higher than the 10% in the case of *Helianthus;* there was much less difference with wheat, the stems of which were also much more erect. By June 27 the height values were reversed and the 50% sunflowers were now almost twice as tall as those in the 10% light, while the other values were approximately 50% greater. As before, the differences in wheat were much less considerable.

Final results in the light series—The plants were removed on July 2, with the results given in the following table:

TABLE 103—*Final values for Helianthus and Triticum*

Criteria	Helianthus		Triticum	
	50 %	10 %	50 %	10 %
Ave. ht. cm.	72	42	28	24
Ave. diam stem, mm.	8.3	5.9	2.5	1.5
Ave. no. leaves	9	8	10	2.7
Ave. size 2nd pair lvs.	12.4 × 9.5	9.3 × 6.1	24.1 × 6.8	23.3×4.6 (largest leaf)
Ave. no. tillers	...	...	3.4	0
Ave. leaf-area 1 side, sq. in.	76.5	49.5	...	...
Ave. dry wt. gm.	6.13	1.7	0.342	0.124

In the case of *Helianthus*, the plants in 50% light were a third to a half better in all stem and leaf measurements, while the dry weight was more than thrice that for the 10% shade-tent. By this time the values for wheat had become more widely different, the number of leaves and tillers being three times greater in the stronger illumination. This was in close correspondence with the dry weight, which was practically thrice larger also. The explanation of these differences is supplied by starch-tests of the leaves in the three light intensities; starch was abundant in the greenhouse light, small in amount in the 10% shade-tent and none at all in the 1%.

WATER SERIES

Installation—Container, soil and the values for water and nitrate-content were the same as in the light series. Three containers each of *Helianthus, Xanthium* and *Triticum* were planted on June 3, in order to furnish one of each for the three variations in the holard. The initial values for the latter were 8.4% in the dry set, 13.5% in the medium and 25.2% in the wet. As in all similar experiments, the containers were placed in the same conditions in the greenhouse, but far enough apart not to influence each other.

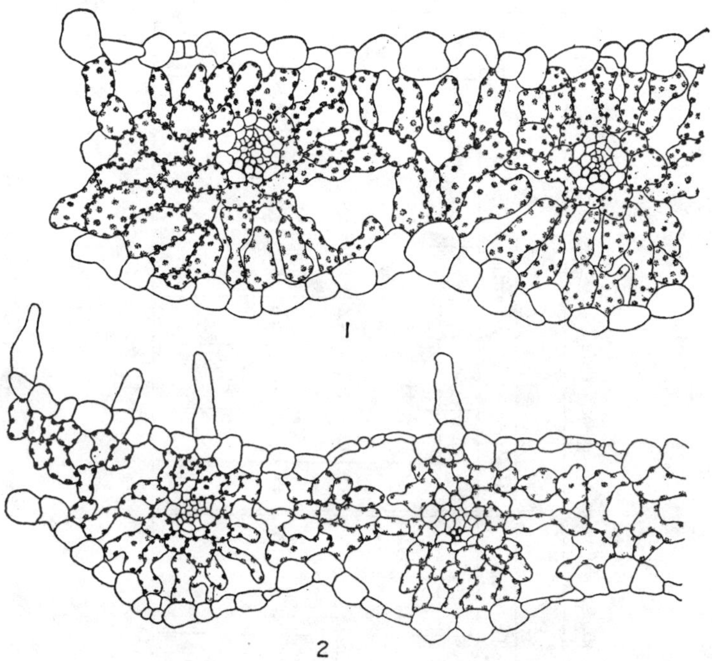

FIG. 26.—Structure of leaf of greenhouse wheat, 1925: (1) full light; (2) 10% light.

The soil in the dry set was kept at as low a holard as possible, resulting frequently in causing wilting or semi-wilting in all three species but especially in the sunflowers with their much larger demands. The medium containers used twice to thrice as much water as the dry ones, and the wet ones somewhat more than the medium, with the exception of the cockle-bur.

Comparative development—Counts and measurements were made on June 19 and 27 with the following results (table 104, p. 282):

The values on June 19, when the plants were about two weeks old, were consistently higher with increased water-content. This was much more striking in the sunflowers with their greater utilization of water, the stem height being twice as great for the wet as the dry, and the other values about a half again as large. By June 27 the effect of high holard in reducing the air-content appeared to be in evidence, since practically all values were now slightly larger for the medium containers.

Final results in the water series—The final measurements obtained on July 2 were as follows (table 105, p. 284):

Without exception, the final results were somewhat better in the medium set than in the wet one and 2–3 times better than in the dry set for the sunflowers. The stature was practically three times as great, and the stem diameter twice as great, while the product of the leaf dimensions was more than twice as large. The differences for stature were less in the

TABLE 104—*Development under different water-contents*

Criteria	June 19									June 27								
	Helianthus			Xanthium			Triticum			Helianthus			Xanthium			Triticum		
	dry	med.	wet	dry	med.	wet	dry	med.	wet	dry	med.	wet	dry	med.	wet	dry	med.	wet
Ave. ht. cm.	10.7	17.7	20.2	7.5	8.3	9.5	...	...	...	17.7	42	41	11.4	15.1	14.6	...	...	...
Ave. ht. cotyls, cm.	2.5	4.5	5.7	4.0	4.5	5.0	...	...	...	...	...	...	...	...	...	...	...	...
Ave. length 1st internode, cm.	7.0	8.9	10.0	...	...	...	...	...	...	9.5	12.2	11.9	4.4	4.6	4.9	...	...	...
Ave. diam. stem, mm.	3.0	4.0	4.5	2.6	3.0	3.4	...	...	...	3	5	5	2.3	3.9	3.7	...	...	...
Ave. no. leaves	2.7	4.3	4.3	3	3	3	3	3	3.5	4	7	6	3.5	5.0	4.0	3	5.3	5
Ave. size largest leaf, cm.	5.9 × 2.9	6.9 × 4.0	7.8 × 4.6	...	...	...	4 mm. wide	4	4.5	...	...	...	...	...	...	...	...	...

cockle-bur and much less in the wheat, but for the other criteria they wer usually as large and sometimes larger. The approximate sequence for leaf area was 1:5:4 for the sunflower and 1:5:2 for the cockle-bur. Wit respect to dry weight, there was little difference between the maximur in the medium and the wet value for sunflower, and both were more tha

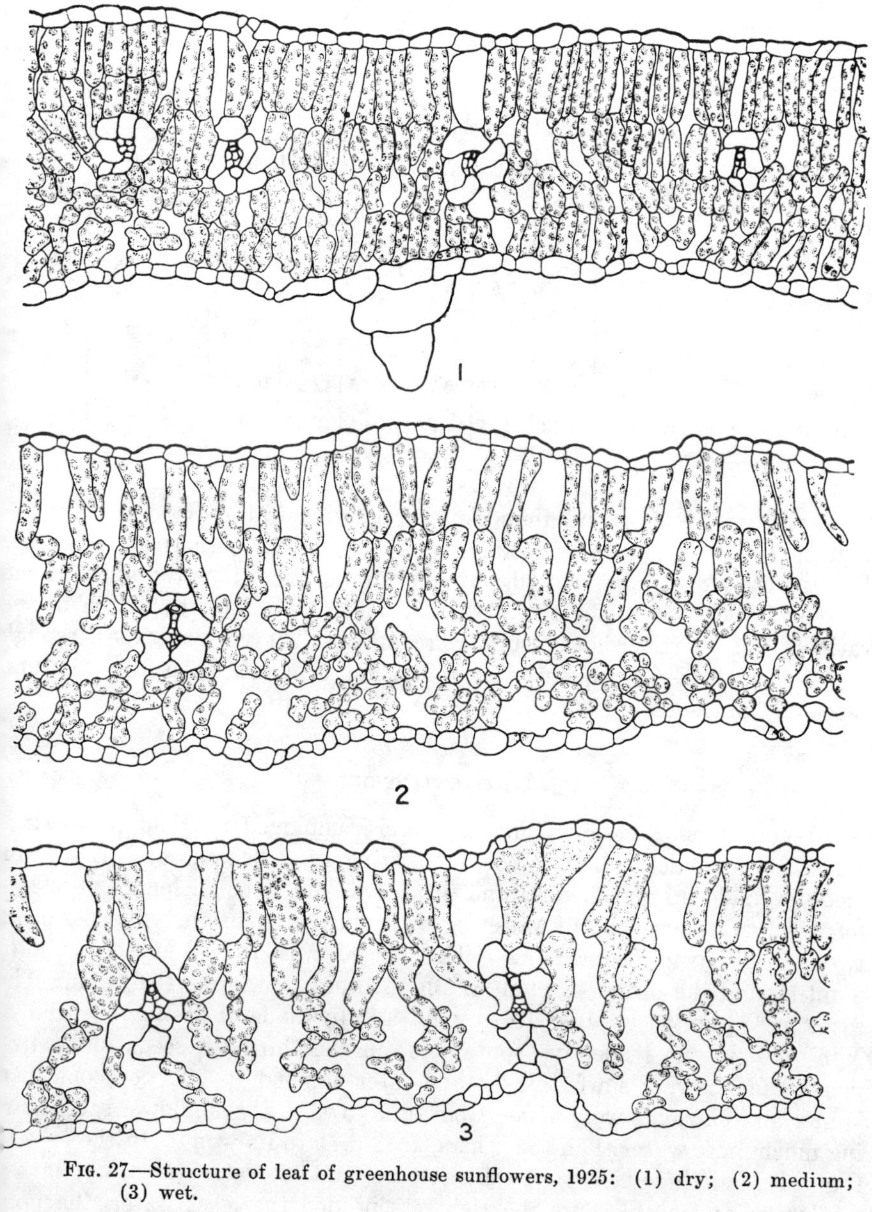

FIG. 27—Structure of leaf of greenhouse sunflowers, 1925: (1) dry; (2) medium; (3) wet.

wice that for the dry. In the case of the cockle-bur the medium value vas five times that of the dry and almost three times that of the wet. Vheat resembled sunflower in showing little difference between medium .nd wet, but the values for these were approximately thrice that for the lry (fig. 27, 28).

These growth results were reflected in the behavior of the stomata nd in the starch-content. Sunflowers in the wet container exhibited stom-

TABLE 105—*Final values for Helianthus, Xanthium and Triticum*

Criteria	Helianthus			Xanthium			Triticum		
	Dry	Med.	Wet	Dry	Med.	Wet	Dry	Med.	Wet
Ave. ht. cm............	22.0	64.0	59.5	11.9	23.1	18.9	17.0	23.0	22.0
Ave. diam. stem, mm.....	3.3	6.4	5.5	3.0	5.4	4.1	1.5	2.5	2.0
Ave. size largest leaf, cm.	6.5 × 4.0	9.6 × 7.6	8.0 × 5.6	7.0 × 4.7	10.1 × 9.1	7.2 × 5.7	14.9 × 0.5	21.0 × 0.7	23.0 × 0.7
Ave. no. full-grown lvs.	3	8	6	2.1	5.7	3.9	2.7	10.	7.0
Ave. leaf-area 1 side, sq. in................	8.59	47.17	36.7	8.65	42.22	16.84	...	..	..
Ave. dry wt. gm........	0.63	2.15	1.97	0.61	3.14	1.17	0.08	0.25	0.23

atal apertures of 17.5μ and abundant starch, while in the dry one the openings ranged from 3μ–4μ and starch was lacking at those times when the leaves were more or less wilted. Under similar conditions the apertures in the cockle-bur were respectively 10 and 2μ–3μ wide and the starch in the ratio of 2:1. Concordant results were obtained at various times, though the wet containers frequently exhibited more starch than the medium, especially in the earlier period.

Nutrient Series

Installation—Three soil mixtures were employed to yield the desired range of nutrients, namely, poor consisting of 2/3 sand and 1/3 loam medium made up of 1/3 sand and 2/3 loam, and rich, all loam. The latter was further enriched by watering with nutrient solution, while the other two sets received water alone. The holard was 16.5% at the time of planting and the nitrate-content of the medium containers 45.6 ppm. The seeds were planted on July 10 to yield 8 individuals of *Helianthus* and of *Xanthium* in the respective containers; in the third set seeds of *Andropogon nutans* were sown thickly and later thinned to 200 per container. Water and nutrients were added from time to time, the sunflower set requiring much more water than the others, according to its wont. In the case of the grass, all three sets used the same amount of water (12 lbs.); in the sunflowers and cockle-burs the rich sets needed the least water, the poor slightly more, while the medium used considerably more in the former and somewhat more in the latter.

Comparative development—Measurements were taken on July 2[illegible] for the sunflowers and cockle-burs, with the results indicated in table 10[illegible]. The grass at this time exhibited little or no difference between the several containers.

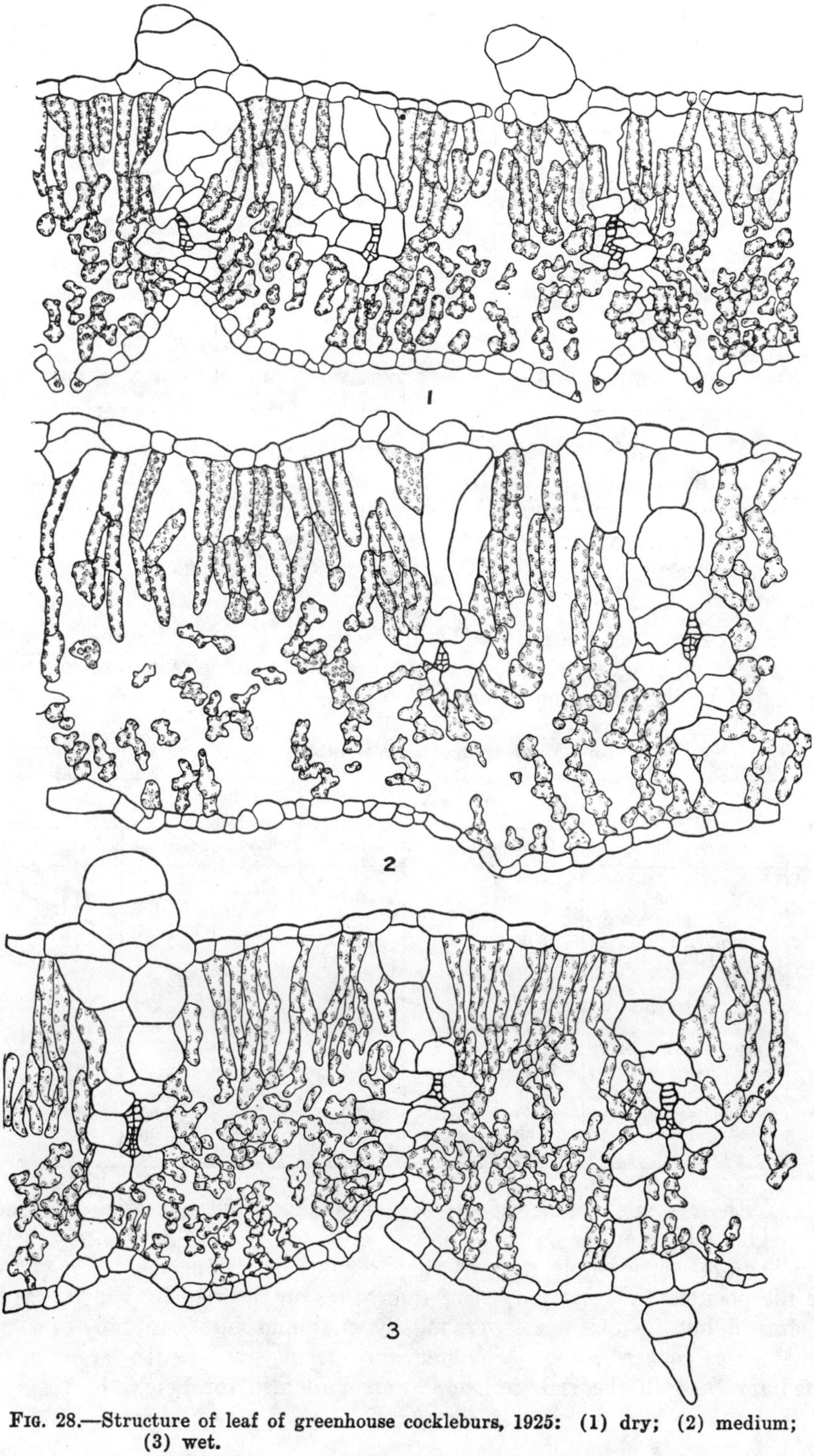

Fig. 28.—Structure of leaf of greenhouse cockleburs, 1925: (1) dry; (2) medium; (3) wet.

TABLE 106—*Development with different nutrient contents*

Criteria	Helianthus			Xanthium		
	Poor	Medium	Rich	Poor	Medium	Rich
Ave. ht. cm.......	29	33	28.4	9.4	10.5	10
Ave. length 1st internode, cm.....	2.5	3.3	3.4	0.9	1.0	0.8
Ave. diam, stem, mm.	6	7	6	3.2	3.4	3.3
Ave. no. full-grown leaves	7	7	7	4.2	4.5	4.5
Ave. size largest leaf, cm.........	10.5 × 8	11 × 9	11 × 8.5	7.7 × 6	8.3 × 6.1	7.7 × 5.6

In the case of both species, the best development took place consistently in the medium container, though the differences were small throughout; those between the poor and rich were still smaller and entirely inconsistent. The most significant fact is the failure of the rich set to give the maximum growth, a result evidently arising from too high a concentration of the soil solution.

Final results in the nutrient series—The plants were removed from the containers on August 8 and final measurements made with the outcome indicated in the following table:

TABLE 107—*Final values for Helianthus and Xanthium*

Criteria	Helianthus			Xanthium		
	Poor	Medium	Rich	Poor	Medium	Rich
Ave. ht. cm.......	80.5	90.7	76.7	27.6	30.2	25.4
Ave. diam. stem, mm.	10.6	12.4	10.4	6.7	7.4	6.9
Ave. length leaves, cm.	15.1	16.1	13.5	13.4	13.9	12.9
Ave. width leaves, cm.	12.4	13.5	11.6	12.6	13.6	13.1
Ave. no full-grown leaves	10.1	11.6	11.5	6.9	6.7	6.9
Ave. leaf-area 1 side, sq. in......	124.60	211.45	123.0	72.86	87.56	81.96
Ave. dry wt. gm...	5.69	7.14	5.81	2.85	3.06	2.79

With a single slight exception, the mediums were better developed throughout than either the poor or rich cultures, the differences now being much more considerable than at the previous determination. The plants in the poor containers were better than those in the rich in 8 instances as against 5, but the differences were mostly small and significant only of somewhat great concentration. The stem dimensions were a fifth larger in the medium than in the rich for sunflowers and also for height in the case of cockle-burs. The largest differences were found in the case of leaf-area and dry weight. The former was approximately three-fourths greater for the medium sunflowers than for poor or rich, these being essentially alike; the relation was much less striking in cockle-burs, the medium being about a fifth higher than the poor, with the rich intermediate. With respect to dry weight, the poor and rich cultures were practically the same for both species; the value for the medium sunflowers was a fourth greater, but for the cockle-burs hardly a tenth. Determinations of stomatal opening and starch-content gave either no differences for the three conditions in either species or only insignificant ones.

The consistent advantage exhibited in the medium containers indicates that nutrients do affect growth response, while the fact that the rich set yielded smaller values as a rule than the poor set shows that the increase in nutrients soon becomes a limiting factor by virtue of the consequent great concentration of water-content and perhaps also of cell-sap. Upon the basis of the most significant criterion, dry weight, it appears that nutrients are much less important to growth than either water or light, and that in consequence they play a much smaller part in competition. The dry weight of *Helianthus* under medium or optimum nutrient content was but a fourth greater than for the poor set, while the differences between the values under medium and deficient holard were three times for *Helianthus* and *Triticum,* and five times for *Xanthium.* Likewise, the differences in dry weight due to growth in 50% and 10% light intensity were nearly four times for the sunflower and three times for the cockle-bur.

GENERAL SUMMARY

Six series of control experiments were conducted to determine the relative importance of the direct factors in competition, and to confirm the results obtained upon the effect of density in the series described in the previous chapters. These included two each of *Helianthus* and *Triticum,* usually with several sets, and one of *Xanthium* and of *Andropogon nutans.* With infrequent slight exceptions, the effect of the various densities upon the different parts was consistent with the demand, and hence in accord with all the similar results previously obtained. The comparison of the series with respect to the paramount factor was rendered more difficult by the effects of density, and the special criteria were somewhat less consistent as a consequence. The relative control of the three factors was naturally revealed much better in the two integrating criteria, leaf-area and dry weight. With practically no exception, these values were highest when both water and nutrients were adequate and lowest when they were both deficient. They were nearly as low or occasionally lower when water alone was deficient, and as a rule a deficiency in water was much more decisive than one in nutrients. The control of light was much less important until the reduction became relatively much greater than with the other two factors. When water and nutrients were abundant and competition was for light alone, the values were usually the highest, except in the densest culture where the reduction in light intensity was greatest. The summaries for the various species and series are to be found on the following pages: 245, 254, 258, 267, 270, 277, 280, 281, 287.

In the two underplanted series designed to permit a direct comparison between competition for light and for water, as well as between no competition and that for both, the values were regularly lower for water minus than with deficient light, with a single exception for dry weight in the second series. The latter is readily explained by the fact that the light value had dropped to 10%, or ten times, while the drop for water was very much less. Thus, though the actual effects were not very different, they were accomplished by a two or three-fold reduction in water by contrast with a tenfold decrease in light. As to the presence or absence of competition, leaf-area was four times and dry weight two and a half times greater in the latter case.

The results of the experiments with respect to the direct response to the three primary factors verify the conclusions already drawn. With the light reduced five times, the leaf-area was decreased a third, and the dry weight three times in *Helianthus* and *Triticum*. When the water-content was diminished from a half to once, the leaf-area for *Helianthus* and *Xanthium* was reduced five times, the dry weight for *Helianthus* and *Triticum* three times and that for *Xanthium* five times. The differences in the nutrient contents were less definitely known in this series, but the medium was approximately twice the poor and half of the rich. As to leaf-area, the medium was about two-thirds greater than either poor or rich in the case of *Helianthus* and only about a tenth greater for *Xanthium*, while the dry weights of the medium were a fourth and a tenth greater respectively.

The first experimental studies of the nature of competition and the rôle of the three primary factors led to the tentative conclusion that water was paramount, light next and nutrients third in importance (Clements, 1905; cf. p. ----). It was recognized, however, that this depended primarily upon the relative quantity of each, and that any one of the four factors, water, light, nutrients or soil-air, would become controlling when the relative supply of it was lowest. It is clear, moreover, that the rate of demand and supply will sometimes be more important than the total amount available. While these conclusions were necessarily preliminary, they now seem to be well warranted by the present much more extensive and varied studies. Furthermore, they appear to be supported by theoretical considerations drawn from the physiological rôle of the several factors. Thus, water is not only a necessary raw material for photosynthesis, but also an indispensable solvent; it serves for diffusion and conduction, as a regulator of leaf temperatures, and as the mechanical agent in growth, in addition to many secondary rôles. Indispensable as light is, only a small fraction or even an apparently inconsiderable one of less than 1% is employed in photosynthesis. As a consequence, it appears that the reduction of the light intensity can proceed very much further than that of the other three factors without untoward effects. The importance of nutrients is determined chiefly by their rôle in chemosynthesis and assimilation, but in nature this is less than that of water, or even of light in dense or layered communities. In the case of cereals or similar field crops especially, where a large yield is the desideratum on the basis of high-producing strains, it easily becomes the critical factor in humid climates. Finally, the significance of soil-air is great in bogs, swamps and water-logged soils in general, particularly agricultural ones, but the direct attack upon its rôle in competition is still to be made.

8. FUNCTIONAL STUDIES IN CONTROL CULTURES

Object and plan—Control series similar to those described in the preceding chapter were established during 1926 for the study of functional behavior in the competitors. This was naturally directed at two major objectives, the modification of function under varying degrees of competition as determined by density, and the compensating effect of increasing the supply of a particular factor. The water relation afforded by far the best sequence of functions and attention was in consequence concentrated upon this, though starch-content, photosynthate and the relation of stomatal opening to light were also taken into account

The respective series differed from those of 1925 by the omission of the 4's and 16's, being limited to the 2's, 8's and 32's in the endeavor to obtain larger differences. In addition, cans with central cylinders were employed for insert phytometers. The installation was made on April 29 and the sets were treated as previously described. Additional series were started from time to time in order to provide a constant supply of material for study and measurement. In the case of *Helianthus annuus*, measurements of function were made at intervals of a few days for nearly two months.

FUNCTIONAL BEHAVIOR OF HELIANTHUS UNDER COMPETITION

Stomatal behavior and starch-content during May—The rate of evaporation under the three degrees of density was determined on May 15 and June 2 for the bright sunny day, with the following results for the W+ series:

2's	12.3 cc.	11.9 cc.
8's	11.2	10.8
32's	9.1	8.8

The rate of loss decreased consistently and quite regularly with increasing density, the atmometers under the 2's losing approximately a third more than those under the 32's. At noon on May 15 the stomata of the 2's were wide-open, those of the 8's ranged from wide-open to almost closed, while in the 32's they were closed or nearly so on both turgid and wilted leaves. In the two higher densities, the behavior was the same in the series with water plus and water minus, indicating that the compensation was inadequate on a warm sunny day. At 3 p.m. the stomata of the 2's were half-open and of the 8's somewhat less open on the average. On May 19 the average width of the pore, as determined from 20 measurements on 3 or more leaves in each density, was respectively 10.5, 5.8, 3.7, and 3.1μ for the 2's, the 8's with water plus, and the 32's with water plus and water minus. Further agreement between the degree of opening and the density was obtained on May 25. The pore ranged from 4μ–12μ in width in the 2's, and 3μ–9μ in the 8's; half of the stomata were open 1μ–6μ in the 32's W+, but in the 32's W−, all were closed.

On the other hand, the number of stomata per unit area (0.1 sq. mm.) increased consistently as the density became greater; on May 17 the lower leaves of 2's gave an average of 8, of the 32's W+, 9 and 32's W—, 10 stomata. Middle leaves on May 25 gave an average of 12.9 for the 2's, 13.4 for the 8's with water plus and 16.4 for those with water minus; the corresponding values for the 32's were 12.7 and 16 stomata. With rare exceptions there were more stomata in denser cultures where there was less water per plant to maintain turgidity and to produce mechanical stretching of the leaves, and the same rule obtained in the series with water minus and for similar reasons. The size and amount of opening decreased with the greater demand for water in the denser plantings, and likewise in the series with the water supply not maintained.

The starch-content on May 27 for leaves at different levels in the several densities was as follows:

TABLE 108—*Starch-content in relation to density and level*

Leaves	2's	8's W+	8's W—	32's W+	32's W—
Lowest pair....	—: little	0:none	0	—	0
Second pair....	++:much	—	—	....	....
Topmost pair in full sun......	+++:very much	++	++	++	+:moderate

The correspondence with position of leaf was good throughout, since this affected the light values more decisively than density; in every case there was a marked decrease from the topmost to the lowest leaves. The 2's contained more starch throughout the several levels than did the 8's or 32's. The addition of water had practically no effect on the 2's and 8's, but made a definite increase in starch-content in the 32's, where the competition for water was much keener.

Stomatal behavior and starch-content during June—On June 1, 10–15 of the uppermost leaves in each container were examined with respect to stomatal opening. The average width of pore was secured for each leaf and the results averaged; closed stomata were not taken into account in this operation. The values were least consistent at 8 a.m., when water and temperature relations favored low transpiration, but even at this time all the stomata of the 2's were open with an average width of 7.1μ. The number of leaves with closed stomata was 2 out of 10 for the 8's W+ by contrast with 5 out of 10 for the 8's W—; for the 32's the respective numbers were 5 and 7 out of 10. By 11 a.m., the pore of the 2's had decreased to 4.2, but all the stomata were still open; this was also true now for the 8's W+ with pore of 3.2, while in the 8's W— the pore was 2.3 and was closed on 7 of 10 leaves. The results for the 32's were not consistent at this time owing to shading in the water minus set, but the higher temperature and lower humidity at 2 p.m. were reflected in an average of 1 for

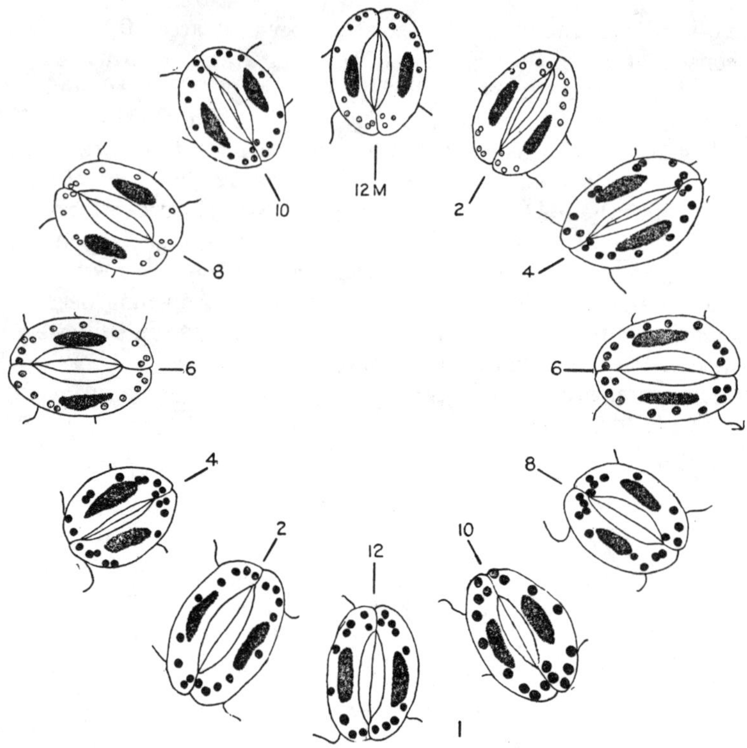

FIG. 29.—Stomatal behavior in lower epidermis of 2's,

32's W— and 3.2μ for 32's W+. The 8's W+ gave a value of 3μ and the 2's, both water plus and minus, of 4.5μ; one leaf of the 10 showed closed stomata, while for the 8's the number was 6 in 10. At 5:15 p.m., the 2's, 8's W+ and 32's W+ exhibited closed stomata on about half the leaves, the 8's and 32's with water minus having them all closed. This was practically the condition at 7 p.m. likewise. With unimportant exceptions, the extent and degree of opening decreased with the density, but increased with the holard throughout most of the day (fig. 29).

On June 19 counts were made of the number of stomata per 0.1 sq. mm. and the extent of opening for different leaves and the width of the pore was measured, with the following results. All determinations were made on the upper epidermis of the uppermost mature leaf; one reading was made from each leaf as to extent and an average of 5 was employed for pore width.

The number of stomata per 0.1 sq. mm. increased with the density, as well as with lower holard. In both cases, the immediate explanation was to be found in the reduced water supply, which resulted in less mechanical expansion and consequent smaller size of the leaf. This took place after the primordia of the guard-cells were formed, with the consequence that the number per unit area was nearly twice as great for the 32's W+ as for the 2's, and more than twice as great for the 32's W—. For both the 8's and 32's the number was approximately 50% greater with water minus than with water plus. These increases in number with increasing density and decreasing holard furnish the explanation of the discord-

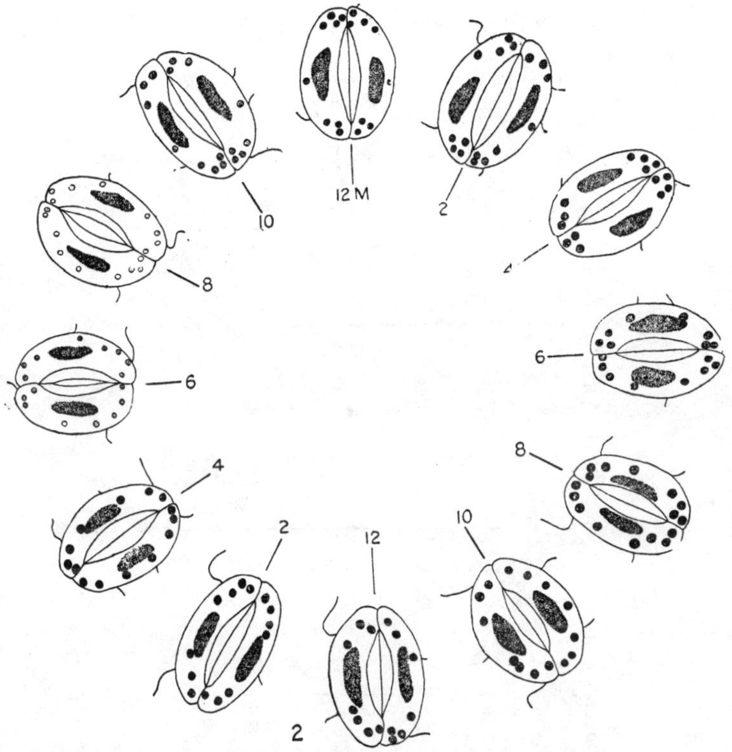

water plus and 8's, water minus, greenhouse sunflowers, 1926.

ant results sometimes obtained in the transpiration in different cultures.

The extent to which the stomata were open on the various leaves was signally affected by the water available, the two containers with water minus exhibiting closed stomata throughout the day. It was also in general accord with the density, the 2's showing much more opening than the 32's and the 8's being intermediate. The size of pore also decreased with much consistency as the density increased, being approximately half as great throughout the day in the 32's as in the 2's.

Starch tests made on the same day at 9:30 a.m. showed a moderate amount in the uppermost mature leaf of the 2's; this decreased through the 8's and 32's with water plus to practically none in the 32's W—. The paired leaf was tested at 6:45 p.m. and gave nearly uniform results for the five containers, the 32's W— showing a slight reduction. This uniformity was to be explained by the fact that the plants were turgid and the leaves selected were fully exposed to bright sunshine.

The results obtained by starch tests on June 2 were essentially consistent and were in close agreement with those of May 27.

The decrease from the top to the lowest leaf was consistent and regular in all the containers. It was less consistent for density and drouth, owing largely to the fact that the 32's W— was more open and the light relation consequently disturbed the sequence.

On June 21 stomatal opening was determined at three different periods; tests were made with cobalt-chloride paper at 2 p.m. and starch tests at 10:45 a.m. and 4 p.m. The condition of the stomata was observed for

TABLE 109—*Number of stomata and size of pore in relation to density*

Container	8:10 a.m.		10:00 a.m.		11:30 a.m.		1:15 p.m.		2:45 p.m.	4:30 p.m.		6:45 p.m.
	Ave. no.	Pore size	Ex-tent	Pore size	Ex-tent	Pore size	Ex-tent	Pore size	Extent	Ex-tent	Pore size	Extent
2's	26	4μ	Many clsd Watered	3μ ...	2 of 5 clsd	4μ	...	3μ	0 closed	0 clsd	3.5μ	All closed
8's W+....	34	3.3	Closed to 2μ	...	3 of 5 clsd Watered	...	0 clsd	3.6	0 closed	1 of 4 clsd	2	All closed
32's W+...	47	2.4	Most closed	1 ...	Watered	...	2 of 5 clsd	2.3	0 closed	Most clsd	2	All closed
8's W—....	50	Clsd wilt-ing	All closed	...	All closed	...	All closed	...	All closed	All closed	...	All closed
32's W—...	64	”	”	...	”	...	”	...	”	”	...	”

both surfaces of the leaf, but is not recorded since it was either identical for both or only slightly greater on the lower surface. The light intensity for the lower leaves was approximately 3%, while the top ones were in full sunshine.

TABLE 110—*Starch-content in relation to density and level, June 2*

Leaf	2's	8's W+	8's W—	32's W+	32's W—
Lowest	+	+	0	—	+
Middle	++	++	+	+	...
Top	+++	+++	++	+++	++

The correspondence between stomatal opening and both density and position was close throughout. This was true also for starch-content and leaf position, but there was rather less agreement with density. The stomata of the top leaves were regularly more open than those of the lowest ones, and they were also usually more open in the less dense cultures. The degree of opening as measured by cobalt-chloride paper was

TABLE 111—*Stomatal behavior and starch-content, June 21*

Leaf	10:45 a.m.		2 p.m.	3 p.m.	4 p.m.	
	Stomata	Starch	Stomata	Cobalt paper sec.	Stomata	Starch
2's Top	All open	++	Open	Up. sur. 7 L. sur. 6	½ open	++
Lowest	Partly open	—	1/3 open	Up. sur. 11 L. sur. 9	½ open	+
8's Top	All open	++	½–2/3 op	Up. sur. 8 L. sur. 7	½ open	++
Lowest	Partly open	+	1/3 open	Up. sur. 16 L. sur. 13	½ open	+
32's Top	All open	+	Mostly closed	Up. sur. 10 L. sur. 9	½ open	++
Lowest	Most closed	—	....	Up. sur. 17 L. sur. 16	Mostly closed	—

thoroughly consistent with leaf position and with density, the values being less for the lower surface in every case and rising for both surfaces of topmost and lowest leaf from the 2's to the 32's. In other words, lower surfaces transpired more than upper, top leaves than lowest, and 2's more than 32's. Similar determinations on July 7 gave equally consistent results for the three densities at three intervals, namely 4, 6, 11; 6, 8, 10; and 7, 8, 16 seconds for 2's, 8's and 32's respectively.

Relation of opening to starch-content of stomata—The relation between the degree of opening and the starch in the chloroplasts of the guard-cells was determined by means of epidermal strips taken at 2-hour intervals on June 23. In addition, the number of chloroplasts per stoma was ascertained by means of counts on 10 stomata. The condition of the stomata is indicated first, the amount of starch second in each column; the extremes are shown by 0 and +, a small amount by —.

TABLE 112—*Relation of pore to starch-content of stoma*

Container	Ave. no. plastids	8:30 a.m.	10:30 a.m.	2:30 p.m. (watered)	2:30 p.m.	4:30 p.m. (cloudy)
2's	16	— —	+ —	— +	+ —	0 +
32's W+	13	— —	+ 0	— +	+ 0	0 +
32's W—	11	— —	0 +	0 +	0 +	0 +

The number of chloroplasts decreased with the density and with the holard, evidently as the direct outcome of a larger number of stomata per unit area and a correspondingly smaller size of stoma. The inverse relation between the degree of opening and the amount of starch in the chloroplasts obtained nearly throughout the series, the only exceptions occurring at 8:30 a.m. Closure with correspondingly high starch-content was found at all times after the first in the 32's with the water lacking; the 2's and 32's W+ exhibited mid-day closure with increase of starch, after which the relation was reversed at 2:30 p.m. and restored at 4:30 p.m.

The relation between stomatal opening and starch was also traced in the 32's when they were wilting on June 25. At 8:30 a.m. the top leaf was turgid, with some stomata open and a small amount of starch; a flaccid leaf below exhibited few open pores and contained a moderate amount of starch, while in a badly wilted one the stomata were closed and starch was abundant. The plants were watered at noon and a new series of strips made at 1:30 with the following results. In the topmost leaf, most of the stomata were closed and starch was present, the open ones containing no starch; in the next leaf below, there was less starch and most of the pores were open, while in the next leaf, there was practically no starch and the stomata were wide open.

Results on July 7–8—Starch tests were made on sunflowers in the W—series with the following results (table 113, p. 297):

The values for starch-content were essentially consistent throughout for both density and leaf position, decreasing downward and in the direction of greater density; wilting was correlated with both features. The amount of starch naturally increased likewise to late afternoon, except in the leaves where wilting was increasing, in which it fell off.

The effect of wilting upon transpiration and food-making was determined by means of four containers, two of which contained 3 and two 4 plants each; one of each was allowed to wilt and the other maintained with full turgor. At the end of the first 24-hour period, direct examination with the microscope as well as the use of epidermal strips showed that all the stomata were closed on the wilted plants (table 114, p. 297).

TABLE 113—*Starch-content of W— sunflowers, July 8*

Plant and Leaf	9:00 a.m.	1:00 p.m.		4:00 p.m.	
	Starch	Leaf cond.	Starch	Leaf cond.	Starch
2's					
Upper	++	Turgid	+++	Turgid	+++
Lower	0	Turgid	++	Turgid	++
8's					
Upper	+	Turgid	++	Turgid	+++
Lower	0	Nearly turgid	++	Turgid	++
32's					
Upper (turgid) ..	—	Slightly wilted	+	Semi-wilted	—
Lower (wilting)	0	Much wilted	—	Fully wilted	— (some at base only)

TABLE 114—*Transpiration of turgid and wilted plants*

Plant	Total loss, gm.		Loss per sq. dm. gm.	
	1st day	2nd day	1st day	2nd day
3's				
Turgid	150	175	25.96	30.27
Wilted	26	23	4.50	3.98
4's				
Turgid	159	203	24.12	30.77
Wilted	35	31	5.31	4.70

The water-loss agreed in every instance with the condition of the plants and the consequent behavior of the stomata. The closure of these as a result of wilting reduced the transpiration from 5 to 10 times by comparison with the turgid individuals.

Summary—With practically no exceptions the degree of stomatal opening decreased regularly with the density of the culture, and with the reduction of the holard independently of density as well. The number of stomata per unit area exhibited the inverse correlation, increasing with the density and from the water plus to minus cultures. This was evidently due to lessened turgor under both sequences and to the consequent reduction in the mechanical stretching of the tissues. Iodin tests of the starch-content of the mesophyll showed a general reduction in amount as the density rose, and a marked decrease with lowered holard. The most striking gradient was associated with leaf level, top leaves often exhibiting twice as much as middle ones, and these twice or thrice that of the lowest. This was largely due to the greater light intensity above, but in the case of lower leaves especially it was also a consequence of age and correspondingly reduced activity. When exceptions to the rule occurred, they were confined to density, and were mostly the result of the interaction of competition and protection at the extremes. The number of chloroplasts decreased with the density, owing to the fact that the stomata were smaller when more numerous. The degree of stomatal opening and the amount of starch in the guard-cells were in inverse relation throughout the day, with rare exceptions.

The effects of competition upon root pressure and conduction are summarized at the end of the chapter.

Root pressure in relation to density—The root pressure of plants from the several densities was measured by means of a manometer on four different occasions. The values obtained on May 29 were as follows:

TABLE 115—*Root pressure in different densities, May 29*

Plant	Stature cm.	Diam. stem mm.	No. leaves	Max. rise mercury, cm.
2's W+	64	10	15	26
8's W+	77	10	11	12
32's W+	59	8	9	7
32's W—	55	5	4	0

On June 2 the rate of rise was recorded at successive intervals during a 2-hour period; the manometers were set up in the order of density, the operation requiring 20 minutes.

TABLE 116—*Root pressure in different densities, June 2*

Plant	Stature cm.	Diam. mm.	Maximum rise mercury, cm.				
			2:14 p.m	2:30 p.m.	2:50 p.m.	3:10 p.m.	3:50 p.m
2's W+	99	11	3.0	4.5	4.8	5.6	6.5
8's W+	105	10	3.0	3.6	2.6	1.9	1.1
32's W+	83	8	1.1	1.7	1.6	1.1	0.8
32's W—	42	5	0	0	0	0	0

The results were entirely consistent and in complete accord between the two experiments. The rise of the column was much more rapid and the maximum much the highest in the 2's; these were not competing for the chresard and the addition of water produced no effect upon the root pressure. The 2's gave a consistent rise during the 2-hour period, while the 8's and 32's with water plus rose to a low maximum and then fell off rapidly, while the 32's W— produced no pressure in either set. As would be expected, there was a fairly close correlation between root pressure and number of leaves, and in the second experiment with the diameter of the stem. The maximum of stature fell between the extremes of density, in accordance with the rule that competition promotes elongation until it reaches a certain degree of intensity.

TABLE 117—*Root pressure in different densities, June 17*

Plant	Stat. cm.	Diam. mm.	No. lvs.	9:05 a.m.	9:12 a.m.	9:40 a.m.	10:00 a.m.	10:10 a.m.	10:27 a.m.	10:45 a.m.	11:15 a.m.	11:45 a.m.	1:30 p.m.
's W+	22.5	3	4	0	0	4.7	7.8	9.4	9.4	9.4	9.4	6.2	....
's W+	25.0	4.7	6	...	...	0	4.7	9.4	13.0	19.0	17.0	17.0	4.0
's W+	25.0	3	4	...	0	4.7	0	0	0	0	0	0	....
's W+	24.0	2.3	4	...	...	0	0	0	0	0	0	0	....

The relation of rise to number of leaves and diameter of stem is well demonstrated by the behavior of the plants from the 2's. The second plant with an advantage of 50% in these respects gave twice the maximum of the first. On the other hand, the 8's with the same values as the latter rose quickly to the initial point and then fell as quickly to zero, while the 32's with but slightly lower measurements showed no rise. In both cases this was evidently due to the greatly increased demand for soil-water in the face of an inadequate supply.

A similar experiment on July 7 gave an initial positive pressure for the 2's of 95 mm. at 9:20 a.m., which fell off to 29 mm. at 1:10 p.m.; the 8's gave a pressure of but 3 mm., which dropped to —30 mm. at the close. One plant of the 32's showed an initial rise of 2 mm., but the column fell to —15 mm. in an hour and to —90 mm. in the next two hours; another rose from —25 to —12 mm. and dropped again to —20 mm.

Conduction and osmotic pressure—Sections of stem 8 cm. long were taken from the first internode, and the amount of water conducted determined at half-hour intervals, with the following results:

TABLE 118—*Conduction in different densities, July 7*

Plant	First Set		Second Set		Stem diam. mm.
	1st half-hour cc.	2nd half-hour cc.	1st half-hour cc.	2nd half-hour cc.	
2's	1.8	1.1	5.5	4.2	6.5
8's	1.5	0.9	3.4	2.8	6.3
32's	0.8	0.5	1.2	0.9	3.7

The difficulties involved in securing good root-hairs from plants rooted in the soil, as well as lack of time, limited determinations of the concentration of the cell-sap to a very few. However, these were consistent with density and holard; the average osmotic pressure for root-hairs of the 2's was 4.5 atmospheres, for the 32's W+, 8.2 and for the 32's W— it was 11 atmospheres. Root-hairs from turgid plants gave a value of 4.2 and from wilting ones, 6.3 atmospheres.

FUNCTIONAL BEHAVIOR OF XANTHIUM UNDER COMPETITION

Stomatal behavior and starch-content—Water was added to all the containers just before this study was begun on July 12, but the 32's W—

were slow in recovering, owing to their wilted condition. The degree of opening was determined by direct microscopical examination of both the lower and upper epidermis. Observations were made at intervals of two hours for the most part, except at night when they were longer, and the series extended from 10 a.m., July 12 to midnight July 13. The + sign indicates wide-open, the — sign slightly open or a few open, fractions intermediate degrees, and 0 is employed for closed. The colon indicates a difference in the behavior of upper and lower epiderm, the former being given first.

TABLE 119—*Stomatal behavior of Xanthium in relation to density*

Time	Light	2's W+	8's W+	8's W—	32's W+	32's W—
July 12						
10 a.m....	Sun	+	+	+	+	½
12 m......	"	+	+	+	+	½
2 p.m.....	Occ. cloud	+	+	+	+	½
4 p.m.....	Cloudy	—	—	—	—	—
6 p.m.....	Cloudy	0	0	0	0	0
July 13						
4 a.m.....	Dark	0	0	0	0	0
5 a.m.....	In shade	0	0	0	0	0
6 a.m.....	Brighter	—:0	—:0	—:0	—:0	—:0
7 a.m.....	Sun	—:0	—:0	—0	—:0	0
8 a.m.....	"	½:—	½	½:—	½	½
10 a.m....	"	+	+	+	+	0
12 m......	"	+	+	½	+	—
2 p.m.....	"	+	+	0 (turgid)	+	0 (wilting)
4 p.m.....	"	+	¾	0 "	½	0 "
6 p.m.....	Shadow	—	—	0	0	0
8 p.m.....	Twilight	0	0	0	0	0
12 midn...	Dark	0	0	0	0	0

As a rule, there was a close correspondence in stomatal behavior between the three densities with adequate water supply; however, closure usually lasted longer and began earlier in the 32's, followed by the 8's. A reduced holard brought about a striking difference, the effects being felt even when the plants still appeared turgid (fig. 30).

On July 13, epidermal strips were taken from the several containers in the water-plus and water-minus series, from which determinations were made of the number of stomata per square millimeter on the lower surface of the topmost mature leaves, their size and the number of chloroplasts per stoma. The average was based upon 50 counts for the stomata and 40 for the chloroplasts.

The values were essentially consistent with density and holard throughout. The number of stomata increased with greater density, being more than a third higher in the 32's W+ than in the 2's; it increased still more with deficient water, rising approximately 10% for the 8's and 20% for the 32's. Size of stoma increased in the inverse relation; it was greatest in the 2's and was diminished about a fourth in the 32's W—. The num-

TABLE 120—*Number and size of stomata in different densities and holard*

Plants	No. per sq. mm.	Length μ	Width μ	Ave. no. plastids
2's (both)......	281	30.3	15.9	14.6
8's W+........	336	27.8	14.2	13.5
32's W+........	394	24.8	13.9	12.0
8's W—........	375	25.0	13.9	12.7
32's W—........	450	22.2	12.5	11.7

ber of chloroplasts varied directly with the size, decreasing also about a fourth from the 2's to the 32's W—.

Determinations of stomatal opening by means of cobalt-chloride strips and of the starch-content on the same day yielded the following results:

TABLE 121—*Stomatal opening and starch-content on July 13*

Plant	No. stomata per sq. mm.	Seconds to color strip		Starch-content	
		Top leaf	Lowest leaf	Top leaf	Lowest leaf
2's W+........	231	13	14	++	....
8's W+........	327	13	17	++	+
32's W+........	381	16	33	++	....
8's W—........	354	23	33	+	....
32's W—........	531	32	52	—	—(base only)

Summary—The values obtained were all in essential harmony with each other and with those from previous experiments. The number of stomata per unit area rose as the density increased and the expansion of the leaf decreased; it was more than half again as great in the 32's than in the 2's of the water-plus series. The 8's with water minus gave a number less than 10% larger than those with water plus, but in the 32's the corresponding increase was nearly 50%, owing to the much greater demand, while the number of stomata for the 32's W— was more than twice that for the 2's W+, both density and water being concerned in this. As measured by the cobalt test, the pores of the stomata were wider in top than in lower leaves, the time required for the latter being twice as great in the maximum density. The difference due to density alone was not great for top leaves, but was more than twice for the lowest leaves of the 2's and 32's. With an inadequate supply of water, the time involved was half again as long for the 32's as the 8's; it was more than twice as long for top leaves in the 32's than in the 2's and nearly four times for the lowest leaves.

The slight difference in stomatal opening for the top leaves of the three containers with adequate water explains why the starch was about equally abundant in them. With deficient water-supply there was a marked drop from the 8's to the 32's, as there was likewise from the 8's and 32's with water plus to those with it minus. The

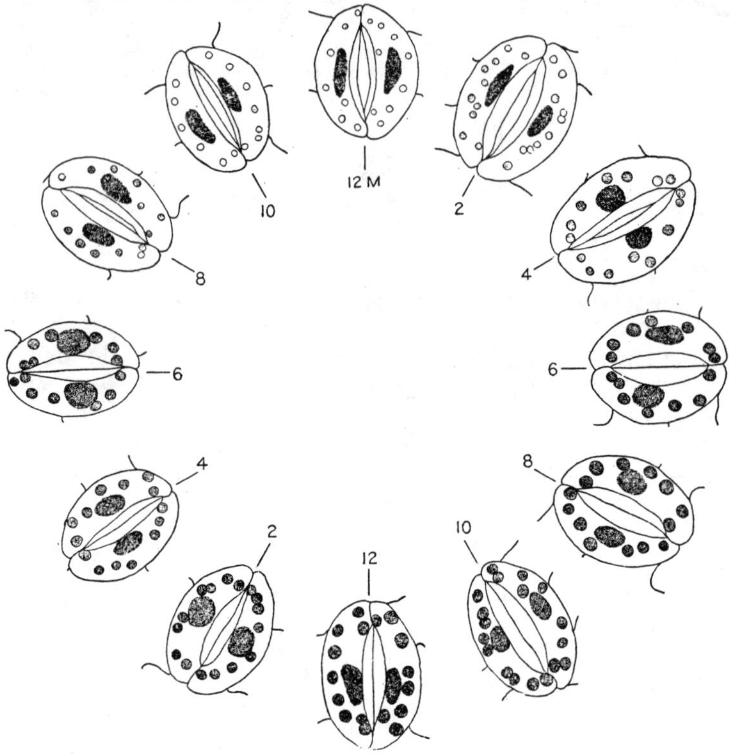

FIG. 30.—Stomatal behavior in lower epidermis of 2's, wa

lowest leaf in both cases contained a smaller amount of starch than the topmost one. Concordant results were secured from the measurement of the photosynthate by means of the picramic method. The amount was practically twice as great for the 2's W+ as for the 32's W—, namely, 0.1248 and 0.0634 gm. per 100 sq. cm.; it was respectively 0.091 and 0.068 gm. for the 8's with water plus and minus.

Root pressure in relation to density and holard—Root pressure was determined in the usual manner for the 8's and 32's with adequate and deficient water-supply, with the results given below:

The rise of the mercury column in millimeters is consistent throughout; in spite of fluctuations during the period due to the relation between water absorption and loss, the difference between the 8's and 32's is uniformly striking. The smallest ratio between the two when water was adequate was 2:1, or 82 to 41; the largest was 10:1, or 70 to 7, which occurred near the peak of transpiration, when the supply was much less adequate in the greater density. The relation in the water minus containers was the same, though the results are less comparable, since they were nearly all positive in the 8's and negative in the 32's. The direct lack of water produced a greater departure than that due to an insufficiency arising from competition. The maximum for the 8's W+ was 153, for the W—, 24 mm., while the respective minima were 63 and —7 mm., both occurring in the period of

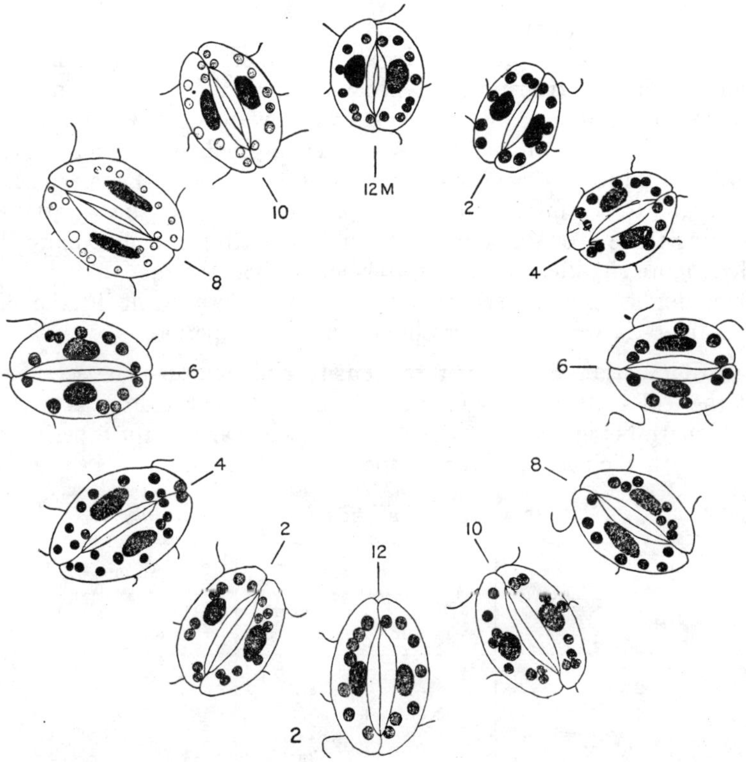

s and 32's, water minus, greenhouse cockle-burs, 1926.

TABLE 122—*Root pressure in Xanthium, July 13–14*

Time	8's W+	8's W—	32's W+	32's W—
a.m.				
8:36	....	....	....	Set at 0
8:44	....	....	Set at 0	....
8:52	Set at 0	....	14 mm.	....
9:04	46 mm.	....	38	—5 mm.
9:14	....	Set at 0	....	....
9:20	82	....	41	—19
10:10	83	5	32	—26
10:40	79	19	28	—27
11:00	78	22	27	—27
11:45	78	24	25	—28
p.m.				
1:35	75	12	23	—27
2:45	70	2	7	—27
3:30	65	—3	14	—29
4:40	63	—7	17	—28
6:10	78	16	25	—33
7:20	78	17	25	—30
8:30	78	18	24	—30
9:10	78	18	22	—30
11:45	116	18	22	—30
a.m.				
1:00	98	17	22	—30
8:00	153	20	22	—30

greatest transpiration. For the 32's the corresponding maxima were 41 and —5 mm., the minima, 7 and —33 mm.

The maximum pressures in all containers occurred at night or in the morning before transpiration became high; the minima on the contrary fell during middle or late afternoon when water-loss was the highest. The demand for water in the 32's W— was so great at all times that the pressure was uniformly negative, fluctuating but 14 mm. from the maximum at 9:20 a.m. to the minimum at 6:10 p.m. On the other hand, after the initial reading the 8's W+ fluctuated from 63 to 153, a range of more than twice. Even more significant of the water-supply was the fact that the values were much higher for the second day than the first, while in the other three containers it was lower for the same general period.

Conduction in relation to density and holard—The sections were cut at the soil-surface and consequently bore the cotyledons as well as leaves, the length being 5.5 cm. Each set of stems was run for a period of 30 minutes, with the results given in the following table:

TABLE 123—*Conduction in Xanthium, July 13*

Plant	First set		Second set	
	Diam. stem mm.	Water cond. cc.	Diam. stem mm.	Water cond. cc.
2's W+......	6.0	4.2	...	...
8's W+......	5.0	2.2	5.3	6.0
8's W—......	4.7	1.5	4.2	4.2
32's W+......	4.4	2.0	3.7	2.9
32's W—......	3.4	0.5	3.2	1.1

The results were consistent throughout, in spite of the fact that the second set of stems gave much larger values than the first. The conduction and the stem diameter increased with decreasing density, and with increasing holard. In the water series, the 2's conducted nearly twice as much water as the 8's, and the 8's twice as much as the 32's in the second set. The difference between the W+ and W— plants of 8's was approximately 50% of the latter in both sets, while for the 32's the W+ plants conducted thrice as much as the W— ones.

FUNCTIONAL BEHAVIOR OF TRITICUM UNDER COMPETITION

Stomatal behavior—The degree of opening in relation to density and holard was determined on May 15, 25, and 28, and June 11; the results were in essential agreement throughout, as indicated by the two tables that follow. On May 28 the stomata of the upper epidermis of the topmost mature leaf were examined in situ, the average being derived from 10–15 readings on several leaves.

The course of stomatal opening through the day was again determined on June 11, a hot sunny day; the leaf below the topmost was examined in each case, the average being derived from 10 leaves in each container.

TABLE 124—*Stomatal opening in Triticum, May 28*

Container	7:45 a.m.	10:45 a.m.	1:30 p.m.	4:30 p.m.
4's W+........	Open, 7.0μ	Open, 5.3μ	½ clsd, 2.3μ	½ clsd, 4.7μ
16's W+.......	" 7.0	½ clsd, 2.8	Most clsd, 2–3	Most clsd, 2–3
16's W—.......	" 4.4	¾ clsd, 2.5	All clsd	All clsd
64's W+.......	" 4.0	Most clsd, 2–3	" "	" "
64's W—.......	½ clsd, 3	All clsd	" "	" "

TABLE 125—*Stomatal opening in Triticum, June 11*

Container	7:30 a.m.	10:30 a.m.	2:30 p.m.
4's W+...........	Open, 6.2μ	Some clsd, 4.0μ	All clsd
16's W+...........	" 6.8	10% clsd, 5.0	90% clsd, 3μ
64's W+...........	" 5.3	Open 3.9	" " 1
16's W—...........	½ clsd, 3.2	90% clsd, 1.0	All clsd
64's W—...........	2/3 clsd, 3.0	All clsd	" "

Almost without exception the degree of opening decreased as the density increased and the holard diminished, varying naturally with the time of day likewise, in accordance with the usual stomatal cycle. As noted previously, the water supply was maintained throughout in the 4's, though in the second table, the opening was less than in the 16's W+, owing to the effect of the hot day upon this open culture.

The number of stomata per unit area differed much less and was also less consistent than in the case of *Helianthus* and *Xanthium*, owing to the characteristic growth and structure of the grass leaf. On May 25 the range in number per 0.1 sq. mm. was from 4.8 in the 4's to 4.3 in the 64's W—, while on May 19 it was from 7.4 in the 4's to 4.2 in the 16's W+. The relation between number and size is shown in the following table:

TABLE 126—*Number and size of stomata in Triticum*

Container	Upper Epiderm		Lower Epiderm	
	No. per 0.22 sq. mm.	Size μ	No. per 0.22 sq. mm.	Size μ
4's W+........	13	52.8×9.8	10.8	49.4×9.0
16's W+........	15	48.0×9.5	13.5	45.2×9.8
64's W+........	15.6	46.9×9.8	12.3	46.5×9.4
16's W—........	14.9	45.7×9.5	13.3	44.6×9.5
64's W—........	20.6	39.6×7.5	15.7	37.6×9.7

The smallest number and the largest size were found in the 4's in accord with minimum density; the largest number and smallest size occurred in the 64's W—, where the density was greatest and the holard lowest. The differences between the other containers were neither considerable nor consistent.

Physical factors and water relations—The holard and nitrate-content of the containers on June 8 were as follows:

TABLE 127—*Holard and nitrate-content in wheat series, June 8*

Depth	4's	16's W+	64's W+	16's W—	64's W—
Holard %					
0– 6″	13.4	12.1	10.4	8.2	9.3
6–12	16.2	17.0	14.5	9.0	7.0
12–22	18.2	17.4	21.8	8.6	6.6
Nitrate-content ppm.					
0– 6″	44.7	25.7	7.2	23.4	16.4
6–18	57.4	46.1	24.1	29.8	24.1

With one or two exceptions the holard increased consistently with depth and decreased with increasing density, and was necessarily much lower in the containers not watered. The higher value in the upper level of the 64's W— as compared with the 16's W— is to be explained by the deeper shade, the two lower layers being markedly less. The maximum value in the water series was found at the third level in the 64's W+, owing to the higher position of the working roots. In general, the nitrate-content decreased with density from a maximum of 44.7 and 57.4 ppm. in the 4's to 16.4 and 24.1 ppm. in the 64's. It fell in accordance with lowered water-content, but rose with depth.

The light values decreased with the density, but were somewhat greater in the containers with water minus than those with water plus, owing to the better growth in the latter. The sequence was as follows: 4's, 13.1%; 16's W—, 4.2%; 16's W+, 3.8%; 64's W—, 3.2%; 64's W+, 1.7%. Leaf temperatures taken on June 17 by means of leaf-thermometers with bulbs 2 mm. in diameter gave readings 2°–5° C. lower for green leaves than for dead dry ones.

The evaporation in the three densities was determined for a two-day period, beginning at 10:30 a.m., May 19. The white porous-cup atmometers were placed so that the evaporating surface was at 5″–8″ above the soil. The open 4's with an average stature of 6″ lost 31.2 cc., the 16's with 8″, 25.2 cc., and the 64's with 10″, 18.9 cc. With the set of five containers, the respective losses for an 8-hour period were 9.8, 9.5 and 6.3 cc., while in the 16's and 64's without additional water, they were 15.4 and 10.5 cc., in correspondence with poorer growth.

Cobalt-chloride tests—A series of tests of stomatal opening was made on June 5, 9, and 11 by means of cobalt-chloride paper. On June 5 the test was applied to the topmost leaf of several plants in each container, the averages being based upon three leaves. The time required to produce the tint was essentially the same for the 4's and 16's W+, viz. 13.2 and 13 seconds; it was 14 for the 16's W—, 16.5 for the 64's W+ and 29 seconds for 64's W—.

The results obtained on June 9 and 11 are indicated in the following table; on the first date the averages were based upon 3–5 readings on different leaves, on the second from 4–6 readings, each from a different plant.

TABLE 128—*Cobalt-paper tests of wheat stomata*

Container	June 9		June 11	
	8:30 a.m. sunny	3:45 p.m. cloudy	8:30 a.m. sunny	3:00 p.m. sunny
	sec.	*sec.*	*sec.*	*sec.*
4's W+......	16.0	7.0	8.3	13.2
16's W+......	18.0	9.0	9.0	18.0
64's W+......	20.0	10.0	8.0	13.5
16's W—......	24.4	24.5	24.3	23.2
64's W—......	29.0	28.3	30.5	25.2

As a rule, the time required increased and the size of the pore correspondingly decreased with the density and with low holard. The chief exception occurred in the 64's W+ on June 11, and was caused by the deep shade of this densest planting. The effect of cloudiness in increasing the opening and decreasing the time is clearly indicated, but this was naturally much less in the denser cultures. On the other hand, the opening decreased during a sunny afternoon in the case of the water series, but conditions were reversed for the dry containers.

The condition of the stomata from the lowest to the uppermost leaf was determined on the same days by the cobalt-paper method. The upper epidermis of each leaf of one plant was employed, and averages obtained from these.

TABLE 129—*Stomatal opening in relation to leaf position*

Leaf	4's		64's W+		64's W—	
	9:30 a.m.	3 p.m.	9:30 a.m.	3 p.m.	9:30 a.m.	3 p.m.
Lowest	152 (dead)	79 (half-dead)	57 (dead)	56 (dead)	74 (dead)	60 (dead)
2nd	11.7	31.5	26	53	37	63
3rd	8.3	13.5	21.7	32.5	83	33
4th	8.0	18.0	17.0	18.0	28.5	25.5
5th	35.0 (top)	....	35.5 (top)	14.0	53.5	31.5
6th (top)..	....	....	....	16.5	126.5	....

With occasional exceptions, usually to be explained by the condition of the leaf, the time required decreased from the dead or inactive leaf at the base to the most active leaf, which was the fourth in all cases but one.

It was 2–3 times as long for the active leaves in the 64's W+ as in the 4's, and half again as long in the 64's W— as in the W+.

Transpiration of insert phytometers—During the period of June 10–18, the transpiration in the various containers was measured by means of insert phytometers. These were planted on May 5 in cylindrical cans with ⅛ the surface of the container and with the respective densities and holard of the series, thus serving as a measure of the transpiration of the corresponding culture. The first column of losses represents the water transpired from June 10–14 before some of the leaves were removed, the second the losses after the removal. In one series the 3 top leaves alone were left on the plants, in the other the 3 top leaves were removed, thus permitting an analysis from the standpoint of position and corresponding intensity of competition. The cut surfaces were sealed with wax after the leaves were excised.

TABLE 130—*Transpiration of wheat cultures as measured by insert phytometers*

Culture	Container	No. pl.	All leaves present: Ave. loss gm.	All leaves present: Loss per plant, gm.	Some leaves removed: Loss per cont. gm. Top lvs.+	Some leaves removed: Loss per cont. gm. Top lvs.—	Some leaves removed: Ave. loss gm. Top lvs.+	Some leaves removed: Ave. loss gm. Top lvs.—
	1				317.5		317.5	
4's	2	1	294.8	294.8		226.8		226.8
	3					226.8		
	1	2			158.6		158.6	
	2				158.6			
16's W+	3	2	176.9	88.4		113.4		113.4
	4					113.4		
	1	2			79.4		85.0	
	2				90.7			
16's W—	3	2	131.5	65.7		79.4		68.1
	4					56.7		
	1	4			68.0		68.0	
64's W+	2	8	77.1	11.5		62.1		56.4
	3					50.8		
	1	4			38.0		32.4	
	2				26.7			
64's W—	3	8	45.3	7.6		22.7		
	4					11.3		17.0

The average transpiration from each container and the water-loss per plant decreased consistently and more or less regularly as the density increased from the 4's to the 64's. There was likewise a marked decrease in the same culture in consequence of a decreased holard; with all the leaves present the phytometers in the W— cultures of the 16's and 64's transpired respectively 74% and 66% as much as those in the W+ cultures. Similar results were obtained after the top leaves or the lower leaves had been removed. In both cases there was a consistent decrease of transpiration with greater density and with reduced holard. The transpiration of the phytometers with top leaves alone present was nearly 10 times as great in the 4's as in the 64's, while in the case of those with the top leaves removed, the difference was greater still. With respect to water-content, the loss was approximately twice as much in the cultures with water plus, under both sets of conditions as to leaves, and in the case of the 64's W+ it was more than thrice as much.

The greater vigor and activity of the upper leaves in comparison with the lower, especially under competition, are well indicated by the differences in transpiration. At the minimum the water-loss was 25% more in the phytometers with upper leaves alone, and the maximum was practically twice as great.

Summary—The degree of stomatal opening in *Triticum* varied inversely as the density but in correspondence with a diminishing water-content. This was equally true whether the determinations were made by means of epidermal strips, direct observation with the microscope, or by means of cobalt-chloride paper. The use of the latter also confirmed the rule with respect to stomatal behavior at different levels, the time required decreasing and the opening correspondingly increasing from the lowest to the topmost leaf. The number and size of stomata followed the rule less closely, but the exceptions were only occasional. They had no visible effect upon the transpiration of insert phytometers, which exhibited large and consistent differences in accordance with density and amount of holard.

FUNCTIONAL BEHAVIOR OF OTHER SPECIES UNDER COMPETITION

Cultures of *Kuhnia glutinosa* were grown in large containers, which permitted the use of insert phytometers, and a mixed culture of *Andropogon nutans* and *Arctium lappa* was employed for similar determinations. In addition, large quadrats of *Andropogon* and *Kuhnia* were grown in the low prairie with insert phytometers to permit the study of transpiration of a dominant and subdominant under different densities.

Stomatal opening and starch-content—Cobalt-chloride paper was employed to determine the degree of opening on the various leaves of insert

TABLE 131—*Stomatal opening in insert phytometers of Kuhnia*

Leaf	Plant 1	Plant 2	Plant 3
	sec.	*sec.*	*sec.*
Lowest	34	41	53
2nd	38	36	34
3rd	21	19	38
4th	24	20	23
5th	17	15	24
6th	....	9	17
7th	....	....	16

plants of *Kuhnia;* the readings were made on June 30, which was partly cloudy. All the leaves used were green, except the lowest pair of plant 3, which were yellow on July 1 when they were utilized.

As a rule, the time required for the tint decreased from the lowest to the topmost leaf; it was longest in the yellow inactive leaf of plant 3. Stomatal opening decreases as the activity of the leaf diminishes, and this is related to age and hence position.

The results obtained with *Arctium* and *Andropogon* indicate the stomatal behavior of suppressed and dominant plants, and the effect of light as well as that of position.

Arctium

	sec.
Upper leaves in the sun	16
Upper leaves in the shade	21
Lower leaves of dominants	23
Lower leaves of suppressed plants	41–59
Large leaf at top in sun	20
Large leaf at top in shade	33
Large leaves at medium level	34
Suppressed plant, leaf at lowest level	45

Andropogon

	sec.
Leaf of dominant in sun	12
Leaf of small suppressed plant	26
Long leaf, lower part in dense shade, upper in sun	
Upper part in sun	18
Lower shaded part	24
Similar leaf	
Upper part in sun	16
Lower shaded part	22

The differences were striking as a rule and essentially consistent throughout. They were practically all due to light, which is usually the paramount factor in suppression. As measured by the degree of stomatal opening, the upper sun leaves of dominants were approximately 2–4 times as active as the leaves of badly suppressed individuals, corresponding closely with the other measures of competition obtained in earlier experiments. There was also a fair degree of agreement between the two species, in spite of great ecological and morpohological differences.

Starch-tests of the *Kuhnia* culture at 10 a.m. on a sunny morning showed a striking difference between dominant and suppressed individuals. In the case of a dominant 6″ tall, starch was fairly abundant to 1¾″, abundant to 3″ and above this very abundant; a suppressed plant 4″ tall and with leaves but half as large contained little starch to 2″, but this was fairly abundant in the upper half. On June 25, determinations were made by both the iodin test and the picramic method, the plants having been kept in the dark the previous day. The light value above the 5″ level in the cultures was approximately 20%; below this level it ranged from 1%–3% at noon. Suppressed plants of *Kuhnia,* in which all the leaves were below 5″, gave no test for starch; the crowded dominants contained little starch below the 5″ level, but an abundance in the leaves above this. Suppressed individuals of *Arctium* contained a moderate amount, while dominant ones gave a uniformly heavy test. The results obtained with respect to the amount of photosynthate were also consistent with the suppression and level.

Transpiration of Kuhnia phytometers—Insert phytometers of *Kuhnia* were employed to determine the transpiration within cultures of this species in relation to that outside. On June 29 a 2-hour period outside the culture gave a loss of 33 cc. by contrast with 22 cc. inside, while a repetition of the experiment gave 34, 13 and 24 cc. for three 2-hour periods, the middle reading being in the culture. A day later the lower older leaves were removed from the inserts and the experiment repeated. The alternative losses beginning with the outside were 29, 11, 38 and 28 cc., the amount rising toward noonday. In all cases the transpiration was much greater on the outside, sometimes 2–3 times as great, the outcome being a decisive demonstration of the value of community grouping in reducing water-loss and of the existence of mutually beneficial reactions.

Insert phytometers in prairie cultures—For the purpose of measuring transpiration and growth in different densities in competition cultures in the low prairie, denuded quadrats with insert containers in place were sown to *Andropogon nutans*. This was planted in three densities, thin, medium and thick, with two containers for each. Similar phytometers were installed for *Kuhnia glutinosa*, except that two densities, thin and thick were used, and the "thin" phytometers were installed in a clipped area of low prairie. A control phytometer of *Andropogon* was placed in the greenhouse, and there were in addition the usual controls without plants as checks on the efficiency of the sand-gravel mulch.

TABLE 132—*Relation of transpiration to density*

Container	Density	Location	Loss cc.	No. plants	Loss per 100 plants	Average
			Andropogon nutans			
1	medium	Greenhouse	74	61	121.31	121.31
2	thin	Thin quad.	64	45	142.22	138.88
3	"	" "	61	45	135.55	
4	medium	Med. quad.	67	64	104.68	117.15
5	"	" "	70	54	129.83	
6	thick	Thick quad.	92	92	100.00	103.67
7	"	" "	73	68	107.35	
			Kuhnia glutinosa			
A	thin	Clipped	124	24	516.66	418.33
B	"	grass area	96	30	320.00	
C	thick	" "	97	29	334.48	318.90
D	"	Kuhnia quad. " "	91	30	303.33	

Although the results in the individual containers are not entirely in accord, the averages exemplify the rule that transpiration is highest in the most open culture and hence bears an inverse relation to density. As already suggested, this fact has a direct bearing upon the relation between competition and reaction, and upon the rôle of community functions (p. 316).

This experiment was repeated for *Andropogon nutans* from July 2 to 10; in addition to transpiration, determinations were also made of dry weight and photosynthate. Light values were read at noon in the middle level; the material for photosynthate was taken at 4 p.m., after a clear day.

TABLE 133—*Transpiration, dry weight and photosynthate in Andropogon*

Container	Density	Loss cc.	Number plants	Loss per plant cc.	Dry wt. gm.	Ave. loss per gm. dry wt.	Photo-synthate gm.	Light value %
1 (Gr.) ..	medium	679	60	11.31	...	....	.....	..
2	thin	484	30	16.13	1.84	263.0	0.1943	30
3	thin	510	36	14.16	...	....	.....	..
4	medium	729	62	11.76	...	....	.....	..
5	medium	633	55	11.51	3.13	202.2	0.1620	23
6	thick	773	95	8.14	5.30	145.8	0.1285	19
7	thick	767	70	10.96	...	....	.....	..

The agreement between the containers of each density was fairly good and there was no overlapping between the three sets. The average loss per plant was approximately 50% greater for the "thins" than for the "thicks," which was in fair correspondence with the results for the preceding set. The water requirement was 30% greater for the "thins" than the "mediums," and 80% greater than in the "thicks." Here again the most effective use of water is in the density with the most marked community reaction, though this is necessarily offset to some degree by the relation to light. The amount of photosynthate for the "thins" was 50% larger than for the "thicks," the sequence being in agreement with that of density and light values.

GENERAL SUMMARY

The stomatal features of the various species were regularly determined from epidermal strips plunged immediately into absolute alcohol for fixing, but this method was frequently checked by means of direct microscopic observation and tests with cobalt-chloride paper. The results obtained were essentially concordant throughout the six genera employed, viz., *Helianthus, Xanthium, Triticum, Kuhnia, Andropogon* and *Arctium,* which represented a considerable range of life-forms. The rate and degree of opening showed a consistent and often regular decrease from the least to the most dense culture, from the higher to the lower water-content, and from topmost to lowest leaves. The number of stomata, as would be expected, increased per unit area with increasing density and decreasing holard, as a consequence of reduced turgor and mechanical expansion, acting upon the epidermal cells. The number of chloroplasts diminished with the density and the holard, in direct relation to the size of the stomata, which was in inverse ratio to the number. The starch-content of the mesophyll behaved essentially like stomatal opening; it decreased with greater density, reduced holard, and with lower leaf-levels.

The water-loss from turgid sunflowers was found to range from 5 to 10 times as much as from wilted ones. This is to be chiefly explained by the fact that all the stomata were closed on the latter, as shown by direct examination as well as by epidermal strips. The transpiration of wheat cultures as measured by insert phytometers was in the closest accord with density and water-content, both with respect to phytometer and individual plant, decreasing with much regularity from 4's to 64's and from plus to minus holard. The protective value of the community grouping was demonstrated in the case of *Kuhnia,* when the water-loss of insert phytometers was alternately determined inside and outside the culture. In every instance, the transpiration was much higher on the outside, sometimes amounting to 2–3 times as much. Similarly, in the case of both *Andropogon* and *Kuhnia* the transpiration of phytometers was found to be greatest in the more open quadrats or areas.

With respect to root pressure, both *Helianthus* and *Xanthium* gave the same results. The root pressure was consistently highest in the least dense culture and in the higher holard; it fell off more or less regularly from this extreme to reach a minimum in the 32's W—. The conduction exhibited by the stem followed the same rule. It dropped consistently in both *Helianthus* and *Xanthium* from the 2's to the 32's, and similar differences were found when the comparison was between W+ and W— cultures.

9. NATURE AND ROLE OF COMPETITION

CONCEPTS

Nature of the community—The concept of the plant community as an organic unit with development and structure was first advanced in "Research Methods in Ecology" (1905), and has served as the basis for the later treatments (Clements, 1907, 1916, 1920, 1928). The term, "complex organism," was designed to emphasize the nature of the community as an organic entity and at the same time to distinguish it from the simple or individual organism, so familiar that it alone seems to bear title to the name. The significance of this distinction has often been overlooked or ignored, with the consequence of failing to understand the intrinsic and universal interplay of development and structure in the life-history of the community. Most of the opposition to the view has come from inadequate knowledge of the cryptogams among plants and the invertebrates among animals, or a lack of acquaintance with the propagative behavior of perennials. Another cause of objection has been too implicit a faith in the dictionary and the failure to realize that in science at least dictionaries are necessarily remote from the frontier of advance.

With the object of avoiding misunderstanding, Tansley has substituted the term, "quasi-organism" (1920:123), which may well serve as an effective compromise until the recognition of the several basic types of organism becomes general. Moreover, he approves of the practice of considering human communities as organisms, and this becomes particularly significant in view of the new concept of the community as a biotic rather than a plant unit merely. While it is unnecessary to press the point at this time, it is evident that the inclusion of animals in the organic entity raises pertinent questions as to the place and rôle of man in it. Spencer has discussed the concept of the social organism with signal clarity, and the student of community development can still turn with great profit to his treatments of this theme (1858, 1864). It is both interesting and suggestive to find that he anticipated certain axioms of plant succession by the statements: "Societies are not made but grow" and "Man may disturb, he may retard or he may aid the natural process of organization (development), but the general course of this process is beyond his control."

A comprehensive discussion of the nature and kinds of organisms belongs elsewhere, and here it is only desirable further to point out that the view of the plant community as an organic entity exhibiting cooperation and division of labor has been anticipated in some degree by several investigators, notably Humboldt and Warming. The latter has expressed this view as follows (1895, 1909): "There are certain points of resemblance between communities of plants and those of human beings or animals; one of these is the competition for food which takes place between similar individuals and causes the weaker to be more or less suppressed. * * * In plant communities there is, it is true, often (or always) a certain natural dependence or reciprocal influence of many species upon one another; they give rise to definite organized division of labor such as is met with in human and animal communities, where certain individuals or groups of individuals work as organs, in the wide sense of the term, for the benefit of the whole community." (p. 94.) "Living beings forming a community have their lives linked and interwoven into one common existence in so manifold, intricate and complex a manner that change at one point may bring in its wake far-reaching changes at other points. In this direction a wide field lies open for investigators." (p. 336.)

The community bond—The evidence as to the closeness of the bond between the individuals of a community is co-extensive with vegetation, but may all be summed up in the interaction of habitat and organism, and the coaction of the individual organisms, whether plants or animals. The bond of association is so strict, its dependence upon habitat and upon aggregation so exact that the same seral stage may recur around the globe in hundreds of thousands of examples, and with the same dominants and subdominants. The correspondence is so close that the experienced ecologist knows in advance what community will be found in fresh-water, swamp, salt-marsh, inland sandhills or coastal dunes, within the limits of the great floras concerned. This correlation is as true in time as it is in space, and upon it have been built the new methods of reconstruction in paleo-ecology ("Plant Succession"; cf., also Clements, 1918). Applying these to the fossil flora of the John Day Basin, Chaney has followed the clue of *Sequoia* through the maze of misnamed genera to the point of reconstructing the ancient community in terms of the dominants and subdominants of the modern one (Chaney, 1925:4, 39).

This principle finds its major exemplification in the bond between climax and climate; the climax is not merely the expression of its climate, but also the best indicator of it. They are not only co-extensive in space, but likewise in time, again within the age of the flora involved. The time relation is best shown in the mass migration of climaxes in the face of a great climatic shift, such as those of the Pleistocene. Not only do the dominants, subdominants and secondary species move as a community, but the various formations preserve their relative position. The most striking evidence of movements today are to be found in the relict rear-guards of forest and grassland, which persist far beyond the limits of the present climax. The postclimaxes of trees known as the "Cross-Timbers" of Texas, and the tall-grass communities of sandhill regions demonstrate the tenacity of the bond that holds the group together, since they recur again and again not merely with their membership intact but also essentially with that of the parent climax. In contrast to this stands the "individualistic" concept of the community, which has been proposed by Gleason (1917, 1926). This appears to involve a confusion of ideas as well as a contradiction of terms; it has been adequately characterized by Tansley (1920:126) and requires no further consideration here.

The functional bond within the community is less apparent than the structural, but it is readily demonstrated by experiment. It is concerned chiefly with competition, reaction, protection, ecesis, and with the coactions between organisms. All of these are processes that owe their very existence to the aggregation of individuals into a community, and hence provide the only adequate understanding of the behavior and development of the latter.

Community functions—The functions of the plant community were first distinguished and analyzed in "Research Methods in Ecology" (1905:199), and they were further discussed in "Plant Physiology and Ecology" (1907:237). The simple functions comprise aggregation, migration, ecesis, competition, adaptation, and reaction while these interact in the complex processes known as invasion and succession. More recent is the recognition of the process termed coaction, which has to do with the direct action of the associated organisms upon each other. The first functions are concerned with movement, within and about the community, as well as that of the latter itself. Ecesis is the complete process of establishment; in bare areas it comprises merely the response of seed or propagule, together with that of the resulting plant, but in the community it involves competition and reaction to some degree, and sometimes adaptation as well. Reaction is the effect of the community upon the habitat; it is obviously an outcome of the response of each individual, but this becomes effective only as the consequence of the cumulative action of all the individuals. Competition is inseparably connected with all the reactions that have to do with the supply of energy or raw material in the habitat. As the name suggests, it is the cardinal function of the community, in which species and individuals seek together to obtain a proper supply of these essentials. It is so universal and often so controlling as to obscure completely or in large measure the mutual protection of individuals, which regularly results from certain reactions, such as those upon wind, sun, etc.

The present study is the second of a series designed to deal experimentally with all the functional processes in the community. The first had for its theme the process of ecesis and the results were embodied in "Experimental Vegetation" (1924). The third is intended to treat of reaction, and it is hoped to devote others to adaptation, migration, and coaction.

NATURE OF COMPETITION

The nature of competition—The nature of competition was considered in much detail in the "Development and Structure of Vegetation" and in "Research Methods in Ecology" (cf. pp. 10, 12, 21), but perhaps the most exact concise characterization is to be found in "Plant Physiology and Ecology" (p. 252). "Competition is purely a physical process. With few exceptions, such as the crowding up of tuberous plants when grown too closely, an actual struggle between competing plants never occurs. Competition arises from the reaction of one plant upon the physical factors about it and the effect of these modified factors upon its competitors. In the exact sense, two plants, no matter how close, do not compete with each other as long as the water-content, the nutrient material, the light and heat are in excess of the needs of both. When the immediate supply of a single necessary factor falls below the combined demands of the plants, competition begins."

The essential feature of the process is disclosed in the last statement; competition in short is a combined need in excess of the supply. An isolated plant may require more energy or material than it can obtain from the habitat, but as the word indicates, competition exists only where two or more individuals toegther seek more than the space they occupy affords. This idea of a common need implies more or less of an equality, in one or more respects and for a certain period at least. Even in the case of dominant tree and secondary herb, competition takes place between the latter and the tree seedling when too close, as also between seedlings of both. As a function, it is necessarily a dynamic process, which may lead to equilibrium, suppression, subordination, or extinction.

In view of the nature of the habitat and the effect of reactions upon it, it is easy to understand the basis for the statement that plants compete for room or space. This has considerable justification in the fact that they can not compete as a rule until roots or shoots come to occupy the same block of soil or air, in whole or in part, and that "spacing" or rate of planting is a primary consideration in growing crops. As a figure of speech, competition for space is probably not objectionable, but it must not be overlooked that space really stands for the raw materials, energy and working factors that it contains. Much more undesirable is the phrase, "sphere of influence," which may have a proper reference to the space about a plant in which roots or shoots exert a reaction, but has sometimes been employed to denote some mysterious or vitalistic relation. The open spacing of desert shrubs in particular suggests some indirect influence in explanation, but studies of the root systems demonstrate that this is a result of competition for water where the deficit is great. Moreover, this is strikingly confirmed in a number of instances by the wealth of grasses and forbs that develops after the shrubs are destroyed. The best example of a "sphere of influence" is to be found in the shadow cast by a tree, or row of them, or in the zone of protection afforded by a wind-break. However, these are both definite and universal reactions, and it is much better to treat them as such.

The real nature of the so-called mechanical competition presents somewhat greater difficulty. Warming has been considered an advocate of the view that such competition occurs, but apparently he held no fixed opinion upon this point, as may be seen from the following: "A consideration of the means by which plants oust each other from habitats raises such questions as to whether roots and rhizomes grow so close together as to bar the way to other plants in a purely mechanical way, or as to rob them of water or nutrients." Sherff inclines to accept mechanical competition between rhizomes as part of the explanation for the layering in the soil of reed-swamps, though the smaller part. However, it seems probable that the actual competition is that of the roots for nutrients, air and water, and that it is desirable to distinguish in this case between the process and the outcome of it in terms of growth and consequent position. Light is thrown on this whole question by the behavior of tuberous-rooted vegetables such as the radish and carrot, in which dense sowing often causes the plants to force each other up and out of the row. In such cases the plants are usually more or less equal in competitive ability, and the "heaving" is the consequence of mutually rapid growth, in which the suppressed individuals have no share.

Course of competition—The general details of the process itself have been given by Clements (1904, 1905, 1907; cf. p. 10), and to some extent by Yapp also (1925). The sequence of smaller details has been described with much fullness in the preceding chapters, especially those dealing with cultures in the prairie. For the purpose of more exact analysis, the course of competition may be divided into (1) incidence (2) cumulation and (3) outcome. The beginning of competition is due to reaction, when the plants are so spaced that the reaction of one affects the response of the other by limiting it. The initial advantage thus gained is increased by cumulation, since even a slight increase in the amount of energy or raw material is followed by corresponding growth, and this by a further gain in response and reaction. A larger, deeper or more active root system enables one plant to secure a larger amount of the chresard, and the immediate reaction is to reduce the amount obtainable by the other. The stem and leaves of the former grow in size and number, and thus require more water; the roots respond by augmenting the absorbing surface to supply the demand, and automatically reduce the water-content still further, and with it the opportunity of a competitor. At the same time, the correlated growth of stems and leaves is producing a reaction on light by absorption, leaving less energy available for the leaves of the competitor beneath it, while increasing the amount of food for the further growth of absorbing roots, taller stems and overshading leaves. Similar reactions are taking place upon all the other direct factors, nutrients in particular, soil-air and in some instances perhaps carbon dioxid as well. The most striking example of cumulation is to be seen in the case of grasses and grains, as the experimental results of the preceding chapters indicate. Not only does a small advantage in root, culm or leaf improve the response and hence the chances of the parent plant, but it also encourages the production of tillers, which are potential individuals and constitute what is virtually a reinforcement in the contest.

The course and degree of competition are constantly indicated by the outcome, at any particular time or at the close of a season or life-cycle. In the absence of an experimental study of the cumulation in reaction and response, the outcome as indicated by the number, size and form of the plants concerned is a good measure of the intensity of competition and the merits of the contestants. However, it is necessary to take into account at all times two other influences, namely, the direct action of physical factors and the effect of coactions. Quite independently of the reactions involved in competition, unusual conditions in the habitat may decide the outcome in terms of growth, survival or adaptation. Not only may they overrule the competitive reaction, but they may augment it or subtract from it. This is even truer and more visibly so of the coactions exerted by parasitic plants or animals, including in the latter all those that consume plant material. Such a consequence is universal in grazing and especially in overgrazing, by which the tall-grasses are hampered in competition with the short-grasses, grasses in the struggle with weeds, and the palatable in that with the unpalatable. The onset of stem-rust greatly increases the need of grains for water and correspondingly handicaps them in securing an adequate amount of a limited supply. Defoliating caterpillars decrease the reaction on light as well as the leaf-area for food-making, bark-beetles interfere with the transport of materials, and root-parasites of all kinds with the absorption of water and solutes. Finally, as has been shown previously, the amount of heat and water during the resting period will determine the toll taken by winter-killing.

Outcome of competition—The effect of competition finds expression in the function, structure or form of the individual or of the community. This expression is continuing or cumulative throughout the process, until final equilibrium or extinction is reached; the period involved may be a season, a year, a sunspot cycle or the life-cycle of individual or community. However, equilibrium is never complete or permanent, since each returning season injects new seedlings and shoots into the situation and each turn of the cycle changes the direct effect of the physical factors. As to the nature of equilibrium itself, this may involve essential equality when the competitors are more or less evenly matched, or any degree of inequality in accordance with their respective abilities. But it never represents the complete cessation or absence of competition, since the associated individuals of both dominant and subordinate species are constantly competing among themselves, and the adults and young of the latter with the seedlings of the former.

With respect to individual and species, the dynamic outcome of competition is expressed by suppression, adaptation, or extinction, the static end-results of which are seen in dominance and subordination. Evasion of competition to some degree and often a decisive one may be secured by separation in space or in time, in the former case usually dealing with root or shoot alone. From the standpoint of the community, the outcome of competition is to be seen in ecesis and migration, in aspection, annuation, and succession, and, in another form, in the coactions between organisms.

Competition and dominance—The regular outcome of competition between the individuals of one species, or between two or more species is dominance. The successful competitors come to control the habitat more or less completely through their reactions and in consequence are termed dominants. This is a term that applies equally well to individual or species, though its significance is somewhat more dynamic in the first case. Dominance is primarily a matter of the controlling reaction upon light or water, or both in the case of natural communities, or of nutrients also in the case of artificial ones. Active or effective competition would appear to cease when dominance is attained, but to a small degree at least it probably always persists (Clements, 1904:169; 1907:251; cf. Yapp, 1925:692), disappearing completely only as individual or species vanishes. Competition evidently reaches the lowest point in a climax forest with only a ground layer of herbs, so far as the relation between dominant and subordinate is concerned. In the absence of adequate measurements and experiments upon the reactions of each, it becomes largely a matter of definition, as Yapp has stated (1925:699). His distinction between competition and priority serves a useful purpose in analysis, though, as he himself suggests, it seems better to regard the latter as a special kind, a partial competition.

As the present experimental studies have shown, successful competition may be due to a shallow root system under certain conditions of chresard, to a deep-seated one in other cases. Success in either instance goes to the plant or species that secures its requirements *first*, as is always typical of competition. Even in the case of light where priority is most marked, tall herbs in particular may hold the advantage for several or many years, herb and sapling then meet on more or less equal terms for some years longer, and finally the tree assumes an increasingly dominant position. During this entire period their roots may remain in competition, as is demonstrated in the many instances where the removal of undergrowth or of trees has stimulated the development of the other. In fact, competition is so complex and often so partial, incomplete, or changing in character, degree or effect that it is best to retain it as the term for the sum total of relations between demand and supply and to seek the necessary distinctions within it.

Perhaps the best statement of the relation between dominant and subordinate is to say that competition decreases as dominance increases, which is a corollary of the rule that plants and species compete the less the more unlike they are. All degrees of competition must then be expected, from the intense struggle between like individuals of the same species to the transient contest between moss and seedling or other life-forms of the greatest dissimilarity. Likewise, differences in the nature, intensity, or completeness of competition are universal, and experiment alone can determine these in any particular community or culture. Such differences are exemplified in the complementary communities of Woodhead (1906; cf. Sherff, 1912), a distinction that helps in the analysis of complex competitive relations. All such communities are competitive in some degree, though with a change in nature or emphasis. This fact is well illustrated by seasonal societies or aspects; the species of vernal societies grow and bloom in spring, thus evading in considerable degree harmful competition with their later taller associates. However, the majority of them persist throughout the growing season in vegetative condition and necessarily play a part in competition.

Concomitant with increasing dominance but opposite to it in effect is suppression. It is first felt by the least successful individuals, but their ranks are constantly recruited by others as the reaction of the dominants becomes greater and more controlling. It may gradually involve all the plants of one or more species, and lead to one of three fates, namely, subordination, adaptation, or extinction. The first two are often associated, since it is only by modification that certain species are able to survive, especially in the case of overshading. Extinction regularly plays a part in the development of all communities, from the family of a single species to the most complex society or association. As the final term of suppression, it is most in evidence between similar life-forms, but especially when one enjoys a marked advantage in the reaction upon light or water-content. Both suppression and extinction are determined in the first instance by the so-called weapons of species, which are considered in a later section.

The assumption of dominance by one or more species does not necessarily lead to the complete subordination of all the others associated, and there are in consequence various degrees of subordination, such as are to be found in layered communities. The most important distinction in this connection is that of subdominance, which has been defined as dominance within a dominance (Clements, 1916:130). The control exerted by the dominant grasses of the prairie is far from exclusive and hence permits a characteristic development of herbaceous species, which at the time of blooming often seem to dominate the grasses. The relative rank of such subdominants is best recognized in forest and woodland, where their subordination to the taller dominants is obvious, their grouping into the various layers being an expression of further competitive relations among themselves. A thorough and comprehensive analysis of layering from the standpoint of competition, adaptation and the reactions concerned is still to be made for forest and other woody communities. As the following abstract indicates, Grevillius has made such a promising beginning in this field that further studies need only the addition of instrumentation and experiment. This has been done for grassland and the forest edge in the present treatment, to great advantage in the understanding of the competitive relations between grass and chaparral, and between tall, mid and short grasses and their associated subdominants.

Grevillius (1894:147) studied the reciprocal struggle for existence in certain Swedish woodlands, as well as the end-results when the development ceased and the constituent forms had reached equilibrium. He stated that a correlation is very often present between the ecological characters of the species concerned, which expresses itself in a definite relation between the manner of exposition of the floral system, the form and exposition of the assimilating organs, the mode of innovation (shoot formation and migration), abundance, the position in the layers of the community, and the time and abundance of flowers and fruits. Three types were distinguished, characterized as follows:

1. The floral system exposed laterally and more or less extended in the vertical direction. The leaves placed closely and regularly on simple or sparse upright main axes with more or less erect branches, more or less elongated, with most typical lateral exposition in the upper region. Sprout formation marked, migration slight.

Occurrence frequent and abundant. Normally constitutes the upper layer of the undergrowth, and blooms mostly in late summer, e. g., *Campanula latifolia.*

2. The floral system exposed upwards and more or less extended in the horizontal direction. The leaves placed in a horizontal plane, more or less separated by elongate oblique organs for exposition, and lobed or divided, with the exposition upward. Sprout formation not marked, migration small or moderate. Occurrence scattered or sparse, placed in the middle layer. Blooms mostly in midsummer, e. g., *Geranium silvaticum.*

3. The floral system variously exposed, but never widely extended in any particular direction. Leaves with more or less regular outline, solitary or arranged in a common surface, exposed upwards by means of petioles more or less elongated in the upward direction. Commonly solitary; migration by stolons. Constitutes the lower layers, down to the lowest ground layer. Blooms chiefly in spring, e. g., *Oxalis acetosella, Paris quadrifolia.*

Competition and adaptation—Adaptation or modification is the regular if not universal outcome of competition. This is the inevitable consequence of changes in the amount of energy or material as a result of reaction. Such is the situation everywhere in nature, except in initial communities where the struggle has not yet begun, and hence competition and adaptation are practically concomitant, unless man steps in. Thus, while competition leads to adjustment of functions on the one hand with the consequent modification of structure and form, it is itself essentially adaptation to direct physical factors as modified by reaction. The functional response to competition naturally increases with the intensity of the latter, and sooner or later concerns all the functions of the plant. In the preceding chapters, it has been shown to involve absorption and root pressure, conduction, stomatal movement, transpiration and starch-content, as well as such features as turgidity, osmotic concentration, etc. The histology of the three organs is correspondingly modified, the leaf in particular showing marked changes in number and size of stomata, thickness of epiderm and mesophyll, the relation of palisade to sponge tissue, and so forth. All of this is reflected in the size and form of the plant body, as well as in the size, number, and structure of flower and fruit.

The final outcome of intense competition both in nature and in culture, is such thorough-going adaptation as to produce forms ranking as species in the more recent manuals. However, there is no proof that such changes are fixed and hence none at present that Darwin was correct in thinking that competition is the mechanism for accumulating minute variations. On the other hand, the experiments with denuded quadrats (p. 130) demonstrates that the forms of species normally found in prairies are to a striking degree a response to the existing competition. This relation is so marked in the bunch-grass prairie of California and the related seral communities that it was made the special object of study in the "Herbarium Ecadium Californiae." (Clements and Clements, 1915.)

The competitive equipment of plants—The success of species in competition with each other is determined primarily by their life-forms. Their phylogenetic relationship is of little or no importance, except as it may be based upon life-form or habitat-form. It is these that bear the ecological impress of climate and habitat, and hence determine the response to the direct physical factors, either with or without the reaction due to competition. The most important feature of the life-form is duration or perennation, owing to its effect upon occupation and to a large degree upon height as well. After this follows the rate of growth, the most effective expression of which is found in the expanse and density of the shoot and root systems, and the depth of the latter. Rate and amount of germination determine the initial advantages to be secured by shoot and root, and hence their success in making the most of the advantages afforded by the life-form. Vigor and hardiness play rôles of great importance also, since they have much to do with survival under reduced water-content or light intensity, or under the stress of winter conditions. These may reside to some extent in the nature of the protoplasm itself, but for the most part they are directly related to growth, structure and such processes as the "ripening" of tissues, especially woody ones.

To enumerate the great variety of forms in which these major advantages appear seems superfluous; many of them have been considered in chapters 2 and 3 especially, as well as summarized at the close of the latter. In grouping species with respect to their competitive ability, practically all the systems of life-forms are of value, but perhaps the most usable are the simplest and most consistent, that of Raunkiaer (1905; cf. Clements 1920, 1928) and of Clements (l.c. 63:270). The one uses the position of the bud-shoot for the primary division, the other emphasizes duration, but both take into account the major features of life-forms. The subdivision is carried much further by Warming (1909) and Drude (1913), but the systematic basis of many of these renders them of somewhat less value for the present purpose (Clements, 1928:67). It is evident from all of these that the species with the best equipment for competition supply the dominants of plant communities, while climax dominants are drawn almost wholly from the three groups, trees, shrubs, and grasses. Subdominants may be woody as well as herbaceous in nature, but the vast majority are forbs of the widest possible range of life-form within the groups concerned.

The factors in competition—The present treatment has dealt entirely with competition in relation to the direct physical factors of the habitat, which control the development of the plant body. Species and individuals also compete for pollinators, though less definitely and with less visible and striking effects. This subject has been considered separately in "Experimental Pollination" (1923), and needs no discussion here in consequence. This is probably the only coaction in which competition enters as a distinct feature, though many others have a bearing upon the course and outcome of competition, as shown later. Plants with migration devices for attachment can not properly be said to compete for carriers, since the relation is entirely incidental and accidental. The situation is slightly different in the case of the carriage of fruits and pollen by the wind. The height of plants does play a part in effective transport, especially in *Taraxacum, Agoseris* and similar genera where the peduncle elongates before the fruits are freed, but this bears little or no relation to competition.

So much has already been said with respect to competition for the various physical factors and their relative importance that it seems desirable merely to summarize this and to point out certain general relations. It appears certain that plants can compete with each other only for those factors of the habitat that constitute energy or raw material, and not at all for the indirect factors, such as humidity, wind, pressure, etc. Since temperature is the one factor that produces important effects of both sorts, it presents some difficulty but there is no evidence to suggest that plants compete for heat, apart from radiant energy. The five direct factors for which plants may be said to compete are light, water, nutrients, oxygen and carbon dioxid. This is strictly true only of holophytes; parasites and saprophytes exhibit little or no competition other than for water and food. While there is no direct experimental proof of competition for carbon dioxid, the small amount present and the effects of enriching the air artificially suggest the possibility. Moreover, the much larger amounts in the air near the ground in forests further suggests that the lower layers of herbs may possess a certain priority in this respect. The large amount of oxygen in the air makes certain that it is not concerned in competition, but the case is quite different in wet or waterlogged soils, or in water. In all saturated soils, but especially in bogs and swamps, the supply of oxygen in the soil-air is regularly below the needs of the plants present, and competition for this gas usually assumes the first importance. Direct studies of competition in the presence of a limited oxygen content are still to be organized, but Bergman (1919) has shown the possibilities of work in this field.

Much evidence has been secured of the relative importance of the three primary factors, water, light and nutrients, in the various aspects of the present investigation. The results are in essential agreement throughout to the effect that water stands first, light next, and nutrients last in native communities, with the order of light and nutrients reversed in the case of many intensive field-crops. The general and probably universal rule is that the factor present in the smallest amount relative to the demand will be paramount. However, the ratio between supply and demand must be determined experimentally with respect to the action of each factor (p. 288), and not merely with reference to the amount of each present. A reduction of the chresard to one-half will produce the most striking effects, while the light intensity must usually fall below 1/10 to have results at all comparable in degree.

Water is the controlling factor in the uppermost layer of a community, since this receives full sunlight and is exposed to full isolation. The grouping in all subordinate layers is primarily a consequence of light, but *within* each layer it is probable that water is again more important than light. Both factors play a part in all these instances, however, and the task as ever is to properly evaluate the rôle of each. Here again the most important principle to keep in mind is the functional demand on each in proportion to the supply. The failure to do this has so far led to little more than contradictory results in the study of tolerance in the forest, and it seems evident that this problem must be approached in a much more comprehensive manner from the standpoint of experimental competition, with both field and laboratory contributing their best methods to the solution.

The effect of a contributing factor such as lime is of great historical interest, as well as physiological importance, though the studies of Nägeli, Christ, Tansley, Pearsall and Wray, and others make it clear that this is a factor operating upon competition rather than directly. Rayner's statement that *Calluna vulgaris* will not grow on calcareous soils, because the seedlings germinating in such soils would be immediately eliminated by competition, probably represents the rule, but the exceptions are not infrequent, and they constitute the most promising points for further attack. Since this is essentially a problem that must be solved in Europe, it is perhaps gratuitous to repeat the statement that physiological and competition studies alone, and mainly in the field, can lead to a complete solution (Clements, 1913:76; Tansley, 1913:16; Massart, 1913:46; Graebner, 1913:70).

Effect of factors and coactions—It has already been pointed out that physical factors may assume a decisive part in competition, quite aside from the "reacted" factor which represents the normal range of chresard or light within the competitive community. It is equally clear that this will take place only when the usual limits are approached or exceeded, and that water will be the factor generally concerned. When the growing season is unusually cloudy, light may play a small part, but this is much overshadowed by the greater influence of a reduced oxygen supply as a result of increased rainfall and lessened water-loss. As many of the experimental cultures have demonstrated, drouth is the great factor that overrules or otherwise modifies the usual course of succession, its effect being almost as decisive in winter as in summer. Whether temperature has a share in winter-killing as a direct factor is still uncertain, but there appears to be no question of its importance in cooperation with water. Naturally what is true of seasonal extremes of climate applies to edaphic extremes as well; the excess or deficit of any direct factor will practically always favor one species more than the other. This has been repeatedly exemplified in those cultures containing one species each from high and low prairie, and in the transfer of sods from one community to the other.

Although the process is very different, the actual effect of a coaction is much the same as that of a factor extreme in throwing the balance in favor of one competitor or another. A coaction may favor one species directly and thus give it a corresponding advantage, but in the great majority of cases it injures one species alone or more than others, to the benefit of the latter. Such a symbiosis as that of the nodule bacteria is an example of the former, though the advantage derived from it by legumes in grassland seems finally to be appropriated by the grasses themselves. The injurious coactions of the greatest moment are due to parasitic fungi and insects, and to herbivorous mammals in the widest sense of the term. In terms of community competition, by far the best understood are the coactions of grazing animals, cattle and rodents (Clements, 1920:293–310, 1928:369–386; Taylor and Loftfield, 1923). Wherever grazing occurs, the most palatable or most nutritious species at any particular time are grazed more heavily, the demand of the plants thus reduced, and the supply to the others correspondingly augmented. In the many instances of slow but progressive overgrazing, the tall-grasses suffer first and most, mid-grasses next, and short-grasses least, the final outcome being pasture or range dominated by the latter alone. The next phase is the overgrazing of the latter to the point where woody undershrubs gain the advantage, and the range may become dominated by *Gutierrezia sarothrae* or *Artemisia frigida,* while the pasture may be filled with annual ruderals or subruderals (Clements, 1920, 1928; Sampson, 1919).

Kinds of competition—While it is undesirable to attempt a complete analysis of competition with respect to kinds at this time, it is helpful to consider the various bases for this. At the outset it is obvious that plant competition differs fundamentally from animal competition, while the latter passes gradually into competition in human society. It is moreover important to recognize that terrestrial holophytes can rarely if ever be said to compete with animals, though the possibility exists to a small degree at least in the case of aquatic forms. On the other hand, though we know little of the details, it is evident that a rust and an aphid may compete with each other for food-stuff from the same leaf or plant, and this may be true of all associated parasites of the same host. The succession of moulds in cultures, as well as on dead or decaying material in nature, indicates that saprophytes exhibit a similar relation to the food-supply. No experimental and quantitative studies are known to have been made as yet in animal ecology, aside from those for nectar and pollen (cf. "Experimental Pollination"). A promising field is that of the competitive relations of mammals to the species of grassland and scrub, but an adequate beginning is still to be made in this direction.

As already suggested, competition among plants may be distinguished on the basis of the object as that for pollinators, chiefly insects, and for the physical factors of the habitat. In accordance, it may be further subdivided with respect to the factor chiefly concerned as competition for water, for light, nutrients, or oxygen, as has been frequently done in the text. It is improbable that competition for room exists as such, energy and material being the actual factors in all the instances known. With reference to the plants, it is simple where these belong to the same species, and mixed when two or more are involved. Most cases of the latter are synchronous, but a species or group of species may germinate later or invade an established community as happens regularly in succession, and the competition may then be heterochronous. This may be regarded as one kind of partial or incomplete competition, while the complementary species and communities of Woodhead represent another. Still a third type is exemplified in layered communities, where species compete most intensively with those in their layer, less so with those in the layer above or below, and little with the dominant and the ground-layer.

In accordance with the intimate bond between the complex organism and the individual, competition in the community passes gradually into what seems a very different process in the plant itself. In such compound plants as are represented by the bunch, rosette, mat and similar life-forms, the units compete with each other in varying degree at different stages, not merely for energy and raw materials, but also for elaborated food-stuffs, especially for storage. The competition between the leafy twigs of shrub or tree is not greatly different, though somewhat less simple. In the case of flowers and fruits, competition for food and water predominates, while with seeds and buds that for food is the most important. This has been repeatedly demonstrated by feeding shoots with weak solutions of glucose (Clements, 1922, 1923). Competition between the organs or parts of an individual plant involves the interaction known as correlation of parts and this term may well be retained for it.

RÔLE OF COMPETITION

From the foregoing it seems clear that competition plays the basic rôle in the community that food-making does in the plant. No community escapes its effects; indeed, it can hardly be said to exist as such until the individuals come into this relation with each other. It is the controlling function in successional development, and is secondary only to the control of climate in the case of climaxes. The complex functions of the community are chiefly or entirely dependent upon it, and the cyclic changes of the climax are an expression of it to a large degree. The former comprise ecesis, mass migration or invasion, succession, and stabilization; the latter are aspection and annuation.

Complex functions and cyclic changes—Ecesis or establishment is the direct outcome of response to the physical factors of the habitat only in such open pioneer groupings that competition has not begun. In all communities proper it is determined as much or more by competition and reaction than by the initial factors themselves. Ecesis represents the outcome in terms of survival, of which dominance, suppression and subordination are the several phases, while extinction is the complementary result. The movement of communities under the compulsion of climatic changes constitutes invasion, which regularly takes place into other communities, except where the retreat of ice-masses or seas is concerned. The invading community is in harmony with the changing climate, the one invaded is correspondingly handicapped by it, and is all the more readily replaced as a result of the competition between them. The course of events in edaphic habitats where succession is occurring is much the same, but the advantage to the invaders arises from the reaction of the occupants, which serves as a progressive hindrance to possession. As a result the invaders again are favored by the factors of the seral habitat, in addition to the competitive weapons of a superior life-form. This interaction of competition, ecesis, and reaction ensues with each succeeding invasion, and constitutes the motive force in succession, each wave marking a seral stage. The accumulating reaction promotes stabilization and this terminates finally in the climax.

With the definite constitution of the climax as the mature stage in harmony with the climate, major functional activity ceases, but competition continues in a secondary degree. If the climate were completely uniform or static, the species of each climax community would be in equilibrium and competition would practically cease. On the contrary, the climate is constantly changing in cycles of varying duration and intensity; the factors of the habitat change in correspondence and the process of competition is set in motion, or kept in motion. As the climatologists point out, no two years are alike in their climatic features and even successive ones may show most striking departures, though these are usually most pronounced at the maximum and minimum of the double sun-spot cycle of 22 years. Conforming with the climatic swing, the climax also exhibits an annual departure in growth and numbers, the process in consequence being termed annuation. In this, competition has both a positive and negative effect; the balance between certain species will be disturbed, while equilibrium may be established between others, and still others will be freed from suppression or the degree of subordination reduced. This is especially true of the balance between tall and short grasses, between dominants and subdominants, and in particular where the influence of coaction is strong.

Bearing a definite relation to season but much modified by annuation is aspection, the sequence of growth and flowering in accordance with the seasons. The response of species or communities to the normal round of the four seasons involves a number of competitive relations, of which evasion as expressed in societies complementary in time is perhaps the most important. The effects of annuation will be regularly superimposed upon those of aspection to bring about readjustments in competition, and even greater disturbance may be caused by the displacement or telescoping of seasons.

Rôle of competition in the forest—Though the general course and outcome of competition are much the same in most communities, there are significant differences in such native types as forest, grassland, and desert, and in the culture ones, field and garden. This is primarily a consequence of life-form, the stature and canopy features of trees yielding the maximum of dominance and giving a degree of complexity to competition not to be found elsewhere. This has already been noted as to the competitive relations within and between layers, but it is of the first importance in connection with the reproduction of the species of the dominant layer. The increasing reaction upon light from the forest canopy to the ground layer and a somewhat similar reaction upon the water-content from the deepest to the shallowest root layer renders the few inches above and below the soil surface the most critical level in the forest. It is to this *reaction-level* and the rôle of competition in relation to it that investigation should be primarily directed, though without losing sight of the part played by the layers above and below it.

It has already been stated that tolerance is really nothing more or less than competition, and in its complete relation, in spite of the fact that the word has long been taken to indicate the response to light. This is the vital process in the life-history of the forest as of all other communities, and must serve as the guide to the fuller understanding of it. Essential to this is the experimental study of the weapons of each species as employed during its life-history, of the physical factors, the reactions upon them and the coactions of animals and fungi, and of the actual course of competition in detail. Shifting views as to tolerance have made it seem as though light and water were antagonistic or mutually exclusive in all this, but this is obviously not the case. No approach is adequate that does not take both fully into account, and nutrients and oxygen-content as well wherever the precipitation is excessive or the drainage insufficient. The relative importance of the two chief factors will vary with the climate and the life-form, as between deciduous, evergreen or coniferous dominants, and it can not be supposed that studies in one region or one climax will settle the question for other regions or climaxes.

Rôle of competition in grassland—Competition in grassland is necessarily keener and follows a somewhat different course than in forest, owing to the absence of a canopy. In this respect, scrub is intermediate, since it forms a canopy in some cases and not in others, but in either event does not promote the development of layers. Moreover, the dominance of grasses is chiefly a matter of competition for water, and light habitually plays a rôle of quite secondary importance. The life-forms of prairie and meadow belong typically to a few types of perennial herbs, and hence layering in any degree like that of the forest is further impossible. These die to the ground at the close of the season and the struggle for dominance begins anew each year, but with the result foregone, the grasses taking the control and the forbs becoming subdominant, though not strictly subordinate. The form of the shoot in the latter is distinctly more favorable for the interception of radiant energy, but both root and shoot of the grass are better adapted to the paramount factor of water. The nutrient values of prairie soils are regularly high and the climatic relations prevent soil-air from becoming a limiting factor, so that these enter the process of competition slightly or rarely.

Grassland further differs from forest in respect to the effect of annuation, responding much more promptly and visibly to extremes of rainfall in especial. Hence, the competitive balance is more easily disturbed, and the relative importance of tall or short grass, forb, weed or undershrub correspondingly affected. Finally, while grassland may be greatly modified by overgrazing, it is practically never destroyed to the extent that forest is by clearing or by the combination of lumbering and fire. As a consequence, it reflects to an exceptional degree the effect of climatic variations and changes and of the coactions of animals and man. In short, its composition at any time is the result of competition operating under the influence of annuation, but often profoundly modified by grazing, burning or cutting. The best guide through this maze of interacting processes is supplied by the investigation of competition, based upon the life-history of the species concerned and the outcome in terms of succession.

Rôle of competition in crop communities—Cultivated plants lend themselves so readily to practical as well as scientific experiments in spacing or rate of planting that the knowledge on this point is much the most extensive in the field of competition. However, the largest portion of it by far is empirical, and few even of the scientific studies have dealt with the process itself or the measurement of reaction and outcome, apart from the test of yield. Since the competition involved is usually of the simple type between individuals of the same species, field crops afford unusually favorable materials for the thorough-going analysis of the process itself, on a scale otherwise impossible. Studies, such as have been made in some cases, of the precise nature of the advantages exhibited by one variety in competition with another are of distinct value in the breeding and testing of improved strains. Moreover, the work of Montgomery and of Kiesselbach indicates that the rôle of competition in producing or selecting individual plants with desired qualities still awaits determination, with some preliminary evidence in favor of such selective action.

FURTHER STUDIES OF COMPETITION

The present organization of the field of competition has been made as comprehensive and at the same time as intensive as possible, the first to bring out its unity and suggest the major correlations, the second to furnish the foundation for more specialized studies in its various aspects. It is hoped that the results obtained and the applications made will stimulate an increasing number of ecologists to carry out similar investigations in different regions and climaxes. Apart from the study of the process for itself, there are five major divisions of the field, in all of which further research will be found distinctly profitable. These are as follows: (1) functions of the individual and the physico-chemical processes involved; (2) interaction of competition, adaptation, and correlation; (3) community functions with especial reference to succession; (4) competition among animals and the competitive significance of coactions; (5) rôle of competition in forest, range and field crop.

Competition and functions of the plant—It has repeatedly been shown in the preceding that competition and the functioning of the individual stand in the most intimate reciprocal relation to each other. Absorption, transpiration, photosynthesis, chemosynthesis, and respiration all affect the physical factors of the habitat directly, producing reactions that determine the course, intensity, and outcome of competition. In return, the progressive increase in demand further reduces the supply for each individual, and its functions are correspondingly affected. This cycle of cause and effect continues as long as plants function and grow, and energy, materials, or both are inadequate to the total requirements. This relation is varied in manifold fashion, as shoot or root bears the brunt of the struggle, as individuals become dominant or suppressed, as the death of leaves or plants, their accumulation and decay, etc., affect reactions still more. The results obtained in these respects in this first organizing study are necessarily preliminary, but their concordance as to the various functions and reactions measured indicates their essential soundness.

The correlation of the basic physical and chemical processes with function, and through this with competition and reaction is obviously a matter for cooperative attack by biochemist and ecologist. The latter can apply present methods and technique when necessary, to questions of permeability, imbibition, and diffusion, of surface tension, dielectric constants and osmotic pressure, of the selective absorption of radiant energy and of total energy relations. But in all these the necessary checks and refinements must be applied by the biochemist and he must be chiefly responsible for the adequacy of methods as well as their improvement.

Interaction of competition, adaptation, and correlation—The functional cycle in competition is adjustment ("Plant Physiology and Ecology," p. 4); the structural outcome is adaptation. In the latter are involved also the nutritional adjustments in the plant body expressed in the correlation of organs and parts. This sequence of action and interaction is fundamental and universal, and is an inevitable concomitant of the cause-and-effect relation of competition and functioning already mentioned. It renders adaptation the basic and visible expression of response in the individual, as succession is in the community, and throws into clear relief the threefold nature of developmental ecology, namely, competition, adaptation, and succession. As emphasized above, function is the motive force in all three of these interlocking processes, and hence the measurement of factor and function with respect to the organism as a working unit, in terms of increasing accuracy, is the greatest desideratum of ecology ("Research Methods in Ecology," p. 20).

The outcome of competition in terms of the adaptation known as the growth-form has been much discussed in connection with the results of the experimental cultures, and the response in the presence and absence of competition has been contrasted by means of denuded quadrats. This analysis is being carried much further by means of paired cultures with and without competition, and especially by quadrat sequences ranging through four or more densities, from solitary individuals to the maximum density in which flowering and fruiting, though profoundly modified, are still possible. This permits not only the correlation of function and form with measured factor, but also makes it possible to evaluate the rôle of direct and of reacted factors in adaptation. Moreover, it affords an exceptional opportunity for tracing the operation of selection and its various effects. Finally, as has been suggested, the correlation of organs in the plant itself has been found to explain adaptational relations that seemed obscure or even contradictory on the basis of a straight factor-function response.

Community functions and succession—It has long been recognized that the reconstruction of succession from a relatively brief or recurrent observation of its stages, desirable as it is for many reasons, should be regarded as a temporary expedient, and that continuous record and experiment in the field of community functions constitute the ideal type of investigation ("Plant Succession," p. 423; "Plant Succession and Indicators," p. 186). This method—at once comprehensive, intensive and continuous—is now being employed in the cooperative investigation of grazing influences in Arizona and of forest influences in California, in both of which succession is the major process. The pressing need is to extend the method to other regions and climaxes, for the sake of dealing with a wide range of variation and modification in the process, and being able to formulate principles and corollaries on a broader basis. The best opportunity for this has been realized in the great grassland formation with its six climax associations, in all of which cooperative studies of succession have been carried on for a decade and in the true and subclimax prairies for thirty years. A similar investigation is even more important in the case of the forest climaxes of the continent, and a like cooperative arrangement is indispensable in view of the much greater distances involved.

While the observational study of succession lends itself to field reconnaissance in some measure, experiment and record demand field exclosures and bases. The chief task in consequence is to link up the various stations and bases over a wide territory, and ultimately perhaps over the entire country, and to focus investigation upon such types of succession as promise the larger returns in relatively briefer periods. Such are burns, overgrazed areas, and abandoned farm-lands in the case of the shorter secondary successions, and peat-bog and freshwater swamps in that of primary successions with much longer life-history. In all of these the clue to development is to be found in the community functions, aggregation, migration, competition, reaction, ecesis and adaptation, among which the crucial rôle is taken by competition.

Competition and coaction among animals—In this field the guide-posts are still to be set, though a beginning has already been made in connection with coactions. In fact, a clear understanding of the nature and rôle of competition among animals seems to rest directly upon much greater knowledge of their coaction upon each other and upon plants. The distinction between the two processes is inherent in the two words; competition is a joint demand in excess of the available supply of food, material, or shelter and hence space in some degree, or even of such physical factors as water, heat or light. Coaction, on the other hand, is the direct and usually reciprocal action of organism upon organism, of animal upon animal or upon plant in the present instance. In other words, animals may be said to compete for their coactions. Because of the two types of organisms concerned in the complex of responses, reactions and coactions, this is the most intricate maze of interrelations and stands in urgent need of comprehensive and adequate analysis.

The coactions that have received some attention up to the present have been concerned with the food supply of animals—nectar and pollen in the case of pollinators, forage in that of grazing animals, and seeds in the case of spermophile rodents and birds. The details of the competition are probably best understood in the first ("Experimental Pollination"), but in no case as yet has been made a direct attack upon competition and coaction as the primary objectives. The growing tendency to emphasize the biotic nature of ecology is rapidly changing this situation and aiding in the organization of this new but fertile field.

Rôle of competition in forest, range, and field crop—The fundamental importance of studies of competition in these three great fields of research has already been emphasized (p. 327), and the chief task at present is to carry them into effect in a comprehensive and detailed manner. In the case of the forest, the relative importance of water and light is the major objective, combined with the outcome of competition in terms of growth and survival, and of successional development. In the range, water is the paramount factor, though light is not to be neglected, and the focus of attack is the complicated interaction of competition, coaction and annuation. Finally, with field crops the problem is centered upon the relative merits of species and varieties in terms of competitive adaptation to seasonal and annual cycles, and to the tillage and rotation control of factors and reactions.

The basic principles and methods are essentially the same for all three fields, but the complete outline for competition research is best illustrated in the case of the forest, owing to its greater complexity and the difficulties to be overcome in measuring the functions of trees in the field. Instrumental batteries in the forest are placed with respect to layers as well as areal differences of composition and density, age, slope-exposure, successional stages, etc. With respect to light, they must be super-batteries characterized by instruments for measuring total energy in terms of incidence and absorption, determining the quality of light by spectrophotometers, etc. The phytometers employed are likewise varied in kind and position, level and insert phytometers being of the first importance. Standard plants have been found most convenient and useful for purposes of comparison, particularly between layers, but native phytometers, both sealed and open, are more significant in measuring the responses and reactions of dominant and subdominant. The chief problem has naturally been experienced in connection with the size of shrubs and trees, and the difficulties overcome with seedlings of a few years. However, these are inadequate for the study of adult response, reaction and competition, for which major installations of tanks, scales and cranes are required. For the complete life-history of the individual and species, dynamic studies of structure and form are indispensable, from the growth and absorption of root-hairs through the tissues concerned with diffusion and conduction to conditions in the mesophyll and the behavior of stomata.

The investigation of the forest community demands the fullest equipment of quadrats and transects. While these have been employed chiefly in the study of successional movement, especially after fire, they have proved to be equally valuable for the analysis of community functions and in particular for competition, reaction and adaptation. In fact, the present plan involves these processes as the immediate objectives of quadrat cultures, natural and artificial, with the course of succession and stabilization as the final goal. The utilization of clearing, denuding, planting, habitat inversion, etc., is regular procedure in all of these, the highest values being obtained when the study of forest influences is combined with that of the ecology of the forest community. Finally, while the degree of control and hence of accuracy is steadily increasing with better field instruments and technique, it has become evident that laboratory and nursery conditions are often indispensable for analyzing and supplementing field experiments. However, it must be fully realized that the greater control they afford is secured at the expense of the natural cause-and-effect relations of the organism as a working unit.

BIBLIOGRAPHY

AALTONEN, V. T. 1923. Ueber die räumliche Ordnung der Pflanzen auf dem Felde und im Walde. Act. For. Fenn. 25:7–85.

———. 1925. Ueber die Selbstabscheidung und den Wuchsraum der Bäume in Naturbeständen. 3–20.

———. 1925. Allgemeines über die Einwirkung der Bäume auf einander. Act. For. Fenn. 29:3–19.

———. 1926. On the space arrangement of trees and root competition. Jour. For. 24:627–644.

ADAMSON, R. S. 1922. The woodlands of Ditcham Park, Hampshire. Jour. Ecol. 9:114–219.

BATES, C. G. 1911. Windbreaks: their influence and value. For. Serv. Bull. 86:7–100.

———. 1917. The rôle of light in natural and artificial reforestation. Jour. For. 15:233–239.

———. 1925. The relative light requirements of some coniferous seedlings. Jour. For. 23:869–879.

———, and J. ROESER. 1928. Light intensities required for growth of coniferous seedlings. Am. Jour. Bot. 15:185–194.

———, and R. ZON. 1922. Research methods in the study of forest environment. U. S. Dept. Agr. Bull. 1059:2–208.

BEATTIE, J. H., F. E. MILLER, and R. E. CURRIN. 1927. Effect of planting distances on yield of peanuts. U. S. Dept. Agr. Bull. 1478:1–14.

BERGMAN, H. F. 1920. The relation of aeration to the growth and activity of roots and its influence on the ecesis of plants in swamps. Ann. Bot. 34:13–33.

BERTOG, H. 1921. Die Buche im nordostdeutschen Kiefernwalde. 38–39.

BONNIER, G. 1879. Quelques observations sur les relations entre la distribution des phanerogames et la nature chimique du sol. Bull. Soc. Bot. Fr. 26:338–341.

BRENCHLEY, W. E. 1917. The effect of weeds upon cereal crops. New Phyt. 16:53–76.

———. 1919. Some factors in plant competition. Ann. App. Biol. 6:142–170.

BROWN, B. A. 1924. Spacing of potato hills. Storrs Agr. Expt. Sta. Bull. 119:143–151.

BUFFON, G. L. L. 1745. Mémoire sur la culture des forests. Hist. Acad. Roy. Sci. Year 1742:233–246.

BÜHLER, A. 1922. Der Waldbau nach wissenschaftlicher Forschung und praktischer Erfahrung.

BURNS, G. P. 1914. Development of white pine seedlings in nursery beds. Ver. Expt. Sta. Bull. 178:127–144.

———. 1914. Relation of shade to evaporation and transpiration in nursery beds. Ver. Agr. Expt. Sta. Bull. 181:235–262.

———. 1916. Discontinuous light in forests. Ver. Agr. Expt. Sta. Bull. 193:3–23.

———. 1920. Tolerance of forest trees and its relation to forest succession. Jour. For. 20:610–615.

———. 1923. Minimum light requirements referred to a definite standard. Ver. Agr. Expt. Sta. Bull. 235:3–32.

———. 1927. Leaf efficiency in thrifty and stunted white pine seedlings. Ver. Agr. Expt. Sta. Bull. 267:3–27.

CHANEY, R. W. 1925. A comparative study of the Bridge Creek flora and the modern redwood forest. Carnegie Inst. Wash. Pub. 349:3–24.

———. 1925. The Mascall flora—its distribution and climatic relation. Ib. 25–48.

CHRIST, H. 1879. Das Pflanzenleben der Schweiz.

CLAPP, E. H. 1926. A national program of forest research.

CLEMENTS, F. E. 1897. Peculiar zonal formations of the Great Plains. Am. Nat. 31:968–970.

———. 1904. Development and structure of vegetation. Rep. Bot. Surv. Nebr. 7:5–175.

———. 1905. Research methods in ecology.

———. 1907. Plant physiology and ecology.

———. 1907. The causes of dwarfing in alpine plants. Science 25:287.

———. 1910. The life history of lodgepole burn forests. For. Serv. Bull. 79:1–56.

———. 1912. Some impressions and reflections. New Phyt. 11:177–179.

———. 1916. Plant succession. Carnegie Inst. Wash. Pub. 242.

———. 1918. Scope and significance of paleo-ecology. Bull. Geol. Soc. Am. 29:369–374.

CLEMENTS, F. E. 1920. Plant indicators. Carnegie Inst. Wash. Pub. 290.

———. 1921. Aeration and air-content. Carnegie Inst. Wash. Pub. 315.

———. 1922. Correlation in bud development. Carnegie Inst. Wash. Year Book. 21:342.

———. 1924. Competition and correlation in the plant body. Ib. 23:250–251.

———. 1928. Plant succession and plant indicators.

———, and E. S. CLEMENTS. 1915. Herbaria Ecadium Californiae.

———, and G. W. GOLDSMITH. 1924. The phytometer method in ecology. Carnegie Inst. Wash. Pub. 356.

———, and F. L. LONG. 1923. Experimental pollination. Carnegie Inst. Wash. Pub. 336.

———, and J. E. WEAVER. 1924. Experimental vegetation. Carnegie Inst. Wash. Pub. 355.

———, J. E. WEAVER, and H. C. HANSON. 1924. The nature and rôle of competition. Carnegie Inst. Wash. Year Book. 23:252–255.

———. 1925. Id. Ib. 24:329–333.

COLE, J. S., and A. L. HALLSTED. 1926. The effect of outside rows on the yields of plots of Kafir and Milo at Hays, Kans. Jour. Agr. Res. 32:991–1002.

CSERHÁTI, A. 1890. Versuche über den Grünmais. Jour. Landw. 38:349–374.

DARWIN, C. 1859. Origin of species.

DECANDOLLE, A. P. 1820. Essai élémentaire de géographie botanique.

DRUDE, O. 1913. Die Oekologie der Pflanzen.

DUHAMEL DU MONCEAU, 1760. Des semis et plantations des arbres et de leur culture.

———. 1764. De l'exploitation des bois.

DUKE OF BEDFORD and S. U. PICKERING. 1914. The effect of one crop upon another. Rep. Woburn Expt. Farm. 14:136–151.

DUNNING, D. 1923. Some results of cutting in the Sierra forests of California. U. S. Dept. Agr. Bull. 1176:1–26.

EAST, E. M. 1926. Mankind at the cross roads.

ENGLER, Arn. 1911. Untersuchungen über den Blattausbruch und das sonstige Verhalten von Schatten- und Lichtpflanzen der Buche und einiger anderer Laubhölzer. Mitt. Schweiz. Cent. Versuch. 10:107–177.

FARROW, E. P. 1917. On the ecology of the vegetation of Breckland. Part V. Observations relating to competition between plants. Jour. Ecol. 5:155–172.

FRANKLIN, B. 1751. Observations concerning the increase of mankind and the peopling of countries.

FRICKE, K. 1904. "Licht und Schattenholzarten" ein wissenschaftlich nicht begründetes Dogma. Cent. Gesamt. Forstw. 30:315.

FROMBLING, E. 1913. Welche Rolle spielt die Wurzelkonkurrenz im Haushalte des Waldes? Forstw. Cent. 57:170–175.

FÜRST, H. 1905. Licht- und Schattenholzarten, ein wissenschaftlich nicht begründetes Dogma? Forstw. Cent. 49:1–10.

GARBER, R. J., and T. E. ODLAND. 1925. Influence of adjacent rows of soy-beans on one another. Jour. Am. Soc. Agro. 18:605–607.

GARNER, W. W., W. M. LUNN, and D. E. BROWN. 1925. Effects of crops on the yields of succeeding crops in the rotation, with special reference to tobacco. Jour. Agr. Res. 30:1095–1132.

GLEASON, H. A. 1917. The structure and development of the plant association. Bull. Torr. Club. 44:463–481.

———. 1926. The individualistic concept of the plant association. Ib. 53:7–26.

GRASOVSKY, A. 1928. Some aspects of light in the forest. Yale Univ. Sch. For. Bull. 23.

GREVILLIUS, A. Y. 1894. Biologisch-physiognomische Untersuchungen einiger schwedischen Hainthälchen. Bot. Zeit. 52:147–168.

HARPER, R. M. 1907. Competition between two oaks. Plant World 10:114–117.

HARTWELL, B. L. 1927. How some crops affect succeeding crops. Jour. Am. Soc. Agro. 19:255–259.

———, B. L., and S. C. DAMON. 1918. The influence of crop plants on those which follow. I. R. I. Agr. Expt. Sta. Bull. 175:4–30.

———, F. R. PEMBER and G. E. MERKLE. 1919. Id. II. Ib. 176:4–47.

———, J. B. SMITH, and S. C. DAMON. 1927. Id. III. Ib. 210:3–23.

HAUCH, L. A. 1910. Zur Variation des Wachstums bei unseren Waldbäumen mit besonderer Berücksichtigung des sogenannten Ausbreitungsvermögens. Forstw. Cent. 32:565–578.

HELLRIEGEL, H. 1883. Beiträge zu den naturwissenschaftlichen Grundlagen des Ackerbaus, etc.

HESSELMANN, H. 1904. Zur Kenntnis des Pflanzenlebens schwedischer Laubwiesen. Mitt. Inst. Univ. Stockholm.

HOFMANN, J. V. 1924. The natural regeneration of Douglas fir in the Pacific Northwest. U. S. Dept. Agr. Bull. 1200:1–62.

HOLE, R. S., and P. SINGH. 1916. Seedling reproduction in natural forests and its improvement. Indian For. Rec. 5:43–85.

HOOKER, J. D. 1867. On the struggle for existence amongst plants. Pop. Sci. Rev. 6:131–139.

HOWARD, A. 1916. Soil aeration in agriculture. Pusa Agr. Res. Inst. Bull. 61.

———, and G. L. C. HOWARD. 1915. Soil ventilation. Quetta Fruit Exp. Sta. Bull. 3.

HULBERT, H. W. and J. D. REMSBERG. 1927. Influence of border rows in variety tests of small grains. Jour. Am. Soc. Agro. 19:585–589.

HUME, D. 1752. Political Discourses.

HUME, A. N., O. D. CENTER, and L. HEGNAUER. 1908. Distance between hills for corn in the Illinois corn belt. Ill. Agr. Expt. Sta. Bull. 126:357–374.

JEFFERIES, T. A. 1915. Ecology of the purple heath grass—Part III. Factors in the distribution of Molinia. Jour. Ecol. 3:98–109.

JEFFREYS, H. 1917. On the vegetation of four Durham Coal-Measure Fells. IV. On various other ecological factors. Jour. Ecol. 5:140–146.

KIESSELBACH, T. A. 1917. Studies concerning the elimination of experimental error in comparative crop tests. Nebr. Agr. Expt. Sta. Res. Bull. 13:5–95.

———. 1922. Corn investigations. Ib. 20:5–151.

———. 1923. Competition as a source of error in comparative corn yields. Jour. Am. Soc. Agro. 15:199–215.

———. 1924. Relation of seed size to the yield of small grain crops. Ib. 16:670–682.

———. 1925. Winter wheat investigations. Nebr. Agr. Expt. Sta. Res. Bull. 31:5–149.

———. 1926. The relation of seeding practices to crop. Jour. Am. Soc. Agro. 18:661–681.

———, and C. A. HELM. 1917. Relation of size of seed and sprout value to the yield of small grain crops. Nebr. Agr. Expt. Sta. Res. Bull. 11:5–73.

———, and W. E. LYNESS. 1924. Spring small grains. Nebr. Agr. Expt. Sta. Bull. 201:2–45.

———, and J. A. RATCLIFF. 1917. Oats investigations. Ib. 160:4–48.

———, and H. B. SPRAGUE. 1926. Relation of the development of the wheat spike to environmental factors. Jour. Am. Soc. Agro. 18:40–60.

KNUCHEL, H. 1914. Spektrophotometrische Untersuchungen im Walde. Mitt. Schweiz. Cent. Vers. 11.

———. 1915. Spectrophotometric investigations in the forest. Proc. Soc. Am. For. 10:87–92. Translated by O. L. Sponsler.

KRAEPELIN, K. 1913. Die Beziehungen der Tiere und Pflanzen zueinander. II. 2 Auflage. (Aus Natur und Geisteswelt, Bd. 427).

KULESCHOW, N. 1922. Der Kampf der Sommerweizenrassen. Landw. Versuchs.

LASCHKE, C. 1902. Geschichtliche Entwicklung des Durchforstungsbetriebs in Wissenschaft und Praxis bis zur Gründung der deutschen forstlichen Versuchsanstalten.

LEMMERMANN, O. 1903. Untersuchungen über den Einfluss eines verschieden grossen Bodenvolumens auf den Ertrag und die Zusammensetzung der Pflanzen. Jour. Landw. 51:1–40, 279–285.

———. 1905. Untersuchungen über den Einfluss eines verschieden grossen Bodenvolumens auf die Entwickelung der Pflanzen. Ib. 53:173–177.

LINDMAN, C. A. M. 1913. Some cases of plants suppressed by other plants. New Phyt. 12:1–6.

LIVERMORE, J. R. 1927. A critical study of some of the factors concerned in measuring the effect of selection in the potato. Jour. Am. Soc. Agro. 19:857–896.

LJUBIMENKO, W. N. 1926. O. STSCHEGLOVA, und Z. BULGAKOVA. 1926. Versuch über den Wettbewerb um dem Raum bei Pflanzen. Jour. Russ. Bot. Ges. 10.

LÖNNROTH, ERIK. 1925. Untersuchungen über die innere Struktur und Entwicklung gleichaltriger naturnormaler Kiefernbestände. Act. For. Fenn. 30:1–269.

LYELL, C. 1867–68. Principles of geology. 10th ed.

MALTHUS, T. R. 1798. The principles of population. 2nd ed. 1803; 3rd ed. 1807.

MAYER, A. 1879. Ein Vegetationshaus ohne directes Sonnenlicht. Landw. Versuchs. 23:249–257.

———. 1892. Ueber die Atmungsintensität von Schattenpflanzen. Landw. Versuchs. 40:203–216.

———. 1897. Das Maximum der Pflanzenproduktion. Ib. 48:61–76.

McKeever, H. G. 1924. Spacing experiments with Acala cotton in southern California. Jour. Agr. Res. 28:1081–1093.

Mitscherlich, E. A. and F. Dühring. 1921. Gefässversuche mit Erlen und Fichtenpflanzen. Zeits. Forst. Jagdw. 53:486–494.

Montgomery, E. G. 1910. Oats: variety tests; rate of sowing; cultivation. Nebr. Agr. Expt. Sta. Bull. 113:3–16.

———. 1912. Competition in cereals. Ib. 127:3–22.

Mooers, C. A. 1928. The effect of spacing on the yield of cotton. Jour. Am. Soc. Agro. 20:211–230.

Moore, B. 1922. Humus and root systems in certain northeastern forests in relation to reproduction and competition. Jour. For. 20:233–254.

Musgrave, G. W. 1926. A further note on competition in potatoes. Jour. Am. Soc. Agro. 18:166–171.

Nägeli, C. 1865. Ueber die Bedingungen des Vorkommens von Arten und Varietäten innerhalb ihres Verbreitungsbezirkes. Sitzb. Akad. Wiss. München. 2:367–395. Bot. Mitt. 2:159–187:1866.

———. 1872. Das gesellschaftliche Entstehen neuer Species. Ib. 9:305–344, 3:165–204: 1881.

———. 1874. Verdrängung der Pflanzenformen durch ihre Mitbewerber. Ib. 11:109–164, 3:205–260:1881.

Pearson, G. A. 1923. Natural reproduction of western yellow pine in the Southwest. U. S. Dept. Agr. Bull. 1105:1–143.

———. 1928. Forest types in New Mexico and Arizona. Ms.

Petrus De Crescentiis. 1305. Handschrift.

Pfeiffer, Th. 1918. Der Vegetationsversuch.

———, E. Blanck, and M. Flügel. 1912. Wasser und Licht als Vegetationsfactoren und ihre Beziehungen zum Gesetze vom Minimum. Landw. Versuchs. 76:169–236.

———, ———, and W. S. Simmermacher. 1915. Beziehungen zwischen dem Einfluss von Licht und Stickstoff als Minimumfaktoren auf das Wachstum der Pflanzen. Landw. Versuchs. 86:45–62.

Phillips, J. F. V. 1928. Olea laurifolia Lam. ("Ironwood"): an introduction to its ecology. Trans. Roy. Soc. S. A. 16:169–190.

Pool, R. J., J. E. Weaver, and F. C. Jean. 1918. Further studies in the ecotone between prairie and woodland. Rep. Bot. Surv. Nebr. n. s. 1:1–47.

Pound, R., and F. E. Clements. 1898. The phytogeography of Nebraska. 2nd ed. 1900.

Rayner, M. C. 1913. The ecology of Calluna vulgaris. New Phyt. 12:59–77. Jour. Ecol. 1:59–60.

———. 1921. The ecology of Calluna vulgaris. II. The calcifuge habit. Jour. Ecol. 9:60–74.

———, and W. N. Jones. 1911. Preliminary observations on the ecology of Calluna vulgaris on the Wiltshire and Berkshire Downs. New Phyt. 10:227–240.

Renvall, A. 1912. Die periodischen Erscheinungen der Reproduktion der Kiefer an der polaren Waldgrenze. Act. For. Fenn. 1:1–154.

Robertson, D. W., and A. Kezer. 1927. Residual effects of different irrigation treatments on the crops grown the succeeding year. Jour. Am. Soc. Agro. 19:923–943.

Romell, L. G. 1922. Die Bodenventilation als ökologischer Faktor. Mitt. Forst. Versuchs. Schwed. 19:125–359.

Ruber, K. 1921. Oekologische und pflanzengeographische Lichtfragen. Forstw. Cent. 65:327–345.

———. 1924. Die pflanzengeographischen Grundlagen des Waldbaus.

Russell, E. J. 1914. The effect of one growing crop on another. Rep. Woburn Expt. Farm. 14:51–68.

Sachs, J. 1860. Bericht über physiologische Thätigkeit an der Versuchstation in Tharandt. II. Wurzelstudien. Landw. Versuchs. 2:1–31.

Salisbury, E. J. 1916. The oak-hornbeam woods of Hertfordshire, parts I and II. Jour. Ecol. 4:83–117.

———. 1917. The ecology of scrub in Hertfordshire: a study in colonization. Trans. Hert. Nat. Hist. Soc. 17:53–64.

———. 1918. The oak-hornbeam woods of Hertfordshire, parts III and IV. Jour. Ecol. 6:14–52.

———. 1922. Stratification and hydrogen-ion concentration of the soil in relation to leaching and plant succession with special reference to woodlands. Ib. 9:220–240.

———. 1924. The effects of coppicing as illustrated by the woods of Hertfordshire. Trans. Hert. Nat. Hist. Soc. 18:2–21.

SALISBURY, E. J. 1925. The incidence of species in relation to soil reaction. Jour. Ecol. 13:149–160.

———. 1929. The biological equipment of species in relations to competition. Jour. Ecol. 18.

SAMPSON, A. W. 1919. Plant succession in relation to range management. U. S. Dept. Agr. Bull. 791:1–76.

SCHMIDT, —. 1923. Ueber neue Untersuchungen zum Kohlensäure-problem. Zeit. Forst-Jagdw. 55:534–542, 715–717.

———. 1924. Vegetationsversuche zum Ertragsfaktor Licht. Ib. 56:461–472.

SEELHORST, C. and I. P. PANAOTOVIC. 1899. Einfluss der Standweite auf die chemische Zusammensetzung der Pflanzen. Jour. Landw. 46:379–389.

SIEGLINGER, J. B. 1923. Grain-sorghum experiments at the Woodward field station in Oklahoma. U. S. Dept. Agr. Bull. 1175:1–65.

SHANTZ, H. L. 1913. The effects of artificial shading on plant growth in Louisiana. Bur. Plant Ind. Bull. 279:7–29.

SHERFF, E. E. 1912. The vegetation of Skokie Marsh, with special reference to subterranean organs and their interrelationships. Bot. Gaz. 53:415–435.

SHOW, S. B., and E. I. KOTOK. 1924. The rôle of fire in the California pine forests. U. S. Dept. Agr. Bull. 1294:1–80.

SPENCER, H. 1858–1860. Essays.

———. 1864. Principles of biology.

———. 1876–80. Principles of sociology.

STEWART, F. C. 1919. Missing hills in potato fields: their effect upon the yield. N. Y. Agr. Expt. Sta. Bull. 459:45–69.

———. 1920. Experiments on the spacing of potato plants. Ib. 474:3–32.

STRINGFIELD, G. H. 1927. Intervarietal competition among small grains. Jour. Am. Soc. Agro. 19:971–983.

SUKATSCHEW, W. 1927. Einige experimentelle Untersuchungen über den Kampf ums Dasein zwischen Biotypen derselben Art. Zeits. Indukt. Abst. Vererb. 47:54–74.

TANSLEY, A. G. 1917. On competition between Galium saxatile L. (G. hercynicum Weig.) and Galium silvestre Poll. (G. asperum Schreb.) on different types of soil. Jour. Ecol. 5:173–179.

———. 1920. The classification of vegetation and the concept of development. Jour. Ecol. 8:118–149.

———. 1922 Studies of the vegetation of the English chalk. II. Early stages of redevelopment of woody vegetation on chalk grassland. Jour. Ecol. 10:168–177.

———, and R. S. ADAMSON. 1925. III. The chalk grasslands of the Hampshire-Sussex border. Ib. 13:177–223. IV. A preliminary survey of the chalk grasslands of the Sussex Downs. Jour. Ecol. 14:1–32.

THOMPSON, W. 1915. Population: a study in Malthusianism.

TISDALE, H. B. 1928. Effect of spacing on yield and size of cotton bolls. Jour. Am. Soc. Agro. 20:298–301.

TOUMEY, J. W. 1928. Foundations of silviculture upon an ecological basis.

———, 1929. The vegetation on the forest floor—light versus moisture. Proc. 4th Inter. Cong. Pl. Sci.

VARENNE DE FENILLE, P. C. M. 1790. Mémoires sur l'administration forestière.

VINALL, H. N. 1922. The sunflower as a silage crop. U. S. Dept. Agr. Bull. 1045:1–32.

WAGNER, C. 1907. Die Grundlagen der räumlichen Ordnung im Walde.

WARMING, E. 1895. Plantesamfund. Grundträk af den ökologiske Plantegeografi.

———. 1896. Lehrbuch der ökologischen Planzengeographie. Translated by Knoblauch.

———. 1909. Oecology of plants. Translated by Groom.

———, and P. GRAEBNER. 1914–1918. Lehrbuch der ökologischen Planzengeographie.

WEAVER, J. E. 1919. The ecological relations of roots. Carnegie Inst. Wash. Pub. 286.

———. 1920. Root development in the grassland formation. Ib. 292.

———. 1924. Plant production as a measure of environment. Jour. Ecol. 12:205–237.

———, and A. F. THIEL. 1917. Ecological studies in the tension zone between prairie and woodland. Rep. Bot. Surv. Nebr. n. s. 1:3–59.

———, H. C. HANSON, and J. M. AIKMAN. 1925. Transect method of studying woodland vegetation along streams. Bot. Gaz. 80:168–187.

WEISER, S. and A. ZAITSCHEK. 1913. Ueber den Einfluss der Saatweite auf den Ertrag und Nährwert des Futtermais. Landw. Versuchs. 81:49–100.

WILSON, M. 1911. Plant distribution in the woods of north-east Kent. I. Ann. Bot. 25:857–902.

WOLLNY, E. 1881. Untersuchungen über den Einfluss des Standraumes auf die Entwicklung und die Erträge der Kulturpflanzen. Jour. Landw. 29:25–62, 217–255, 493–536.

WOODHEAD, T. W. 1906. Ecology of woodland plants in the neighborhood of Huddersfield. Linn. Soc. Jour. Bot. 37:333–406.

YAPP, R. H. 1908. Sketches of vegetation at home and abroad. IV. Wicken Fen. New Phyt. 7:61–81.

———. 1909. On stratification in the vegetation of a marsh, and its relations to evaporation and temperature. Ann. Bot. 23:275–319.

———. 1922. The concept of habitat. Jour. Ecol. 10:1–17.

———. 1925. The inter-relations of plants in vegetation and the concept of "association". Fest. Schröter, Geobot. Inst. Rübel. 3:684–706.

ZEDERBAUER, E. 1907. Das Lichtbedürfniss der Waldbäume und die Lichtmess Methoden. Cent. Ges. Forstw. 33:325–330.

———. 1908. The light requirements of forest trees and the methods of measuring light. For. Quar. 6:255–262. Translated by R. Zon and B. T. Boisen.

ZOLLER, PH. 1867. Vegetationsversuche und agrikulturchemische Untersuchungen. II. Bewurtzelungsversuche. Jour. Landw. II. 2:193–204.

ZON, R., and H. S. GRAVES. 1911. Light in relation to tree growth. For. Serv. Bull. 92:5–59.

HISTORY OF ECOLOGY

An Arno Press Collection

Abbe, Cleveland. **A First Report on the Relations Between Climates and Crops.** 1905

Adams, Charles C. **Guide to the Study of Animal Ecology.** 1913

American Plant Ecology, 1897-1917. 1977

Browne, Charles A[lbert]. **A Source Book of Agricultural Chemistry.** 1944

Buffon, [Georges-Louis Leclerc]. **Selections from Natural History, General and Particular, 1780-1785.** Two volumes. 1977

Chapman, Royal N. **Animal Ecology.** 1931

Clements, Frederic E[dward], John E. Weaver and Herbert C. Hanson. **Plant Competition.** 1929

Clements, Frederic Edward. **Research Methods in Ecology.** 1905

Conard, Henry S. **The Background of Plant Ecology.** 1951

Derham, W[illiam]. **Physico-Theology.** 1716

Drude, Oscar. **Handbuch der Pflanzengeographie.** 1890

Early Marine Ecology. 1977

Ecological Investigations of Stephen Alfred Forbes. 1977

Ecological Phytogeography in the Nineteenth Century. 1977

Ecological Studies on Insect Parasitism. 1977

Espinas, Alfred [Victor]. **Des Sociétés Animales.** 1878

Fernow, B[ernhard] E., M. W. Harrington, Cleveland Abbe and George E. Curtis. **Forest Influences.** 1893

Forbes, Edw[ard] and Robert Godwin-Austen. **The Natural History of the European Seas.** 1859

Forbush, Edward H[owe] and Charles H. Fernald. **The Gypsy Moth.** 1896

Forel, F[rançois] A[lphonse]. **La Faune Profonde Des Lacs Suisses.** 1884

Forel, F[rançois] A[lphonse]. **Handbuch der Seenkunde.** 1901

Henfrey, Arthur. **The Vegetation of Europe, Its Conditions and Causes.** 1852

Herrick, Francis Hobart. **Natural History of the American Lobster.** 1911

History of American Ecology. 1977

Howard, L[eland] O[ssian] and W[illiam] F. Fiske. **The Importation into the United States of the Parasites of the Gipsy Moth and the Brown-Tail Moth.** 1911

Humboldt, Al[exander von] and A[imé] Bonpland. **Essai sur la Géographie des Plantes.** 1807

Johnstone, James. **Conditions of Life in the Sea.** 1908

Judd, Sylvester D. **Birds of a Maryland Farm.** 1902

Kofoid, C[harles] A. **The Plankton of the Illinois River, 1894-1899.** 1903

Leeuwenhoek, Antony van. **The Select Works of Antony van Leeuwenhoek.** 1798-99/1807

Limnology in Wisconsin. 1977

Linnaeus, Carl. **Miscellaneous Tracts Relating to Natural History, Husbandry and Physick.** 1762

Linnaeus, Carl. **Select Dissertations from the Amoenitates Academicae.** 1781

Meyen, F[ranz] J[ulius] F. **Outlines of the Geography of Plants.** 1846

Mills, Harlow B. **A Century of Biological Research.** 1958

Müller, Hermann. **The Fertilisation of Flowers.** 1883

Murray, John. **Selections from *Report on the Scientific Results of the Voyage of H.M.S. Challenger During the Years 1872-76.*** 1895

Murray, John and Laurence Pullar. **Bathymetrical Survey of the Scottish Fresh-Water Lochs.** Volume one. 1910

Packard, A[lpheus] S. **The Cave Fauna of North America.** 1888

Pearl, Raymond. **The Biology of Population Growth.** 1925

Phytopathological Classics of the Eighteenth Century. 1977

Phytopathological Classics of the Nineteenth Century. 1977

Pound, Roscoe and Frederic E. Clements. **The Phytogeography of Nebraska.** 1900

Raunkiaer, Christen. **The Life Forms of Plants and Statistical Plant Geography.** 1934

Ray, John. **The Wisdom of God Manifested in the Works of the Creation.** 1717

Réaumur, René Antoine Ferchault de. **The Natural History of Ants.** 1926

Semper, Karl. **Animal Life As Affected by the Natural Conditions of Existence.** 1881

Shelford, Victor E. **Animal Communities in Temperate America.** 1937

Warming Eug[enius]. **Oecology of Plants.** 1909

Watson, Hewett Cottrell. **Selections from *Cybele Britannica.*** 1847/1859

Whetzel, Herbert Hice. **An Outline of the History of Phytopathology.** 1918

Whittaker, Robert H. **Classification of Natural Communities.** 1962